AF322728

Memorial Volume for
YI-SHI DUAN

Memorial Volume for
YI-SHI DUAN

Edited by

Mo–Lin Ge
Chern Institute of Mathematics, China

Rong–Gen Cai
Chinese Academy of Sciences, China

Yu–Xiao Liu
Lanzhou University, China

 **World Scientific**

NEW JERSEY · LONDON · SINGAPORE · BEIJING · SHANGHAI · HONG KONG · TAIPEI · CHENNAI · TOKYO

Published by

World Scientific Publishing Co. Pte. Ltd.

5 Toh Tuck Link, Singapore 596224

USA office: 27 Warren Street, Suite 401-402, Hackensack, NJ 07601

UK office: 57 Shelton Street, Covent Garden, London WC2H 9HE

Library of Congress Cataloging-in-Publication Data

Names: Ge, M. L. (Mo-Lin), editor. | Cai, Rong-Gen (Physics professor),
    editor. | Liu, Yu-Xiao, editor. | Duan, Yi-Shi, 1927–2016, honouree.
Title: Memorial volume for Yi-Shi Duan / edited by, Mo-Lin Ge (Chern
    Institute of Mathematics, China), Rong-Gen Cai (Chinese Academy of
    Sciences, China), Yu-Xiao Liu (Lanzhou University, China).
Description: Singapore ; Hackensack, NJ : World Scientific Publishing Co. Pte. Ltd., [2018]
Identifiers: LCCN 2017058721| ISBN 9789813237261 (hardcover ; alk. paper) |
    ISBN 9813237260 (hardcover ; alk. paper)
Subjects: LCSH: Gauge fields (Physics) | General relativity (Physics) |
    Duan, Yi-Shi, 1927–2016.
Classification: LCC QC793.3.G38 M46 2018 | DDC 530.14/3--dc23
LC record available at https://lccn.loc.gov/2017058721

British Library Cataloguing-in-Publication Data
A catalogue record for this book is available from the British Library.

For any available supplementary material, please visit
http://www.worldscientific.com/worldscibooks/10.1142/10911#t=suppl

Printed in Singapore

Preface

One of the world renowned pioneers in the study of gauge field theory and general relativity in China, Professor Yi-Shi Duan of Lanzhou University, passed away peacefully in December 2016. Trained in former Soviet Union, Prof. Duan returned to China in 1957 to work in Lanzhou University for 60 years. In 1963, he came up with a general co-variant form of conservation law of energy-momentum tensor in general relativity. In 1979, he suggested that the gauge potential could be decomposed, which has important implications to gauge field theory. He trained a big team of talents in theoretical physics in China. His contributions to theoretical physics in China have earned him praise from both Professor S. S. Chern and Professor C. N. Yang.

In July 2016, friends and students organized a conference at Lanzhou University to celebrate his 88th birthday. With his passing, we are very grateful to Professor K. K. Phua, who has encouraged us to publish a memorial volume for Professor Yi-Shi Duan. With this initiative, I also like to take this opportunity to thank Professor Rong-Gen Cai of the Institute of Theoretical Physics at Chinese Academy of Sciences, and colleagues at Lanzhou University for their great efforts to put this volume together, without them, this testimonial of Yi-Shi Duan will not be possible. I wish Prof. Duan would be consoled when he read this volume in another spacetime.

Mo-Lin Ge
Professor at Nankai University
20 October 2017

Photos of Yi-Shi Duan

Professor Yi-Shi Duan (July 1927 – December 2016)

Early childhood in Beijing (1935)

At Moscow State University (1954)

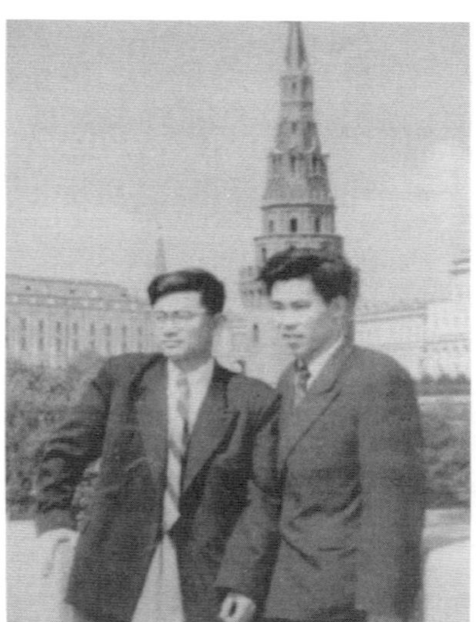

With Kai-Hua Zhao in Moscow (1955)

In a talk (1955)

With talents from all over the world in Moscow State University (1955)

With advisor Prof. Shirokov (1956)

With wife You-Mei Huang (1956)

At Joint Institute for Nuclear Research (JINR), Dubna (1956)

At shoot range of Lanzhou (1971)

In discussion with his student Mo-Lin Ge on gauge field theory in Lanzhou （1974）

With Prof. Chen-Ning Yang and others in Lanzhou University （1977）

During sports meeting of staffs in Lanzhou University　(1980)

Visiting University of Missouri　(1980)

Visiting Stanford University (1983)

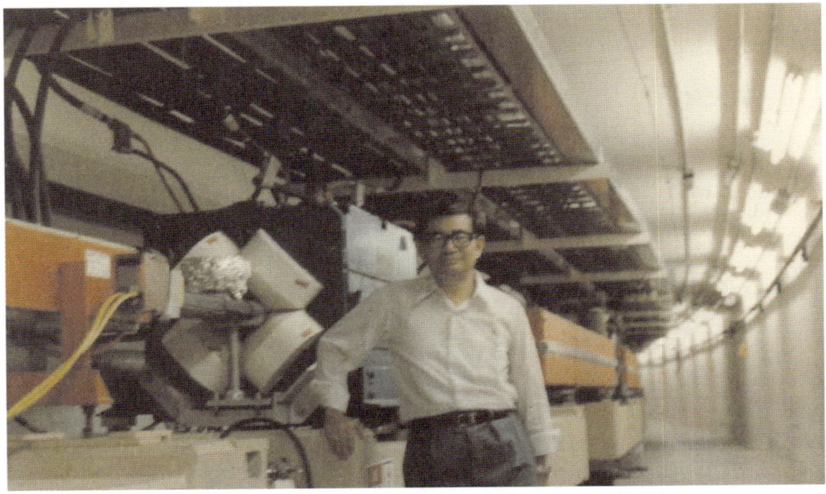

Visiting SLAC National Accelerator Laboratory (1983)

With Prof. Sidney Drell, director of SLAC, at Stanford University （1984）

During the 11th John Hopkins Workshop on Current Problems in Particle Theory in Lanzhou
（1987）

In the library of physics department, Lanzhou University　(1988)

Visiting Moscow State University　(1993)

Visiting Rome, Italy （1994）

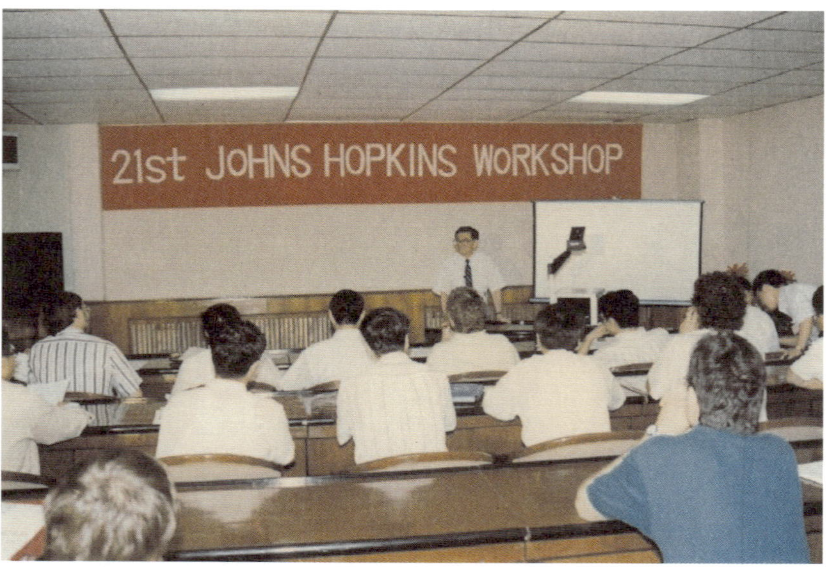

During 21st John Hopkins Workshop in Lanzhou （1997）

Prof. Yi-Shi Duan's 70th birthday celebration at Lanzhou University (1997)

With graduate students in Lanzhou (1998)

With wife in Paris（2001）

At work（2002）

With Prof. Shiing-Shen Chern at Nankai University （2002）

During the International Conference on String Theory （Beijing 2002）

In a talk on topological current theory
at Lanzhou University （2004）

Visit Sichuan University (2005)

With wife in Chengdu （2005）

In a talk of Lecture Series of World Year of Physics at Lanzhou University (2005)

With Nobel Laureate Samuel Chao Chung Ting in Lanzhou (2006)

With Prof. Liao Liu in Lanzhou (2006)

With Prof. Andrew Strominger from Harvard University (2006)

With students at Xiamen University （2006）

With Nobel Laureate David Gross （2007）

Prof. Yi-Shi Duan's 80th birthday celebration at Lanzhou University（2007）

In the live interview of Junyi Shui during the Centennial Celebration of Lanzhou University
（2009）

Workshop on the Frontiers of Theoretical Physics
— In celebration of Prof. Duan in teaching for 65 years (2016)

Contents

Chapter 1

SU(2) Gauge Theory and Electrodynamics with N Magnetic Monopoles

Yi-Shi Duan and Mo-Lin Ge

Lanzhou University, Lanzhou 730000, China

Abstract: We in brief present a viewpoint that the gauge potential B_μ can be decomposed and possesses the inner structure. We point out that the B_μ can be decomposed into two parts: b_μ and Γ_μ, where the b_μ satisfies the adjoint transformation corresponding to a massive vector particle and Γ_μ satisfies the gauge transformation can be composed with other fundamental field. In $SU(2)$ gauge theory, the fundamental field is Higgs field $\phi(x)$. With this decomposition the electrodynamics of N magnetic monopoles system is obtained. In terms of the topological conservation current of magnetic monopoles, one can find that the quantization condition of magnetic charge is an obvious result in our theory. Furthermore, we prove the worldline of zero of $\phi(x)$ corresponding to the trajectory of monopoles, i.e., the electrodynamics of monopoles is described by the property of zeroes of $\phi(x)$. At last, we obtain the sourceless solution of $SU(2)$ gauge field for the N monopoles system, and the Wu–Yang potential is generalized to the N monopoles system.

Note: Originally published in *Scientia Sinica Mathematica* in Chinese.[1]

1.1 Introduction

Yang–Mills field [1], i.e. the $SU(2)$ gauge field, is proposed over twenty years. But with the traditional viewpoints, the gauge potential is always considered as an impartible whole. We think this viewpoint hampered the deep understanding for the essence of Yang–Mills theory and limited the application of gauge field theory in a certain extent.

In fact, for the gauge theory with arbitrary group G, the gauge potential can always be decomposed into two parts b_μ and Γ_μ. The b_μ part satisfies the adjoint transformation of group G, which can be considered as a kind of field with massive vector particle. It is obvious that this kind of mass is independent of Higgs mechanics and symmetry spontaneous broken theory [2]. Γ_μ satisfies the gauge

---

[1] Translated with permission from *Sci. Sinica* **9**(11) (1979) 1072–1081. ©Science China Press.

transformation and possesses the inner structure, i.e., Γ_μ can be composed with other fundamental field. The inner structure reveals the more profound physical meaning of gauge theory. General relativity is such an example. This paper, as well the other our work about magnetic monopole [3, 4, 5, 6], are all based on the inner structure of Γ_μ.

1.2 Decomposition of gauge potential and its inner structure

In the non-Abelian gauge theory with group G, the gauge field

$$G_{\mu\nu} = \partial_\mu B_\nu - \partial_\nu B_\mu - [B_\mu, B_\nu], \tag{1.1}$$

and the covariant derivative of field $\psi(x)$

$$\nabla_\mu \psi = (\partial_\mu - B_\mu)\psi. \tag{1.2}$$

The gauge field $G_{\mu\nu}$ and the covariant derivative $\nabla_\mu \psi$ satisfy

$$G'_{\mu\nu} = SG_{\mu\nu}S^{-1} \tag{1.3}$$

and

$$\nabla'_\mu \psi' = S \, \nabla_\mu \psi, \tag{1.4}$$

$$\psi(x) = S \, \psi(x), \tag{1.5}$$

The gauge potential B_μ satisfies the gauge transformation

$$B'_\mu = SB_\mu S^{-1} + (\partial_\mu S)\left(S^{-1}\right), \tag{1.6}$$

where $B_\mu = B_\mu^a I_a$, I_a is the generator. In the fibre bundle theory, B_μ and $G_{\mu\nu}$ is the connection and curvature of principal fibre bundle $P(M, G)$ with the 4-dimensional spacetime manifold M and structure group G.

The gauge potential B_μ which satisfies the transformation (1.6) can always be decomposed into two parts

$$B_\mu = \Gamma_\mu + b_\mu, \tag{1.7}$$

where the Γ_μ following the same transformation as (1.6)

$$\begin{aligned} \Gamma'_\mu &= S\Gamma_\mu S^{-1} + (\partial_\mu S)S^{-1}, \\ \Gamma_\mu &= \Gamma_\mu^a I_a. \end{aligned} \tag{1.8}$$

But b_μ satisfies the adjoint transformation

$$\begin{aligned} b'_\mu &= Sb_\mu S^{-1}, \\ b_\mu &= b_\mu^a I_a. \end{aligned} \tag{1.9}$$

It is obvious that $\Gamma_\mu + b_\mu$ still satisfies the gauge transformation (1.6). With Eq.(1.9) we find that b_μ is not only a vector in the 4-dimensional spacetime, but also a gauge covariant quantity. As a gauge invariance, $\mathrm{Tr}[b_\mu b_\nu]$ can be considered as a mass term $M^2 \mathrm{Tr}[b_\mu b_\nu]$. So with the decomposition of gauge potential, we can obtain naturally some massive vector particles with b_μ, without the Higgs mechanics and symmetry spontaneous broken theory. For the different group G, b_μ can be intermediate boson, gluon or other new vector particles, which enlarge the application of gauge field theory.

Furthermore, the Γ_μ part in the gauge potential is not definitely an impartible whole. It can be composed by one or several fundamental fields, i.e. Γ_μ is still have its inner structure. If we suppose the fundamental fields are ϕ^A, χ^A, $\cdots$, the inner structure of Γ_μ can be

$$\Gamma_\mu = \Gamma_\mu \left(\phi^A, \chi^A, \cdots, \partial_\mu \phi^A, \partial_\lambda \chi^A, \cdots\right). \tag{1.10}$$

For the gauge group G, these fundamental field ϕ^A, χ^A, $\cdots$ possess their transformation. The above Γ_μ should satisfies the transformation (1.8), and such inner structure should have the concrete physical meaning (or some special differential geometry).

For the gauge theory with $GL(4)$, the gauge potential (connection) B_μ can be decomposed and have the inner structure, which is well known. With the generator I_λ^ν of $GL(4)$, gauge potential $B_\mu = B_{\mu\nu}^\lambda I_\lambda^\nu$, can be decomposed

$$\Gamma_\mu = \Gamma_{\mu\nu}^\lambda I_\lambda^\nu, \quad b_\mu = \Lambda_{\mu\nu}^\lambda I_\lambda^\nu,$$

where $\Lambda_{\mu\nu}^\lambda$ is a strict tensor corresponding to the torsion, and $\Gamma_{\mu\nu}^\lambda$ satisfies (1.8). If we suppose the $\Gamma_{\mu\nu}^\lambda$ possesses the inner structure and its fundamental field ϕ^A is taken as metric field $g_{\mu\nu}$, we can obtain the well-known result

$$\Gamma_{\mu\nu}^\lambda = \frac{1}{2} g^{\lambda\sigma} \left(\partial_\mu g_{\nu\sigma} + \partial_\nu g_{\mu\sigma} - \partial_\sigma g_{\mu\nu}\right)$$

with the condition $\nabla_\lambda g_{\mu\nu} = 0$ and $\Gamma_{\mu\nu}^\lambda = \Gamma_{\nu\mu}^\lambda$.

Once we obtain the inner structure, with Eq. (1.1) and (1.2), the 4-dimensional Riemannian geometry can be derived from the torsion-free $GL(4)$ gauge theory. From this viewpoint, general relativity has close relationship with the fact that the $GL(4)$ gauge potential possesses its inner structure, and the introduction of fundamental field $g_{\mu\nu}$ is necessary to describe the gravity field and inertial force field.

For the $SU(2)$ gauge theory with the electrical charge e as the coupling constant, let gauge potential be $\mathbf{B}_\mu = e\mathbf{W}_\mu$, and suppose the fundamental field is the vector field $\phi(x)$ in the isotropic space (i.e. Higgs field) and electromagnetic potential A_μ, and let $\mathbf{n}(x)$ is the direct vector of $\phi(x)$

$$\begin{aligned} \mathbf{n}(x) &= \phi(x)/\phi(x), \\ \phi(x) &= |\phi(x)|. \end{aligned} \tag{1.11}$$

Let electromagnetic potential $A_\mu(x)$ is the component of $\mathbf{W}_\mu(x)$ along the $\mathbf{n}(x)$, and the covariant derivative of $\mathbf{n}(x)$

$$\nabla_\mu \mathbf{n}(x) = \partial_\mu \mathbf{n}(x) + e\mathbf{W}_\mu(x) \times \mathbf{n}(x), \tag{1.12}$$

then the decomposition of $\mathbf{W}_\mu$ is

$$\mathbf{W}_\mu = \mathbf{\Gamma}_\mu + \mathbf{b}_\mu, \tag{1.13}$$

where

$$\mathbf{\Gamma}_\mu = A_\mu \mathbf{n} + \frac{1}{e}\partial_\mu \mathbf{n} \times \mathbf{n}, \tag{1.14}$$

$$\mathbf{b}_\mu = -\frac{1}{e}\nabla_\mu \mathbf{n} \times \mathbf{n}, \tag{1.15}$$

$$A_\mu = \mathbf{W}_\mu \cdot \mathbf{n}. \tag{1.16}$$

The above decomposition of $SU(2)$ gauge potential and its inner structure is based that the Γ_μ part should describe the electromagnetic theory including magnetic monopoles and the existence of monopole has the close relation with the fundamental field $\phi(x)$. And since b_μ is perpendicular to n, it should corresponding to some electrical vector particle, for example the boson for weak interaction or a new vector particle.

1.3 Electrodynamics of magnetic monopoles and electrical particles system in the SU(2) gauge theory

Without magnetic monopoles, the usual electromagnetic tensor $f_{\mu\nu}$ satisfies the Maxwell equations

$$\partial_\nu f_{\mu\nu} = 4\pi j_\mu, \tag{1.17}$$

$$\partial_\nu {}^* f_{\mu\nu} = 0, \tag{1.18}$$

where

$$\begin{aligned} f_{\mu\nu} &= \partial_\mu A_\nu - \partial_\nu A_\mu, \\ {}^* f_{\mu\nu} &= \frac{1}{2}\varepsilon_{\mu\nu\lambda\rho} f_{\lambda\rho}, \end{aligned} \tag{1.19}$$

and j_μ is the electrical density of electrical particles. With the Lorentz gauge condition $\partial_\mu A_\mu = 0$, A_μ satisfies

$$\nabla^2 A_\mu = -4\pi j_\mu. \tag{1.20}$$

Then except the sourceless electromagnetic potential, A_μ can be determined by $j_\mu(x)$, for example the retarded potential is

$$A_\mu(x) = \int G(x - x') j_\mu(x') d^4 x', \tag{1.21}$$

where $G(x - x')$ is the retarded Green function

$$\nabla^2 G(x - x') = -4\pi\delta(x - x'). \tag{1.22}$$

When there exist the monopoles in the system, usually we add the magnetic current density in the rhs of Eq. (1.18). But since A_μ is determined by j_μ with Eq. (1.17), the electromagnetic potential A_μ can't include the monopole if there is only the electrical current density j_μ described the electrical particle system in the rhs of (1.17). So we should add the effective electrical current in the rhs of (1.17) induced by the movement of monopoles. Let $K_{\mu\nu}$ is the electromagnetic field induced by monopoles, the effective current is $(1/4\pi)\partial_\nu K_{\mu\nu}$. Then the Eq. (1.17) should be revised as

$$\partial_\nu f_{\mu\nu} = 4\pi j_\mu + \partial_\nu K_{\mu\nu}. \tag{1.23}$$

Now the electromagnetic field $f_{\mu\nu}$ and potential A_μ determined by Eq. (1.23) and (1.19) is that of real electrical current and effective current. The equation $\partial_\nu {}^* f_{\mu\nu} = 0$ is still kept.

Since $K_{\mu\nu}$ is the electromagnetic field induced by the movement of monopoles system, let ${}^* j_\mu$ be the magnetic current density, then $K_{\mu\nu}$ satisfied

$$\partial_\nu {}^* K_{\mu\nu} = -4\pi \, {}^* j_\mu, \tag{1.24}$$

where

$${}^* K_{\mu\nu} = \frac{1}{2}\varepsilon_{\mu\nu\lambda\rho} K_{\lambda\rho}.$$

Defining a new electromagnetic tensor

$$F_{\mu\nu} = f_{\mu\nu} - K_{\mu\nu}, \tag{1.25}$$

or

$$F_{\mu\nu} = (\partial_\mu A_\nu - \partial_\nu A_\mu) - K_{\mu\nu}. \tag{1.26}$$

with $\partial_\nu {}^* f_{\mu\nu} = 0$, we can obtain two symmetrical Maxwell equations with Eqs. (1.23) and (1.24)

$$\text{I.} \qquad \partial_\nu F_{\mu\nu} = 4\pi j_\mu, \tag{1.27}$$

$$\text{II.} \qquad \partial_\nu {}^* F_{\mu\nu} = 4\pi \, {}^* j_\mu. \tag{1.28}$$

We call this new electromagnetic tensor $F_{\mu\nu}$ which satisfies the above new Maxwell equations as physical electromagnetic tensor with magnetic monopoles, which can be derived strictly from the invariance of $SU(2)$ gauge theory with couple constant e. In the usual electromagnetic theory, the electromagnetic tensor is $U(1)$ gauge invariance. Generalizing this idea, we suppose $F_{\mu\nu}$ is a $SU(2)$ gauge invariance. It is not difficult to find out this invariance. With (1.26) one can recognize this question is equivalent to find an antisymmetrical tensor $K_{\mu\nu}$.

With the decomposition of $SU(2)$ gauge potential (1.13), the $SU(2)$ gauge field is

$$\mathbf{G}_{\mu\nu} = \partial_\mu \mathbf{W}_\nu - \partial_\nu \mathbf{W}_\mu + e\mathbf{W}_\mu \times \mathbf{W}_\nu, \tag{1.29}$$

then

$$\mathbf{G}_{\mu\nu} \cdot \mathbf{n} = (\partial_\mu A_\nu - \partial\nu A_\mu) - \frac{1}{e}\mathbf{n} \cdot (\partial_\mu \mathbf{n} \times \partial_\nu \mathbf{n}) + \frac{1}{e}\mathbf{n} \cdot (\nabla_\mu \mathbf{n} \times \nabla_\nu \mathbf{n}), \tag{1.30}$$

where the lhs and third term of rhs are $SU(2)$ invariance. Thus the difference between $(\partial_\mu A_\nu - \partial_\nu A_\mu)$ and $\frac{1}{e}\mathbf{n} \cdot (\partial_\mu \mathbf{n} \times \partial_\nu \mathbf{n})$ is also a $SU(2)$ gauge invariance, i.e.

$$K_{\mu\nu} = \frac{1}{e}\mathbf{n} \cdot (\partial_\mu \mathbf{n} \times \partial_\nu \mathbf{n}), \tag{1.31}$$

and

$$F_{\mu\nu} = (\partial_\mu A_\nu - \partial_\nu A_\mu) - \frac{1}{e}\mathbf{n} \cdot (\partial_\mu \mathbf{n} \times \partial_\nu \mathbf{n}). \tag{1.32}$$

With (1.30)

$$F_{\mu\nu} = \mathbf{G}_{\mu\nu} \cdot \mathbf{n} - \frac{1}{e}\mathbf{n} \cdot (\nabla_\mu \mathbf{n} \times \nabla_\nu \mathbf{n}), \tag{1.33}$$

this is just the physical electromagnetic tensor defined by 't Hooft [7].

In the following we discuss the equation of motion of monopoles and electrical particles. In the system without the electrical vector particles b_μ, we have $b_\mu = 0$, i.e. $\nabla_\mu \mathbf{n} = 0$. With Eq. (1.33) the physical electromagnetic tensor

$$F_{\mu\nu} = \mathbf{G}_{\mu\nu} \cdot \mathbf{n}. \tag{1.34}$$

On the other hand, with

$$[\nabla_\mu \nabla_\nu - \nabla_\nu \nabla_\mu]\,\mathbf{n} = e\mathbf{G}_{\mu\nu} \times \mathbf{n}, \tag{1.35}$$

it is clear that $\mathbf{G}_{\mu\nu}$ is parallel to $\mathbf{n}$ with the condition $\nabla_\mu \mathbf{n} = 0$. Thus with (1.34) $G_{\mu\nu}$ is

$$\mathbf{G}_{\mu\nu} = F_{\mu\nu}\mathbf{n}. \tag{1.36}$$

Furthermore, with $\nabla_\mu \mathbf{n} = 0$ one can find

$$\nabla_\nu \mathbf{G}_{\mu\nu} = (\partial_\nu F_{\mu\nu})\,\mathbf{n}, \tag{1.37}$$

$$\nabla_\nu {}^*\mathbf{G}_{\mu\nu} = (\partial_\nu {}^*F_{\mu\nu})\,\mathbf{n}, \tag{1.38}$$

where

$${}^*\mathbf{G}_{\mu\nu} = \frac{1}{2}\varepsilon_{\mu\nu\lambda\rho}\mathbf{G}_{\lambda\rho}.$$

Substituting Maxwell Eqs. I and II into Eq. (1.37) and (1.38), we arrive the two $SU(2)$ gauge field equations:

$$\nabla_\nu \mathbf{G}_{\mu\nu} = (4\pi j_\mu)\,\mathbf{n},$$
$$\nabla_\nu {}^*\mathbf{G}_{\mu\nu} = (4\pi\,{}^*j_\mu)\,\mathbf{n}. \tag{1.39}$$

So the source of two gauge field equations are determined by j_μ and ${}^*j_\mu$, i.e.,

$$\mathbf{J} = j_\mu\mathbf{n}, \quad {}^*\mathbf{J}_\mu = j_\mu\mathbf{n}.$$

For the system possesses N point-like monopoles and N' point-like electrical particles, ${}^*j_\mu$ and j_μ and energy-momentum tensor $t^g_{\mu\nu}$ and $t^q_{\mu\nu}$ can be expressed as

$$^*j_\mu(x) = \sum_{i=1}^{N} g_i \int \frac{dz_i^\mu}{dS}\delta\,(x - z_i(s))\,dS, \tag{1.40}$$

$$j_\mu(x) = \sum_{i=1}^{N'} q_i \int \frac{d\bar{z}_i^\mu}{dS}\delta\,(x - \bar{z}_i(s))\,dS, \tag{1.41}$$

$$t^g_{\mu\nu}(x) = M\sum_{i=1}^{N} \int \frac{dz_i^\mu}{dS}\frac{dz_i^\nu}{dS}\delta\,(x - z_i(s))\,dS, \tag{1.42}$$

$$t^q_{\mu\nu}(x) = m\sum_{i=1}^{N} \int \frac{d\bar{z}_i^\mu}{dS}\frac{d\bar{z}_i^\nu}{dS}\delta\,(x - \bar{z}_i(s))\,dS, \tag{1.43}$$

where g_i is the magnetic charge of i-th monopole and its worldline $x^\mu = z_i^\mu(s)$, and $q_i = \pm e$ is the electrical charge of i-th point-like electrical particle and its worldline $x^\mu = \bar{z}_i^\mu(s)$. M and m are their mass, respectively. With Eqs. (1.42) and (1.43) one can find

$$\partial_\nu t_{\mu\nu}^g = M \sum_{i=1}^{N} \int \frac{d^2 z_i^\mu}{dS^2} \delta\left(x - z_i(s)\right) dS, \tag{1.44}$$

$$\partial_\nu t_{\mu\nu}^q = m \sum_{i=1}^{N} \int \frac{d^2 \bar{z}_i^\mu}{dS^2} \delta\left(x - \bar{z}_i(s)\right) dS. \tag{1.45}$$

The energy-momentum tensor of $SU(2)$ gauge field is

$$T_{\mu\nu} = \frac{1}{4\pi}\left[\mathbf{G}_{\mu\lambda} \cdot \mathbf{G}_{\nu\lambda} - \frac{1}{4}\mathbf{G}_{\lambda\sigma} \cdot \mathbf{G}_{\lambda\sigma}\delta_{\mu\nu}\right], \tag{1.46}$$

and

$$\partial_\nu T_{\mu\nu} = \frac{1}{4\pi}\left[\mathbf{G}_{\lambda\mu} \cdot \nabla_\nu \mathbf{G}_{\lambda\nu} + {}^*\mathbf{G}_{\lambda\mu} \cdot \nabla_\nu {}^*\mathbf{G}_{\lambda\nu}\right]. \tag{1.47}$$

Substituting Eqs. (1.36) and (1.39) into the above expression, we obtain

$$\partial_\nu T_{\mu\nu} = \frac{1}{4\pi}\left[F_{\lambda\mu}\partial_\nu F_{\lambda\nu} + {}^*F_{\lambda\nu}\partial_\nu {}^*F_{\lambda\nu}\right] = F_{\lambda\mu}j_\lambda + {}^*F_{\lambda\mu}{}^*j_\lambda. \tag{1.48}$$

For the total system of magnetic monopoles, electrical particles and gauge field $G_{\mu\nu}$, the energy-momentum should be conservative, i.e.

$$\partial_\nu \left[t_{\mu\nu}^g + t_{\mu\nu}^q + T_{\mu\nu}\right] = 0, \tag{1.49}$$

or

$$\partial_\nu \left[t_{\mu\nu}^g + t_{\mu\nu}^q\right] = -\left[F_{\lambda\mu}j_\lambda + {}^*F_{\lambda\mu}{}^*j_\lambda\right]. \tag{1.50}$$

Substituting Eqs. (1.40), (1.41), (1.44) and (1.45) into the above expression and comparing the singular points of magnetic monopole and electrical particles, we get the equation of motion for the monopoles and electrical particles,

$$M\frac{d^2 z_i^\mu}{dS^2} = -g_i \frac{dz^\lambda}{dS} {}^*F_{\lambda\mu}, \tag{1.51}$$

$$m\frac{d^2 \bar{z}_i^\mu}{dS^2} = -q_i \frac{d\bar{z}^\lambda}{dS} F_{\lambda\mu}. \tag{1.52}$$

The Maxwell equations are

$$\partial_\nu F_{\mu\nu} = 4\pi \sum_{i=1}^{N'} q_i \int \frac{d\bar{z}_i^\mu}{dS} \delta\left(x - \bar{z}_i(s)\right) dS, \tag{1.53}$$

$$\partial_\nu {}^*F_{\mu\nu} = 4\pi \sum_{i=1}^{N'} q_i \int \frac{dz_i^\mu}{dS} \delta\left(x - z_i s\right) dS. \tag{1.54}$$

The four the above equations are the basic equations of electrodynamics system with N magnetic monopoles and N' electrical particles. The difference of these equation from the usual $U(1)$ monopole theory [8, 9] is the physical electromagnetic tensor $F_{\mu\nu}$ defined by (1.32). These equations are derived from the $SU(2)$ gauge theory, more precise, are the $U(1)$ electrodynamics based on the $SU(2)$ gauge theory.

1.4   Features of the solution of monopoles' electrodynamical system and topological conservative current

For the system with N point-like magnetic monopoles, the magnetic current density is

$$j_\mu(x) = \sum_i g_i \int \frac{dz_i^\mu}{dS} \delta^4\left(x - z_i(s)\right) dS. \tag{1.55}$$

In the following we consider the solution of the Maxwell equation II

$$\partial_\nu {}^*K_{\mu\nu} = -4\pi \sum_i g_i \int \frac{dz_i^\mu}{dS} \delta^4\left(x - z_i(s)\right) dS. \tag{1.56}$$

With Eqs. (1.31) and (1.11), ${}^*K_{\mu\mu}$ can be expressed with fundamental field $\phi^a (a = 1, 2, 3)$

$$\begin{aligned}
{}^*K_{\mu\nu} &= \frac{1}{2e} \varepsilon_{\mu\nu\lambda\rho} \mathbf{n} \cdot (\partial_\lambda \mathbf{n} \times \partial_\rho \mathbf{n}) \\
&= \frac{1}{2e} \varepsilon_{\mu\nu\lambda\rho} \varepsilon_{abc} \frac{\phi^a}{\phi^3} \partial_\lambda \phi^b \partial_\rho \phi^c.
\end{aligned} \tag{1.57}$$

Suppose the ϕ^a is smooth, one can prove [5]

$$\begin{aligned}
\partial_\nu {}^*K_{\mu\nu} &= \frac{1}{2e} \varepsilon_{\mu\nu\lambda\rho} \varepsilon_{abc} \partial_\nu \left[\frac{\phi^a}{\phi^3} \partial_\lambda \phi^b \partial_\rho \phi^c \right] \\
&= \frac{1}{2e} \varepsilon_{\mu\nu\lambda\rho} \varepsilon_{abc} \left(\frac{\partial}{\partial \phi^l} \frac{\phi^a}{\phi^3} \right) \partial_\nu \phi^l \partial_\lambda \phi^b \partial_\rho \phi^c.
\end{aligned} \tag{1.58}$$

The vector Jacobi $D_\mu(\phi/x)$ $(\mu = 0, 1, 2, 3)$ can be defined as

$$\varepsilon_{lbc} D_\mu \left(\frac{\phi}{x} \right) = -\varepsilon_{\mu\nu\lambda\rho} \partial_\nu \phi^l \partial_\lambda \phi^b \partial_\rho \phi^c, \tag{1.59}$$

in which

$$D_0 \left(\frac{\phi}{x} \right) = -D \left(\frac{\phi}{x} \right), \tag{1.60}$$

$$D \left(\frac{\phi}{x} \right) = \frac{\partial(\phi^1, \phi^2, \phi^3)}{\partial(x^1, x^2, x^3)}, \tag{1.61}$$

is the usual Jacobi. Substituting (1.59) into (1.58), and with $\varepsilon_{abc}\varepsilon_{lbc} = 2\delta_{la}$ one can find

$$\partial_\nu {}^*K_{\mu\nu} = \frac{1}{e} D_\mu \left(\frac{\phi}{x} \right) \frac{\partial}{\partial \phi^a} \left[\frac{\phi^a}{\phi^3} \right] = -\frac{1}{e} D_\mu \left(\frac{\phi}{x} \right) \triangle_\phi \left(\frac{1}{\phi} \right), \tag{1.62}$$

where

$$\triangle_\phi = \frac{\partial^2}{\partial \phi^a \partial \phi^a}$$

is the Laplace operator of isospin space. According to the generalized function theory, we have

$$\triangle_\phi \left(\frac{1}{\phi} \right) = -4\pi \delta(\phi), \tag{1.63}$$

which is just the "Coulomb theorem" in the isospin space. With Eqs. (1.62) and (1.63) one obtain

$$\partial_\nu {}^*K_{\mu\nu} = \frac{4\pi}{e} D_\mu \left(\frac{\phi}{x}\right) \delta(\phi(x)), \tag{1.64}$$

which means that the zeroes of $\phi(x)$ is the singularities of function $\delta(\phi(x))$.

Suppose there are N moving isolated zeroes of $\phi(x)$, let worldline of i-th zero be

$$x^\mu = z_i^\mu(s), \quad i = 1, 2, \cdots, N, \tag{1.65}$$

i.e. $z_i^\mu(s)$ is the solution of equation

$$\phi(z_i(s)) = 0, \quad i = 1, 2, \cdots, N. \tag{1.66}$$

Thus $\delta(\phi(x))$ can be expressed as

$$\delta(\phi(x)) = \frac{1}{|D(\phi/x)|} \sum_{i=1}^{N} \beta_i \delta(\mathbf{x} - \mathbf{z}_i(s)), \tag{1.67}$$

where β_i is the Hopf index. Under the condition $D(\phi/x) \neq 0$, with (1.66) it is easy to prove along the worldline of i-th zero

$$\frac{dz_i^\mu}{dS} = - \left[\frac{D_\mu(\phi/x)}{D(\phi/x)}\right]_{x=z_i} \frac{dz_i^0}{dS}, \quad \mu = 0, 1, 2, 3,$$

or

$$D_\mu \left(\frac{\phi}{x}\right)\bigg|_{x=z_i} \frac{dz_i^0}{dS} = -D \left(\frac{\phi}{x}\right)\bigg|_{x=z_i} \frac{dz_i^\mu}{dS}. \tag{1.68}$$

With Eqs. (1.67) and (1.68), we can get the exact form of magnetic current density with N monopoles

$$\partial_\nu {}^*K_{\mu\nu} = -\frac{4\pi}{e} \sum_{i=1}^{N} \eta_i \beta_i \int \frac{dz_i^\mu}{dS} \delta^4(x - z_i(s)) dS, \tag{1.69}$$

where

$$\eta_i = [D(\phi/x)/|D(\phi/x)|]_{x=z_i}. \tag{1.70}$$

Comparing the Eqs. (1.69) and (1.56), we can find that there exist solution of Maxwell equation II under the quantization condition of magnetic charge

$$g_i = n_i \frac{1}{e}, \quad n_i = \eta_i \beta_i, \tag{1.71}$$

where n_i is an integer. From the above discussion we arrive the following conclusion:

1. Zeroes of fundamental field $\phi(x)$ is the source of magnetic monopole. The trajectory of monopole is the worldline $x^\mu = z_i^\mu(s)(i = 1, 2, \cdots, N)$ of zero and the equation of motion (1.51) of monopole is just the equation of zeroes of $\phi(x)$.

2. Electromagnetic field $K_{\mu\nu}$ induced by N monopole is determined by fundamental field $\phi(x)$. The electrodynamics of monopole is described by zeroes of $\phi(x)$.

3. The absolute value of magnetic charge must be the integer multiple of dual charge $1/e$. The integer is corresponding to the Hopf index β_i of zero of $\phi(x)$. The basic unit of magnetic charge is $1/e$ (electrical charge e).

4. The sign of magnetic charge is decided by η_i, i.e., the sign of Jocabi $D(\phi/x)$ at the zero of $\phi(x)$.

According to the quantization condition of magnetic charge (1.71), the magnetic current density (1.40) is fixed to the following form

$$^*j_\mu(x) = \frac{1}{e}\sum_{i=1}^{N} n_i \int \frac{dz_i^\mu}{dS}\delta^4(x - z_i(s))dS, \tag{1.72}$$

and the magnetic charge density $^*j_0(x) = \rho(x)$,

$$\rho(x) = \frac{1}{e}\sum_{i=1}^{N} n_i\delta(x - z_i(s)). \tag{1.73}$$

With Eqs. (1.64) and (1.24), the magnetic current density can be expressed with fundamental field $\phi(x)$

$$^*j_\mu(x) = -\frac{1}{e}D_\mu\left(\frac{\phi}{x}\right)\delta(\phi(x)), \tag{1.74}$$

$$\rho(x) = \frac{1}{e}D\left(\frac{\phi}{x}\right)\delta(\phi(x)). \tag{1.75}$$

Furthermore, with the above equation and (1.73) the total magnetic charge can be expressed as an integral in the isospin space

$$G = \int_V \rho(x)dV = \frac{1}{e}\int_T \delta(\phi)(d\phi) = \frac{n}{e}, \tag{1.76}$$

$$n = \sum_{i=1}^{N} n_i = n^+ - n^-, \tag{1.77}$$

where n^+ and n^- are the positive and negative magnetic charges in V. The integral of magnetic charge indicates that there are $k = n^+ + n^-$ coverings when the set in the usual 3-dimension space V is mapped into the isospin space T by function $\phi(x)$. The n^+ and n^- coverings are corresponding to the different zeroes of $\phi(x)$ where $D(\phi/x) > 0$ and $D(\phi/x) < 0$, respectively. The topological property can be obtained in the case of $n^+ = 0$ or $n^- = 0$.

Furthermore from the integral of G one can find the magnetic current density in the isospin space

$$\rho_\phi = \frac{1}{e}\delta(\phi), \tag{1.78}$$

which means that the N moving magnetic monopoles in the usual spacetime correspond to a static magnetic monopole at $\phi = 0$ in the isospin space.

We have to mention an important characteristic of $^*K_{\mu\nu}$ with the above discussion. When the fundamental field $\phi(x)$ possesses N moving zeroes, the divergence of $^*K_{\mu\nu}$ given in Eq. (1.57) can be expressed as the form (1.69) of a 4-dimensional magnetic current density by the zero of $\phi(x)$. The system of N moving magnetic monopoles can be described with this magnetic current density. And one can obtain the quantization of magnetic charge in terms of the topological property of the magnetic current density. Thus we call this conservative current as a topological conservative current. This is an important result based on the inner structure of $SU(2)$ gauge theory with the fundamental field $\phi(x)$.

The electrodynamics of zeroes of $\phi(x)$ can be found with the above topological conservative current and Eqs. (1.51)-(1.54). However, in Refs. [3, 4, 14] only the quantization of magnetic charge was discussed with the topological property defined in those papers.

1.5 Sourceless solution of SU(2) gauge field and magnetic monopole

For the case of $\nabla_\mu \mathbf{n} = 0$, with Eqs. (1.37) and (1.39), the sourceless equations are $\partial_\nu F_{\mu\nu} = 0$ and (1.69), which means the sourceless solution of $SU(2)$ gauge field corresponding to the existence of magnetic monopole. Imposing the Lorentz gauge condition in the (1.55), one can find the retarded solution

$$A_\mu(x) = \int G(x - x')\partial_\nu K_{\mu\nu}(x')d^4x'. \tag{1.79}$$

Substituting the above solution into decomposition (1.13) of W_μ, we obtain the formal solution of sourceless equation with the condition $\nabla_\mu \mathbf{n} = 0$

$$W_\mu(x) = \frac{\phi^a(x)}{\phi(x)} \int G(x - x')\partial_\nu K_{\mu\nu}(x')d^4x' - \frac{1}{e}\varepsilon_{abc}\frac{1}{\phi^2}\phi^b(x)\partial_\mu\phi^c(x). \tag{1.80}$$

For the $\phi(x)$ with N zeroes, the above expression determines the system of N magnetic monopoles. Let consider a special case: magnetic charge density $\rho(r) = (1/e)\delta(r)$ and $\phi(x) = r$. Obviously we have $\partial_\nu K_{\mu\nu} = 0$, the sourceless solution of a static magnetic monopole is

$$W_\mu^a(x) = -\frac{1}{e}\varepsilon_{ab\mu}\frac{x^b}{r^2}, \tag{1.81}$$

which is just the solution obtained by 't Hooft [1]. It is easy to prove $\nabla_\mu \mathbf{n} = \nabla_\mu(\mathbf{r}/r) = 0$. With the Lagrangian introduced by 't Hooft [1], the gauge field equation is

$$\nabla_\nu G_{\mu\nu} = e\phi \times \nabla_\mu\phi = e^2\phi^2\mathbf{n} \times \nabla_\nu\mathbf{n}, \tag{1.82}$$

which is a sourceless equation when $\nabla_\mu\phi = 0$. So 't Hooft's solution is just the simplest sourceless solution independent of ϕ.

Furthermore, it should be pointed out that flat connection $G_{\mu\nu} = 0$ is a kind of sourceless solution [6, 10].

1.6 U(1) gauge potential of electromagnetic tensor $K_{\mu\nu}$

Let $e_a(x)(a = 1, 2)$ be orthogonal moving frame perpendicular to $\mathbf{n}$ in the isospin space. One can prove there exist a $U(1)$ potential, i.e. electromagnetic potential [6]

$$a_\mu = \frac{1}{2e}\varepsilon_{ab}(\mathbf{e}_a \cdot \partial_\mu \mathbf{e}_b), \tag{1.83}$$

which satisfies $K_{\mu\nu} = \partial_\mu a_\nu - \partial_\nu a_\mu$. And it is easy to prove that the potential (1.83) satisfies just the usual electromagnetic gauge transformation when the frame rotates along $\mathbf{n}$. Since $\mathbf{e}_1$ and $\mathbf{e}_2$ are the direction vector tangent to the sphere in the spherical coordinates of the isospin space, let $\mathbf{e}_1 = \mathbf{e}_\theta$, $\mathbf{e}_2 = \mathbf{e}_\varphi$ and rotate them. And let σ_1 be the north part of unit sphere and σ_2 be the south part,

$$\sigma_1 : \begin{aligned} e_1 &= \cos\varphi e_\theta - \sin\varphi e_\varphi, \\ e_2 &= \sin\varphi e_\theta + \cos\varphi e_\varphi, \end{aligned}$$

$$\sigma_2 : \begin{aligned} e_1 &= \cos\varphi e_\theta + \sin\varphi e_\varphi, \\ e_2 &= -\sin\varphi e_\theta + \cos\varphi e_\varphi, \end{aligned} \tag{1.84}$$

where the sign of rotation angle changes in different part. In terms of Eqs. (1.84), (1.83) and the orthogonal relationship, we have

$$a_\mu = \tilde{\mathbf{A}} \cdot \partial_\mu \phi = \tilde{A}_a \partial_\mu \phi^a, \tag{1.85}$$

in which

$$\tilde{\mathbf{A}} = -\frac{1}{e}\frac{\cos\theta - 1}{\phi\sin\theta}\mathbf{e}_\varphi, \quad \text{in} \quad \sigma_1, \tag{1.86}$$

$$\tilde{\mathbf{A}} = -\frac{1}{e}\frac{\cos\theta + 1}{\phi\sin\theta}\mathbf{e}_\varphi, \quad \text{in} \quad \sigma_2. \tag{1.87}$$

$\tilde{\mathbf{A}}$ is just the connect of principal fibre bundle $P(S^2 U(1))$ with manifold S^2 in the isospin space and structure group $U(1)$, or the Wu–Yang potential in the isospin space [12, 13]. For the $\phi^a(x)$ with N zeroes, a_μ is the Wu–Yang potential of system with N magnetic monopole.

Let R_1 and R_2 be two neighborhoods of a 2-dimensional closed surface R in the usual 3-dimensional space, and $\mathbf{n}(x)$ maps them into isospin space as σ_1 and σ_2,

$$R_1 : a_\mu = [\tilde{\mathbf{A}}]_1 \cdot \partial_\mu \phi, \tag{1.88}$$

$$R_2 : a_\mu = [\tilde{\mathbf{A}}]_2 \cdot \partial_\mu \phi. \tag{1.89}$$

Introducing $\partial_a = \partial/\partial\phi^a$ $(a = 1, 2, 3)$, i.e. $\partial_\mu = \partial_\mu \phi^a \partial_a$, we find the tensor $K_{\mu\nu} = f_{ab}\partial_\mu\phi^a\partial_\nu\phi^b$. With $f_{ab} = \partial_a\tilde{A}_b - \partial_b\tilde{A}_a$, immediately one can find

$$f_{ab} = \varepsilon_{abc}H_c, \tag{1.90}$$

$$H_c = \frac{1}{e}\frac{\phi^c}{\phi}, \tag{1.91}$$

which can be regarded as a "static magnetic field" in the isospin space.

On the overlap of R_1 and R_2, it is easy to prove that the two gauge potentials satisfy

$$[a_\mu]_1 - [a_\mu]_2 = \frac{1}{e}2\partial_\mu\varphi, \tag{1.92}$$

$$[\tilde{\mathbf{A}}]_1 - [\tilde{\mathbf{A}}]_2 = \frac{1}{e}2\nabla_\phi\varphi, \tag{1.93}$$

where ∇_ϕ is the gradient operator in the isospin space. Let Γ be a closed curve on the overlap of R_1 and R_2, corresponding to a curve γ on the overlap σ_1 and σ_2

$$\oint_\Gamma [a_\mu]_1 dx^\mu - \oint_\Gamma [a_\mu]_2 dx^\mu = \frac{2}{e}\int_\gamma d\varphi. \tag{1.94}$$

With Eqs. (1.28) and (1.73), one can prove the rhs of (1.94) is just $(4\phi/e)(n^+-n^-)$, i.e.

$$\oint d\varphi = 2\pi(n^+ - n^-). \tag{1.95}$$

The above analysis reveals the following topological property. When a point rotates 2π angle one time along closed curve Γ on the ordinary 2-dimensional closed surface, the corresponding point in the isospin unit sphere will rotate along positive direction of φ n^+ times and negative direction n^- times through the mapping $\mathbf{n}(x)$. This property indicates the close connection between the topology and the Wu–Yang potential.

Acknowledgement

Authors thank Prof. C. N. Yang for his encourage during his visit in Lanzhou University, and thank Y. B. Dai, B. Y. Hou, Y. S. Wu and M. D. Xin for their discussion.

Bibliography

[1] Yang, C. N. and Mills, R. L., Phys. Rev. **96** (1954) 191.

[2] Higgs, P., Phys. Letts, **12** (1966) 132.

[3] Hou, B. Y., Duan, Y. S. and Ge, M. L., Journal of Lanzhou University (Natural Science), **2** (1975) 26; Acta Physica Sinica, **25** (1976) 451. (in Chinese)

[4] Duan, Y. S. and Ge, M. L., Journal of Lanzhou University (Natural Science), **4** (1975) 19. (in Chinese)

[5] Duan, Y. S. and Ge, M. L., Sci. Sinica, **21** (1976) 282. (in Chinsese)

[6] Duan, Y. S. and Ge, M. L., Journal of Lanzhou University (Natural Science), **1** (1977) 33. (in Chinese)

[7] 't Hooft, G., Nucl. Phys. B **79** (1974) 276.

[8] Dirac, P. A. M., Phys. Rev. **74** (1948) 817.

[9] Wu, T. T. and Yang, C. N., Phys. Rev. D **14** (1976) 437.

[10] Kobayashi, S. and Nomizu, K., Foundations of Differential Geometry I, 1963.

[11] Wu, T. T. and Yang, C. N., Phys. Rev. D **12** (1975) 3843.

[12] Hou, B. Y., Acta Physica Sinica **26** (1977) 83. (in Chinese)

[13] Wu, Y. S. and Guo, H. Y., Sci. Sinica, **4** (1977) 308. (in Chinese)

[14] Arafune, J., et al., Jour. Math. Phys., **16** (1975) 433.

Chapter 2

Abelian Decomposition of QCD

Y. M. Cho [1]

Administration Building 310-4, Konkuk University,
Seoul 143-701, Korea
School of Physics and Astronomy, Seoul National University,
Seoul 151-747, Korea
Department of Physics, Florida International University,
Miami, Florida 33199, USA

Abstract: The importance of the Abelian decomposition of QCD and the role of Professor Yi-Shi Duan in this decomposition is emphasised. The Abelian decomposition (known as Cho–Duan–Ge decomposition) decomposes the gluons to the color neutral binding gluons (the neurons and the monopoles) and the colored valence gluons (the chromons), and shows that QCD can be viewed as the restricted QCD (RCD) made of the binding gluons which has the chromons as colored source. This simplifies the QCD dynamics greatly. In the perturbative regime this decomposes the gluon propagater to the neuron propagatoes and the chromon propagaters, and simplifies the Feynman diagram. In the non-perturbative regime this allows us to demonstrate the monopole condensation gauge independently. In particular, this tells that the gauge covariant valence gluons can be treated as the constituents of hadrons, and generalizes the quark model to the quark and chromon model. Moreover, we can apply the Abelian decomposition to simplify Einstein's theory and obtain the restricted gravity which describe the core dynamics of Einstein's theory. As importantly, we show that the Abelian decomposition plays a crucial role to resolve the proton spin crisis problem.

Keywords: Abelian decomposition, binding gluon, valence gluon, neuron, chromon, monopole, monopole condensation, quark and chromon model, glueballs, vacuum decomposition, proton spin crisis problem

---

[1] Email: ymcho7@konkuk.ac.kr

2.1   Introduction

In this article I will discuss about the importance of the Abelian decomposition in QCD and its physical implication in gauge theory, which was developed by Professor Yi-Shi Duan and my dear friend Mo-Lin Ge. I met Professor Duan in 2010 when I visited Lanzhou. To explain how I met him, I have to explain how I have so many Chinese friends.

In 1985 I organized the XIVth International Conference on Group Theoretical Methods in Physics (ICGTMP) in Seoul. It was a big international gathering, and I invited physicists from all over the world, in particular from China and Russia. This was a real challenge, because at that time the world was still in cold war, and Korea had no diplomatic relation with Russia and China. I sent 20 invitations to Russia and another 20 to China. Russia simply refused to receive the invitations, and turned back all invitations.

But for some reason China accepted the invitation, and sent three prominent physicists, Mo-Lin Ge at Nankai University, Jinquan Chen at Nanjing University, and Yuanzhong Zhang at ITP in Beijing, to Korea. This was a big news in Korea, because they were the first Chinese who were officially allowed to visit Korea after the Korean war. Korean government protected them 24 hours a day from the beginning to the end, to make sure them to return safely. During the conference Mo-Lin Ge told me he would like to invite me to China. I thanked him, but really did not believe this would be possible, because Korea and China had no diplomatic relation so that there was no way I could visit China.

I was wrong. To my surprise, several months later I did get the invitation to visit China for one month from a Chinese government official. But my trip had to be unofficial, as there was no diplomatic relation between Korea and China at that time. In July 1987, I entered China through Hong Kong with no trace of my visit to China on my passport, and I traveled with Mo-Lin to Beijing (Beijing University and ITP), Tianjin (Nankai University), Xian (Northwestern University), Shanghai (Fudan University), Hangzhou (Zhejiang University), and Suzhou. Probably I was the first Korean scientist who visited the communist China after the Korean war. After this trip I have had many friends in China, and Mo-Lin and I became real brothers.

This time I could not meet Professor Duan because Lanzhou was too remote, but Mo-Lin did give me the paper written by Duan and him on the Abelian decomposition. Unfortunately I did not pay much attention as it was written in Chinese which I was unable to read. Several years later, in 2004, I invited Pengming Zhang, one of Duan's students, to Seoul National University to work with me. And he also mentioned about this paper. So I asked him to translate this paper in English, and I immediately realized that it was exactly on the Abelian decomposition of Yang–

Mills theory that I have been advocating in my whole life. After that I dubbed the Abelian decomposition Cho–Duan–Ge (CDG) decomposition.

I have visited China several times after my first trip, but have had no chance to visit Lanzhou. But in 2010 Pengming invited me to Lanzhou University after he returned to China, and I met Professor Duan for the first time. He was gentle but alert, and both of us were very happy to meet each other. He explained how he and Mo-Lin had developed this idea during the cultural revolution. It was really remarkable that we had the same idea almost at the same time.

Now let me explain why this Abelian decomposition is so important. It has generally been believed that the non-Abelian gauge symmetry is so tight that it defines the theory almost uniquely. Moreover, the gauge symmetry binds different gauges so strongly that the theory does not allow any simplification. Exactly the same assertion has been made in Einstein's general relativity. Here again the assertion is that the general invariance is so tight that the theory can not be simplified further. The Abelian decomposition tells that this popular wisdom is wrong.

The Abelian decomposition decomposes the gluons (in general the non-Abelian gauge potentials) into two different types, the color neutral binding gluons called the neurons and the colored valence gluons called the chromons, gauge independently. This is very important in hadron spectroscopy because the chromons, just like the quarks, can form the color singlet bound states and become the constituent of hadrons. This generalizes the quark model to the quark and chromon model. In particular, this plays the essential role for us to understand the glueball spectrum and its mixing with quarkonium.

As importantly, it tells that we can have simpler QCD called the restricted QCD (RCD) made of the binding gluons which nevertheless has the full non-Abelian gauge symmetry. Moreover, QCD can be interpreted as RCD which has the chromons as colored source. This greatly simplifies the dynamics of QCD, in general the non-Abelian gauge theory.

Furthermore, the Abelian decomposition separates the non-Abelian monopole gauge independently, and allows us to prove the confining potential in QCD comes from the monopole. In fact it allows us to calculate the effective potential of QCD and demonstrates that QCD vacuum is made of the monopole condensation.

QCD is not the only theory which can be simplified with the Abelian decomposition. Applying the Abelian decomposition to Einstein's theory, we can simplify the Einstein's theory and obtain the restricted gravity which describes the core dynamics of Einstein's theory.

The Abelian decomposition has another important applications. For example, we can easily generalize it to the vacuum decomposition which decomposes the non-Abelian gauge potential to the pure vacuum part and the gauge covariant physical part gauge independently. In particular, in this vacuum decomposition we can write down the most general vacuum potential explicitly in a very simple and compact form. This allows us to study the topological structure of the non-Abelian vacuum.

More importantly, the vacuum decomposition allows us to express the non covariant momentum operator (the plain derivative operator) in the gauge covariant momentum operator (the gauge covariant derivative operator) without changing the physical content. This plays a crucial role for us to resolve the nucleon spin crisis problem, as we will see in the following.

Another important application of the vacuum (and Abelian) decomposition is that we can quantize the non-Abelian gauge theory with the vacuum potential as the background field, and develop a new Feynman rule with only the physical as the dynamical (propagating) fields. The advantage of this is that the calculations of the Feynman diagrams are automatically guaranteed to be gauge independent.

The paper is organized as follows. In Section II we review the Abelian decomposition for later purpose. In Section III we discuss the Abelian decomposition of SU(3) QCD. In Section IV we prove the Abelian dominance and monopole condensation in QCD. In Section V we discuss the quark and chromon model and the quarkonium-glueball mixing in QCD. In Section VI we discuss the monoball, the vacuum fluctuation of the monopole condensation in QCD. In Section VII we review the Lorentz gauge formalism of Einstein's theory. In Section VIII we discuss the Abelian decomposition of Einstein's theory. In Section IX we review the vacuum decomposition. In Section X we discuss the canonical momentum and kinematic momentum. In Section XI we discuss the proton spin crisis problem. Finally in Section XII we make the closing remark.

2.2   Abelian decomposition: a review

Let us review the Abelian decomposition first. Consider the SU(2) QCD for simplicity, and let $(\hat{n}_1, \hat{n}_2, \hat{n}_3 = \hat{n})$ be an arbitrary right-handed local orthonormal basis. To make the Abelian decomposition we choose any direction, to be explicit $\hat{n}$, to be the Abelian direction and impose the isometry to project out the restricted potential $\hat{A}_\mu$ which describes the Abelian sub-dynamics of QCD [1, 2, 3]

$$D_\mu \hat{n} = (\partial_\mu + g\vec{A}_\mu \times)\hat{n} = 0,$$

$$\vec{A}_\mu \to \hat{A}_\mu = A_\mu \hat{n} - \frac{1}{g}\hat{n} \times \partial_\mu \hat{n} = \mathcal{A}_\mu + \mathcal{C}_\mu,$$

$$\mathcal{A}_\mu = A_\mu \hat{n}, \quad \mathcal{C}_\mu = -\frac{1}{g}\hat{n} \times \partial_\mu \hat{n}, \quad A_\mu = \hat{n} \cdot \vec{A}_\mu. \tag{2.1}$$

This is the Abelian projection which projects out the color neutral binding gluons.

Notice the followings. First, $\hat{A}_\mu$ is precisely the connection which leaves the Abelian direction invariant under the parallel transport (which makes $\hat{n}$ covariantly constant). Second, it is made of two parts, the topological (Diracian) $\mathcal{C}_\mu$ which describes the non-Abelian monopole and the non-topological (Maxwellian) $\mathcal{A}_\mu$. Third, the decomposition is gauge independent because $\hat{n}$ is arbitrary. We can rotate $\hat{n}$ to any direction and still get exactly the same decomposition. This is the most important point.

With this we have

$$\hat{F}_{\mu\nu} = (F_{\mu\nu} + H_{\mu\nu})\hat{n} = G_{\mu\nu}\hat{n},$$
$$F_{\mu\nu} = \partial_\mu A_\nu - \partial_\nu A_\mu,$$
$$H_{\mu\nu} = \partial_\mu C_\nu - \partial_\nu C_\mu, \quad C_\mu = -\frac{1}{g}\hat{n}_1 \cdot \partial_\mu \hat{n}_2,$$
$$G_{\mu\nu} = \partial_\mu B_\nu - \partial_\nu B_\mu, \quad B_\mu = A_\mu + C_\mu. \tag{2.2}$$

This tells two things. First, $\hat{F}_{\mu\nu}$ has only the Abelian component. Second, $\hat{F}_{\mu\nu}$ is made of two potentials, the electric (non-topological) A_μ and magnetic (topological) C_μ.

With (2.1) we can recover the full QCD potential adding the non-Abelian (colored) part $\vec{X}_\mu$ [2, 3, 4]

$$\vec{A}_\mu = \hat{A}_\mu + \vec{X}_\mu,$$
$$\vec{X}_\mu = \frac{1}{g}\hat{n} \times D_\mu \hat{n}, \quad \hat{n} \cdot \vec{X}_\mu = 0. \tag{2.3}$$

Under the infinitesimal gauge transformation

$$\delta \vec{A}_\mu = \frac{1}{g} D_\mu \vec{\alpha}, \quad \delta \hat{n}_i = -\vec{\alpha} \times \hat{n}_i, \tag{2.4}$$

we have

$$\delta \hat{A}_\mu = \frac{1}{g} \hat{D}_\mu \vec{\alpha}, \quad (\hat{D}_\mu = \partial_\mu + g\hat{A}_\mu \times),$$
$$\delta A_\mu = \frac{1}{g}\hat{n} \cdot \partial_\mu \vec{\alpha}, \quad \delta C_\mu = -\frac{1}{g}\hat{n} \cdot \partial_\mu \vec{\alpha},$$
$$\delta \vec{X}_\mu = -\vec{\alpha} \times \vec{X}_\mu. \tag{2.5}$$

This tells that $\hat{A}_\mu$ has the full SU(2) gauge degrees of freedom, even though it is restricted. Moreover, this confirms that $\vec{X}_\mu$ becomes gauge covariant. This is a direct consequence of the fact that the connection space (the space of all potentials) forms an affine space.

This is the Abelian decomposition which decomposes the gluons to the color neutral binding gluons and the colored valence gluons gauge independently. We can express the Abelian decomposition graphically. This is shown in Fig. 2.1, where the gluons are decomposed to the binding gluons and the valence gluons in (A), and the binding gluons are decomposed further to the non-topological Maxwell part $\mathcal{A}_\mu$ and the topological Dirac part $\mathcal{C}_\mu$ in (B). In the literature this is known as

Cho decomposition, Cho–Duan–Ge (CDG) decomposition, or Cho–Faddeev–Niemi (CFN) decomposition [5, 6, 7].

From this we can construct RCD which has the full non-Abelian gauge symmetry but is simpler than the QCD

$$\mathcal{L}_{RCD} = -\frac{1}{4}\hat{F}_{\mu\nu}^2 = -\frac{1}{4}F_{\mu\nu}^2 + \frac{1}{2g}F_{\mu\nu}\hat{n}\cdot(\partial_\mu\hat{n}\times\partial_\nu\hat{n}) - \frac{1}{4g^2}(\partial_\mu\hat{n}\times\partial_\nu\hat{n})^2, \quad (2.6)$$

which describes the Abelian subdynamics of QCD. Since RCD contains the non-Abelian monopole degrees explicitly, it provides an ideal platform for us to study the monopole dynamics gauge independently.

From (2.3) we have

$$\vec{F}_{\mu\nu} = \hat{F}_{\mu\nu} + \hat{D}_\mu\vec{X}_\nu - \hat{D}_\nu\vec{X}_\mu + g\vec{X}_\mu\times\vec{X}_\nu. \quad (2.7)$$

With this we can express QCD by

$$\mathcal{L}_{QCD} = -\frac{1}{4}\vec{F}_{\mu\nu}^2 = -\frac{1}{4}\hat{F}_{\mu\nu}^2 - \frac{1}{4}(\hat{D}_\mu\vec{X}_\nu - \hat{D}_\nu\vec{X}_\mu)^2$$
$$- \frac{g}{2}\hat{F}_{\mu\nu}\cdot(\vec{X}_\mu\times\vec{X}_\nu) - \frac{g^2}{4}(\vec{X}_\mu\times\vec{X}_\nu)^2. \quad (2.8)$$

This is the extended QCD (ECD) which confirms that QCD can be viewed as RCD made of the binding gluons, which has the colored valence gluons as its source [2,3,4,8]. Notice that, although mathematically ECD is identical to QCD, physically it acquires a totally different meaning.

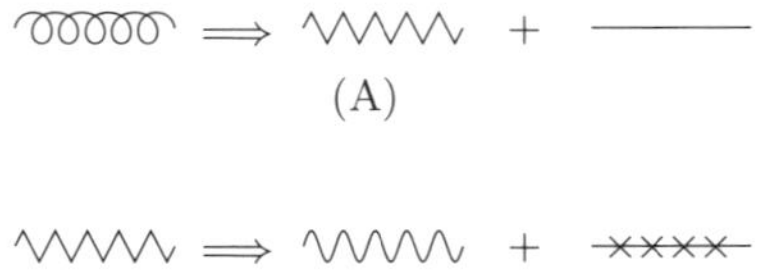

(A)

(B)

Fig. 2.1 The Abelian decomposition of the gluons. The gluon is decomposed to the binding gluon (kinked line) and the valence gluon (straight line) in (A), and the binding gluon is further decomposed to the Maxwell part (wiggly line) and Dirac part (spiked line) in (B).

2.3 Abelian decomposition of SU(3) QCD

The Abelian decomposition of SU(3) QCD is a bit more complicated, but is well known. Since SU(3) has rank two, we have two Abelian subgroups in SU(3). Let $\hat{n}_i$ $(i = 1, 2, ..., 8)$ be the local orthonormal SU(3) basis. Clearly we can choose the Abelian directions to be $\hat{n}_3 = \hat{n}$ and $\hat{n}_8 = \hat{n}'$. Now make the Abelian projection by

$$D_\mu\hat{n} = 0. \quad (2.9)$$

This automatically guarantees [9]

$$D_\mu \hat{n}' = 0, \quad \hat{n}' = \frac{1}{\sqrt{3}} \hat{n} * \hat{n}. \tag{2.10}$$

where $*$ denotes the d-product. This is because SU(3) has two vector products, the anti-symmetric f-product and the symmetric d-product.

Solving (2.9), we have the following Abelian projection which projects out two neutral binding gluons,

$$\begin{aligned}
\vec{A}_\mu \to \hat{A}_\mu &= A_\mu \hat{n} + A'_\mu \hat{n}' - \frac{1}{g} \hat{n} \times \partial_\mu \hat{n} - \frac{1}{g} \hat{n}' \times \partial_\mu \hat{n}' \\
&= \sum_p \frac{2}{3} \hat{A}^p_\mu, \quad (p = 1, 2, 3), \\
\hat{A}^p_\mu &= A^p_\mu \hat{n}^p - \frac{1}{g} \hat{n}^p \times \partial_\mu \hat{n}^p = \mathcal{A}^p_\mu + \mathcal{C}^p_\mu, \\
A^1_\mu &= A_\mu, \quad A^2_\mu = -\frac{1}{2} A_\mu + \frac{\sqrt{3}}{2} A'_\mu, \quad A^3_\mu = -\frac{1}{2} A_\mu - \frac{\sqrt{3}}{2} A'_\mu, \\
\hat{n}^1 &= \hat{n}, \quad \hat{n}^2 = -\frac{1}{2} \hat{n} + \frac{\sqrt{3}}{2} \hat{n}', \quad \hat{n}^3 = -\frac{1}{2} \hat{n} - \frac{\sqrt{3}}{2} \hat{n}',
\end{aligned} \tag{2.11}$$

where the sum is the sum of the Abelian directions of three SU(2) subgroups made of $(\hat{n}_1, \hat{n}_2, \hat{n}^1)$, $(\hat{n}_6, \hat{n}_7, \hat{n}^2)$, $(\hat{n}_4, -\hat{n}_5, \hat{n}^3)$. Notice the factor 2/3 in front of $\hat{A}^p_\mu$ in the p-summation. This is because the three SU(2) binding potentials are not independent.

From this we have the restricted the restricted QCD made of the restricted field strength,

$$\mathcal{L}_{RCD} = -\sum_p \frac{1}{6} (\hat{F}^p_{\mu\nu})^2, \tag{2.12}$$

which has the full SU(3) gauge symmetry.

With (2.11) we have the Abelian decomposition of the SU(3) gauge potential,

$$\begin{aligned}
\vec{A}_\mu &= \hat{A}_\mu + \vec{X}_\mu = \sum_p \left(\frac{2}{3} \hat{A}^p_\mu + \vec{W}^p_\mu \right), \qquad \vec{X}_\mu = \sum_p \vec{W}^p_\mu, \\
\vec{W}^1_\mu &= X^1_\mu \hat{n}_1 + X^2_\mu \hat{n}_2, \quad \vec{W}^2_\mu = X^6_\mu \hat{n}_6 + X^7_\mu \hat{n}_7, \\
\vec{W}^3_\mu &= X^4_\mu \hat{n}_4 - X^5_\mu \hat{n}_5.
\end{aligned} \tag{2.13}$$

Here again $\vec{X}_\mu$ transforms covariantly, and can be decomposed to the three valence gluons $\vec{W}^p_\mu$ of the SU(2) subgroups. But unlike $\hat{A}^p_\mu$, they are mutually independent. So we have two binding gluons and six (or three complex) valence gluons in SU(3) QCD.

From (2.13) we have

$$\hat{D}_\mu \vec{X}_\nu = \sum_p \hat{D}_\mu^p \vec{W}_\nu^p, \quad \hat{D}_\mu^p = \partial_\mu + g\hat{A}_\mu^p \times,$$

$$\vec{X}_\mu \times \vec{X}_\nu = \sum_{p,q} \vec{W}_\mu^p \times \vec{W}_\nu^q,$$

$$\vec{F}_{\mu\nu} = \hat{F}_{\mu\nu} + \hat{D}_\mu \vec{X}_\nu - \hat{D}_\nu \vec{X}_\mu + g\vec{X}_\mu \times \vec{X}_\nu$$

$$= \sum_p \left[\frac{2}{3}\hat{F}_{\mu\nu}^p + (\hat{D}_\mu^p \vec{W}_\nu^p - \hat{D}_\nu^p \vec{W}_\mu^p) \right] + \sum_{p,q} \vec{W}_\mu^p \times \vec{W}_\nu^q, \qquad (2.14)$$

so that we have the following form of SU(3) ECD [10, 11]

$$\mathcal{L} = \sum_p \left\{ -\frac{1}{6}(\hat{F}_{\mu\nu}^p)^2 - \frac{1}{4}(\hat{D}_\mu^p \vec{W}_\nu^p - \hat{D}_\nu^p \vec{W}_\mu^p)^2 - \frac{g}{2}\hat{F}_{\mu\nu}^p \cdot (\vec{W}_\mu^p \times \vec{W}_\nu^p) \right\}$$

$$- \sum_{p,q} \frac{g^2}{4}(\vec{W}_\mu^p \times \vec{W}_\mu^q)^2 - \sum_{p,q,r} \frac{g}{2}(\hat{D}_\mu^p \vec{W}_\nu^p - \hat{D}_\nu^p \vec{W}_\mu^p) \cdot (\vec{W}_\mu^q \times \vec{W}_\mu^r)$$

$$- \sum_{p \neq q} \frac{g^2}{4} \left[(\vec{W}_\mu^p \times \vec{W}_\nu^q) \cdot (\vec{W}_\mu^q \times \vec{W}_\nu^p) + (\vec{W}_\mu^p \times \vec{W}_\nu^p) \cdot (\vec{W}_\mu^q \times \vec{W}_\nu^q) \right]. \qquad (2.15)$$

This shows that the interactions in SU(3) QCD is more complicated than the SU(2) QCD. But what is remarkable about (2.15) is that it is Weyl symmetric, symmetric under the permutation of the three SU(2) subgroups of SU(3).

We can easily add quarks in the Abelian decomposition,

$$\mathcal{L}_q = \sum_k \bar{\Psi}_k(i\gamma^\mu D_\mu - m)\Psi_k$$

$$= \sum_k \left[\bar{\Psi}_k(i\gamma^\mu \hat{D}_\mu - m)\Psi_k + \frac{g}{2}\vec{X}_\mu \cdot \bar{\Psi}_k(\gamma^\mu \vec{t})\Psi_k \right]$$

$$= \sum_{p,k} \left[\bar{\Psi}_k^p(i\gamma^\mu \hat{D}_\mu^p - m)\Psi_k^p + \frac{g}{2}\vec{W}_\mu^p \cdot \bar{\Psi}_k^p(\gamma^\mu \vec{\tau}^p)\Psi_k^p \right],$$

$$\hat{D}_\mu = \partial_\mu + \frac{g}{2i}\vec{t}\cdot \hat{A}_\mu, \quad \hat{D}_\mu^p = \partial_\mu + \frac{g}{2i}\vec{\tau}^p \cdot \hat{A}_\mu^p, \qquad (2.16)$$

where m is the mass, k and p denote the flavor and color of the quarks, and Ψ_k^p represents the three SU(2) quark doublets (i.e., (r,b), (b,g), and (g,r) doublets) of the (r,b,g) quark triplet. With this the colors of the six chromons are given by $(r\bar{b}, b\bar{g}, g\bar{r}, \bar{r}b, \bar{b}g, \bar{g}r)$, which can be denoted by $(R, B, G, \bar{R}, \bar{B}, \bar{G})$.

We can show how the Abelian decomposition refines QCD interaction graphically. This is shown in Fig. 2.2. In (A) the three-point QCD gluon vertex is decomposed to two vertices made of one neuron and two chromons and three chromons. In (B) the four-point gluon vertex is decomposed to three vertices made of one neuron and three chromons, two neurons and two chromons, and four chromons. In (C) the quark-gluon vertex is decomposed to the quark-neuron vertex and quark-chromon vertex.

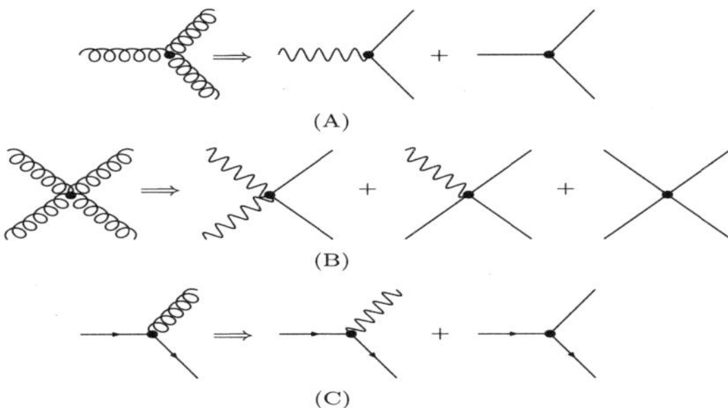

Fig. 2.2   The decomposition of vertices in SU(3) QCD. The three and four point gluon vertices are decomposed in (A) and (B), and the quark gluon verteces are decomposed in (C). Notice that here (and in the followings) the neurons are representedby wiggly lines and the chromons are represented by straight lines.

Notice that here (and in the following figures) the neurons are expressed by the wiggly lines (Maxwell part). This is because the monopole potential (Dirac part) makes the condensation, so that in the perturbative regime (inside the hadrons) it does not contribute to the Feynman diagrams. Also here three-point vertex made of three neurons or two neurons and one chromon, and four-point vertex made of three or four neurons are forbidden by the conservation of color. Moreover, the quark-neuron interaction does not change the quark color, but the quark-chromon interaction changes the quark color.

An important implication of Fig. 2.2 is that there are two types of gluon jets, the neuron jet and chromon jet. In principle we can test this experimentally by studying the gluon jets. Experiments can tell the difference between the photon-quark jet from the gluon-quark jet. If so, by (re-)analyzing the gluon-gluon jets and/or gluon-quark jets more carefully we could confirm that indeed there are two types of gluon jets. This can endorse the existence of two types of gluons.

But what is the most important is that this Abelian decomposition is gauge independent. This is because the decomposition is made without the gauge fixing. This has deep implications. The conventional wisdom is that all gluons are equal because of the gauge invariance, so that one can not differentiate the neutral gluons from the colored ones. This is simply not true. There is a mathematically well defined way to separate the neutral gluons from the colored ones gauge independently.

2.4   Abelian dominance and monopole condensation in QCD

Our analysis tells that, although the Abelian decomposition does not change QCD, it makes many hidden structures of QCD explicit. First, it tells that RCD is responsible for the confinement, because the valence gluons (being colored) have to be confined [12, 13]. This is the Abelian dominance advocated by 'tHooft [12].

But actually the Abelian dominance does not tell what is responsible for the confinement, because the restricted potential is made of two parts, the non-topological Maxwellian potential and topological Diracian potential. The Abelian decomposition tells exactly what is responsible for the confinement [13].

It allows us to prove that the monopole is responsible for the area law in the Wilson loop integral. Implementing (2.9) on lattice, we can performed a truly gauge independent lattice calculation and show that the monopole is responsible for the confinement. Two lattice groups, KEK-Chiba and SNU-Konkuk, made the numerical calculation independently and confirmed the monopole dominance [14, 15]. The SNU-Konkuk result is shown in Fig. 2.3, which shows that the full gauge potential, the restricted potential, and the monopole potential all produce the same linear confining potential in Wilson loop integral. This assures that we only need the monopole potential for the confining force.

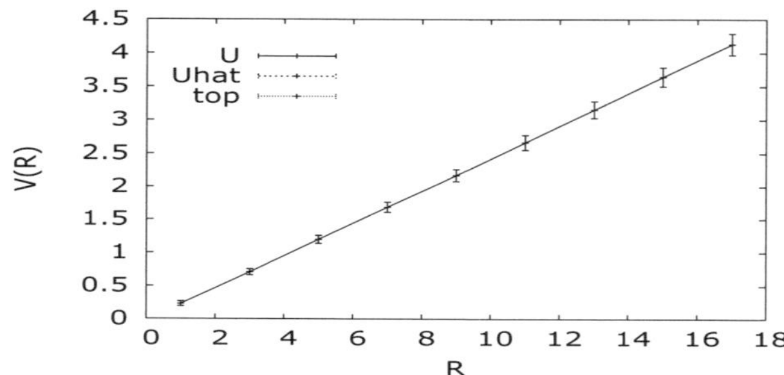

Fig. 2.3   The lattice QCD calculation which establishes the monopole dominance in Wilson loop.

Second, it reduces the complicated non-Abelian gauge symmetry to a simple discrete symmetry called the color reflection invariance. To see this, consider the rotation of basis called the color reflection in SU(2) QCD

$$(\hat{n}_1, \hat{n}_2, \hat{n}) \rightarrow (\hat{n}_1, -\hat{n}_2, -\hat{n}). \tag{2.17}$$

Obviously this is a gauge transformation, so that this must remain a symmetry of QCD. On the other hand, the isometry condition (2.1) does not change under (2.17). This means that, after we select the Abelian direction $\hat{n}$ we have two different but gauge equivalent Abelian decompositions related by the color reflection.

What makes the color reflection symmetry so important is that it is the only remaining symmetry of the full gauge symmetry left over, after we make the Abelian decomposition [3, 4]. So the color reflection invariance plays the role of the non-Abelian gauge invariance after we have chosen the Abelian direction. This greatly simplifies for us to implement the gauge invariance to calculate the QCD effective action [10, 11].

The color reflection symmetry of SU(2) QCD is made of 4 elements, but in SU(3) QCD it consists of 24 element. An important subgroup of the color reflection symmetry is the Weyl symmetry, which is made of the permutation of the colors. This automatically guarantees that QCD is invariant under the permutation of any two colors. In SU(2) QCD we have two colors so that the Weyl symmetry is made of two elements. But in SU(3) QCD it is made of six elements since SU(3) QCD has three colors.

Moreover, the Abelian decomposition automatically puts QCD to the background field formalism [16, 17, 18]. This is because we can view the Abelian part as the classical background and the colored part as the quantum fluctuation. This enlarges and doubles the gauge symmetry to the classical and quantum gauge symmetries, since we can associate the original gauge symmetry to the classical part or to the quantum part.

More importantly this allows us to calculate the effective action of QCD and demonstrate the monopole condensation gauge independently. Savvidy was the first who tried to prove the monopole condensation calculating the effective action of SU(2) QCD, and has "almost" succeeded to do so [19]. However, this monopole condensation called the Savvidy vacuum turned out to be unstable, which is known as the Savvidy-Nielsen-Olesen (SNO) instability [20]. Since then many people tried to restore the stability without much success [21, 22].

Actually there are ways to restore the stability in the Savvidy vacuum [23, 24]. But the real problem with the Savvidy vacuum is not that it is unstable but that the gauge invariance was not properly implemented in the calculation [10]. First of all, the Savvidy background was not the monopole background. At that time people did not know how to separate the monopole background gauge independently. But most seriously, the calculation of the functional determinant of the chromons was not done gauge invariantly, so that the determinant contained the tachyonic mode which made the Savvidy vacuum unstable. This was the critical defect.

The Abelian decomposition allows us to calculate the effective action of QCD gauge independently and gauge invariantly. First, the Abelian decomposition decomposes the gauge potential to the binding potential and the valence potential gauge independently. Second, it separates the monopole part from the binding potential gauge independently. This allows us to separate the monopole background gauge independently. Without the Abelian decomposition this would have been impossible.

Third, in the earlier calculations it was not clear which gluons are integrated out in the functional integral. As a result it was not clear what kind of effective action was calculated. But with the Abelian decomposition we can clearly say that, integrating out only the gauge covariant chromons, we are calculating the effective action of RCD.

Most importantly, imposing the color reflection invariance in the calculation of the functional determinant of the chromons under the monopole background, we can remove the tachyonic mode which cures the SNO instability.

To show how we can do this we start from the Lagrangian (2.8) and fix the quantum gauge by the generalized Lorentz gauge condition

$$\vec{F} = \hat{D}_\mu \vec{X}_\mu = 0, \quad \mathcal{L}_{gf} = -\frac{1}{2\xi}(\hat{D}_\mu \vec{X}_\mu)^2. \tag{2.18}$$

The corresponding Faddeev–Popov determinant can be expressed by

$$M^{FP}_{ab} = \frac{\delta F_a}{\delta \alpha^b} = \Pi_{ac}(\hat{D}_\mu D_\mu)_{cb},$$

$$\Pi_{ab} = \delta_{ab} - \hat{n}_a \hat{n}_b. \tag{2.19}$$

Notice that Π_{ab} is the projection operator which projects out the $\hat{n}$ component.

To proceed we have to choose the background. We choose the gauge independent and parity conserving arbitrary constant Diracian background,

$$\hat{F}^{(b)}_{\mu\nu} = \bar{H}_{\mu\nu}\hat{n}. \tag{2.20}$$

Notice that when $\bar{H}_{\mu\nu} = H\delta^1_{[m}\delta^2_{n]}$, it describes the monopole background. In comparison Savvidy background was the parity violating Maxwellian $\bar{F}_{\mu\nu}\hat{n}$.

With this we can calculate the effective action of RCD integrating out the valence potential [10]

$$\exp^{iS_{eff}(\hat{A}_\mu)} = \int \mathcal{D}\vec{X}_\mu \mathcal{D}\vec{c}\,\mathcal{D}\vec{c}^* \exp\left\{ i \int \left[-\frac{1}{4}\hat{F}^2_{\mu\nu} - \frac{1}{4}(\hat{D}_\mu \vec{X}_\nu - \hat{D}_\nu \vec{X}_\mu)^2 \right.\right.$$

$$- \frac{g}{2}\hat{F}_{\mu\nu}\cdot(\vec{X}_\mu \times \vec{X}_\nu) - \frac{g^2}{4}(\vec{X}_\mu \times \vec{X}_\nu)^2$$

$$\left.\left. - \frac{1}{2\xi}(\hat{D}_\mu \vec{X}_\mu)^2 + \vec{c}^*\hat{D}_\mu D_\mu \vec{c} \right] d^4x \right\}, \tag{2.21}$$

where $\vec{c}$ and $\vec{c}^*$ are the ghost fields which are orthogonal to $\hat{n}$. Here we need only the ghosts which are orthogonal to $\hat{n}$ because they come from the gauge fixing of the valence gluons which are orthogonal to $\hat{n}$.

Integrating out the valence potential under the constant Diracian background imposing the color reflection invariance (the C-projection) in the calculation of the functional determinant of the chromons, we obtain [10]

$$\Delta\mathcal{L} = \lim_{\epsilon \to 0} \frac{1}{8\pi^2} \int_0^\infty \frac{dt}{t^{3-\epsilon}} \frac{abt^2/\mu^4}{\sinh(at/\mu^2)\sin(bt/\mu^2)}$$

$$\times \left[\exp(-2at/\mu^2) + \exp(+2ibt/\mu^2) - 1 \right],$$

$$a = \frac{g}{2}\sqrt{\sqrt{\bar{H}^4 + (\bar{H}\tilde{\bar{H}})^2} + \bar{H}^2},$$

$$b = \frac{g}{2}\sqrt{\sqrt{\bar{H}^4 + (\bar{H}\tilde{\bar{H}})^2} - \bar{H}^2}. \tag{2.22}$$

This is the new integral expression of QCD effective action. Notice that a describes the Diracian magnetic (i.e., monopole) background, but b describes Diracian electric background.

We emphasize that here the C-projection plays exactly the same role as the G-parity in string theory. In NSR string the GSO projection restores the super-symmetry and modular invariance by projecting out the tachyonic vacuum [25, 26]. *Just like the G-projection, the C-projection in QCD removes the tachyonic modes and restores the gauge invariance of the effective action.*

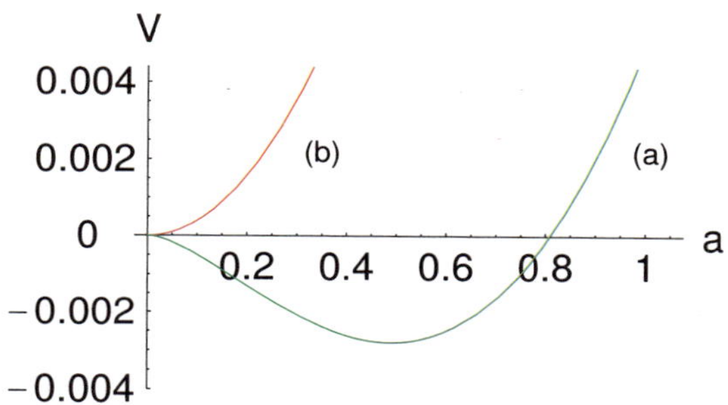

Fig. 2.4   The effective potential of SU(2) QCD in the monopole background. Here (a) is the effective potential and (b) is the classical potential.

From this we have [10]

$$\mathcal{L}_{eff} = \begin{cases} -\dfrac{a^2}{2g^2} - \dfrac{11a^2}{48\pi^2}(\ln \dfrac{a}{\mu^2} - c'), & b = 0 \\ \dfrac{b^2}{2g^2} + \dfrac{11b^2}{48\pi^2}(\ln \dfrac{b}{\mu^2} - c') \\ \quad - i\dfrac{11b^2}{96\pi}, & a = 0 \end{cases} \tag{2.23}$$

The effective action (2.22) generates the much desired dimensional transmutation in QCD. To demonstrate this notice that, with $b = 0$, (2.22) provides the following effective potential

$$V = \frac{a^2}{2g^2}\left[1 + \frac{11g^2}{24\pi^2}(\ln \frac{a}{\mu^2} - c')\right]. \tag{2.24}$$

Defining the running coupling $\bar{g}$ by [23, 24]

$$\frac{\partial^2 V}{\partial a^2}\bigg|_{a=\bar{\mu}^2} = \frac{1}{\bar{g}^2}, \tag{2.25}$$

we have

$$\frac{1}{\bar{g}^2} = \frac{1}{g^2} + \frac{11}{24\pi^2}(\ln \frac{\bar{\mu}^2}{\mu^2} - c' + \frac{3}{2}),$$

$$\beta(\bar{g}) = \bar{\mu}\frac{\partial \bar{g}}{\partial \bar{\mu}} = -\frac{11\bar{g}^3}{24\pi^2}. \tag{2.26}$$

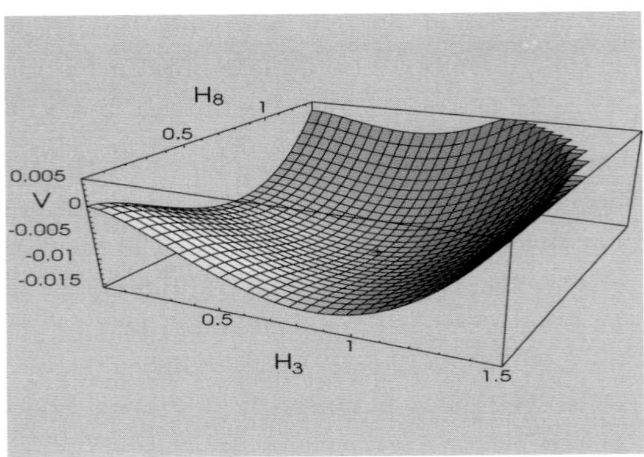

Fig. 2.5   The effective potential of SU(3) QCD.

This is exactly the same β-function in perturbative SU(2) QCD which assures the asymptotic freedom [27].

In terms of the running coupling the renormalized potential is given by [10]

$$V_{\text{ren}} = \frac{a^2}{2\bar{g}^2}\left[1 + \frac{11\bar{g}^2}{24\pi^2}\left(\ln\frac{a}{\bar{\mu}^2} - \frac{3}{2}\right)\right], \tag{2.27}$$

which generates a non-trivial local minimum at

$$\langle a \rangle = \bar{\mu}^2 \exp\left(-\frac{24\pi^2}{11\bar{g}^2} + 1\right). \tag{2.28}$$

This is nothing but the dimensional transmutation by the monopole condensation. The corresponding effective potential is plotted in Fig. 2.4, where we have assumed $\bar{\alpha}_s = 1$ and $\bar{\mu}^2 = 1$.

Using the Weyl symmetry of SU(3) ECD (2.15) we can easily generalize the above result to SU(3) QCD. We have

$$\mathcal{L}_{eff} = \begin{cases} -\sum_p \left(\dfrac{a_p^2}{3g^2} + \dfrac{11a_p^2}{48\pi^2}\left(\ln\dfrac{a_p}{\mu^2} - c\right)\right), & b = 0 \\[2mm] \sum_p \left(\dfrac{b_p^2}{3g^2} + \dfrac{11b_p^2}{48\pi^2}\left(\ln\dfrac{b_p}{\mu^2} - c\right) \right. \\[2mm] \left. \quad -i\dfrac{11b_p^2}{96\pi}\right), & a = 0 \end{cases} \tag{2.29}$$

where a_p ($p = 1, 2, 3$) and b_p are the Diracian chromomagnetic (i.e., monopole) and chromoelectric background of three SU(2) subgroups.

With $b_p = 0$, we can easily show that the effective potential (2.29) has (after the renormalization) the true minimum at

$$\langle a_1 \rangle = \langle a_2 \rangle = \langle a_3 \rangle = \bar{\mu}^2 \exp\left(-\frac{16\pi^2}{11g^2} + \frac{3}{4}\right). \tag{2.30}$$

The effective potential is shown in Fig. 2.5. This demonstrates the monopole condensation which generates the desired mass gap in QCD.

The effective action has two important features. First, it is invariant under the dual transformation [23]

$$a \to -ib, \qquad b \to ia, \tag{2.31}$$

This duality was first discovered in the QED effective action [28]. But subsequently this duality has been shown to be a fundamental symmetry of the effective action of gauge theory, Abelian and non-Abelian.

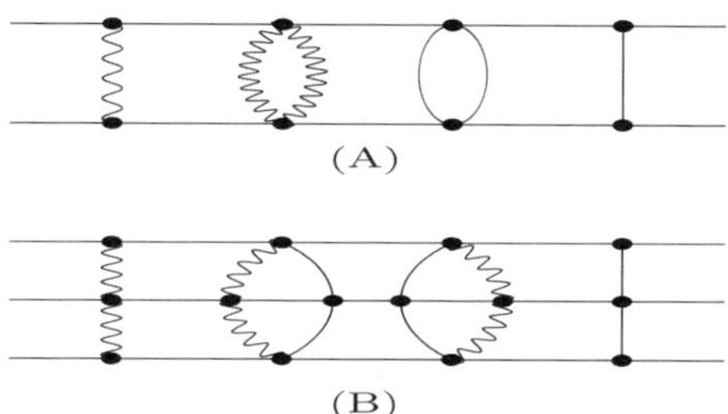

Fig. 2.6　The possible Feynman diagrams which bind the chromons. Two chromon binding is shown in (A), three chromon binding is shown in (B).

Another important feature of (2.22) and (2.29) is that when $a = 0$ (or $a_p = 0$) the imaginary part has a negative signature. This implies the pair annihilation of chromons in chromoelectric background [10, 24, 29]. This must be contrasted with the QED effective action where the electron loop integral generates a positive imaginary part [30, 28]. The positive imaginary part in QED means the pair creation which generates the screening. On the other hand in QCD we have the negative imaginary part, and thus the pair annihilation of chromons.

This is closely related to the asymptotic freedom (anti-screening) of gluons. In QCD it is well known that the gluons and quarks play opposite roles in the asymptotic freedom [27]. In other words the quarks enhance the screening while the gluons overide the quarks and diminish it to generate the anti-screening. We can understand this with the pair creation of the quarks and the pair annihilation of the chromons by the chromo-electric field [29, 10]. And the QCD effective action confirms this with the negative imaginary part.

2.5　Quark and chromon model and quarkonium-glueball mixing

One of the important issues in hadron spectroscopy is the identification of glueballs. The general wisdom is that QCD must have the glueballs made of gluons, and in early days the gauge invariant combinations of the QCD field strength were suggested to make glueballs [31,32]. Later several models of glueballs including the bag model and the constituent model have been proposed [33,34,35,36]. Moreover, the lattice QCD has been able to estimate the mass of the low-lying glueballs [37].

But so far the search for the glueballs has not been so successful for two reasons. First, theoretically there has been no consensus on how to construct the glueballs from QCD. For example, there has been the proposal to make the glueballs with "the constituent gluons", but a precise definition of the constituent gluon was lacking [34]. This has made it difficult for us to predict what kind of glueballs we can expect.

The other reason is that it is not clear how to identify the glueballs experimentally. This is partly because the glueballs could mix with the quarkoniums, so that we must take care of the possible mixing to identify the glueballs. This is why we have very few candidates of glueballs so far, compared to huge hadron spectrum made of quarks listed by Particle Data Group (PDG) [38].

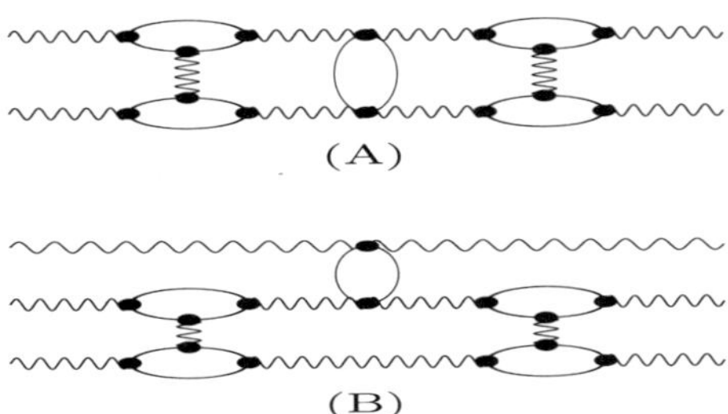

Fig. 2.7   The possible Feynman diagrams of the neuron interaction.

The Abelian decomposition provides a clear picture of glueballs. The fact that the chromons become colored gauge covariant vector particles tells that, just like the quark, they could become the constituents of hadrons. Indeed, Fig. 2.6 shows that the interaction among the chromons is like the interaction among quarks. This means that the color reflection invariant combination of $g\bar{g}$ or ggg could form hadronic bound states called the chromoballs [39,40].

Since we have six gauge covariant chromons $(R_\mu, B_\mu, G_\mu, \bar{R}_\mu, \bar{B}_\mu, \bar{G}_\mu)$, we can construct color reflection invariant chromoballs with two or three chromons,

$$|g\bar{g}\rangle = \frac{|R_\mu \bar{R}_\nu\rangle + |B_\mu \bar{B}_\nu\rangle + |G_\mu \bar{G}_\nu\rangle}{\sqrt{3}},$$

$$|ggg\rangle_d = \frac{\sum_{(RGB)} |R_\mu B_\nu G_\rho\rangle}{\sqrt{6}}, \quad (2.32)$$

$$|ggg\rangle_f = \frac{\sum_{[RGB]} |R_\mu B_\nu G_\rho\rangle}{\sqrt{6}},$$

where the sums in ggg are the totally symmetric (the d-product) and the totally anti-symmetric (the f-product) combination of three colors.

In addition to the above chromoballs, the Abelian decomposition predicts new hybrid hadrons made of quarks and chromons. For example, we can have color reflection invariant $q\bar{q}g$ hybrid meson or $qqqg$ hybrid baryons. This is because both quarks and chromons become the constituents of hadrons [2, 3].

At this point one might ask why the neurons can not be the constituents of hadrons. From group theoretical point of view, there is no reason why they can not form bound states. But dynamically, the binding among the neurons is supposed to be very weak because they do not carry color charge. This can be seen from the Feynman diagrams which describe the interaction among the neurons shown in Fig. 2.7, which is just like the interaction among photons in QED. This strongly implies that, if they form bound states at all, they could only form very loosely bound states.

This sounds all very nice. But this is not the whole story. The problem with the above picture is that this could give us too many chromoball states, while experimentally we have not so many candidates of them. So the real problem with the glueballs is to understand why there are so few candidates of glueballs experimentally, compared to the rich haderon spectrum based on the successful quark model.

As we have already pointed out, one of the reasons (at least partly) is that the chromoballs made of chromons could become unstable. The reason is that, unlike the quarks, the gluons tend to annihilate each other in the chromo-electric background [29, 23, 24]. This is because in the chromo-electric background the effective action (2.29) has negative imaginary part. This must be contrasted with quarks, which remain stable inside the hadrons.

But perhaps a more important reason is that these chromoballs in general may not exist as mass eigenstates, because they could mix with $q\bar{q}$ states unless the conserved quantum numbers prevent it. So we need a clear picture of the mixing mechanism to identify the glueballs [39, 40]. The possible Feynman diagrams of the mixing between the chromoball and quarkonium are shown in Fig. 2.8. The diagram tells that in the mixing the difference between neurons and chromons, in particular the role of the constituent of the chromons, becomes blurred. In fact we can not explain the mixed states in terms of constituents. This makes the experimental identification of glueballs complicated.

Nevertheless the quark and chromon model allows us to discuss the mixing u-nambiguously. For example, we can perform the numerical analysis of glueball-quarkonium mixing in 0^{++}, 2^{++}, and 0^{-+} sectors below 2 GeV, and show that $f_0(500)$ and $f_0(1500)$ in the 0^{++} sector, $f_2(1950)$ in the 2^{++} sector, and $\eta(1405)$ and $\eta(1475)$ in the 0^{-+} sector could be identified as predominantly the glueball states [39, 40].

Although the chromoballs in general mix with the quarkoniums, in particular

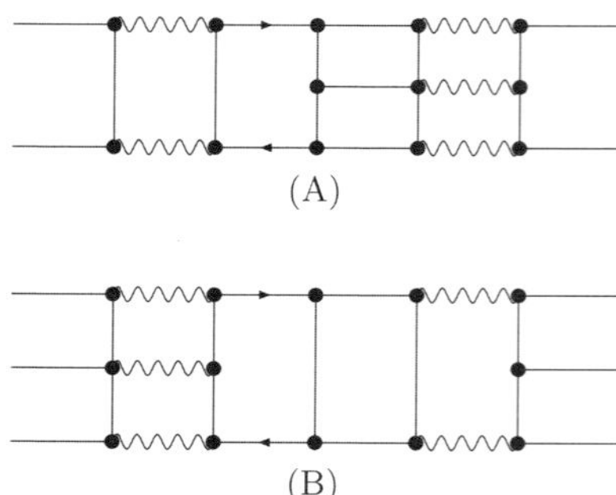

Fig. 2.8   The possible glueball-quarkonium mixing diagrams.

cases the pure chromoballs could exist [34]. This is because some of the $g\bar{g}$ glueballs have the quantum number J^{PC} which $q\bar{q}$ can not have. In the quark model the $q\bar{q}$ states in the natural spin-parity series $P = (-1)^J$ must have spin one, and hence $CP = +1$. So the mesons with natural spin-parity and $CP = -1$ (e.g., $0^{--}, 0^{+-}, 1^{-+}, 2^{+-}$, etc.) are forbidden.

So these particular chromoballs carrying the quantum numbers which $q\bar{q}$ are not allowed to have can not mix with $q\bar{q}$, and they remain as pure chromoballs. They are called "the oddballs" [34]. These oddballs play very important to test the quark and chromon model, so that it becomes an important for us to search for the oddballs experimentally.

2.6   Monoball: vacuum fluctuation of monopole condensation

The Abelian decomposition allows us to demonstrate the monopole condensation (more precisely the monopole-antimonopole pair condensation) which induces the dimensional transmutation and creates the mass gap. If so, one may ask what (if any) is the observable consequence of the monopole condensation. The answer could be the monoball.

To understand this, consider the ordinary superconductor in QED. It is well known that the BCS superconductivity is characterized by two scales, the correlation length of the Cooper pair and the penetration length of the magnetic field. Field theoretically they are represented by two composite fields, a (complex) scalar field for the Cooper pair and a (massive) vector field for the confined magnetic field. And the existence of these modes are the consequence of the BCS superconductivity.

So in QCD we may expect a similar consequence of the monopole condensation. Naively we might think that the monopole condensation creates two mass scales, the correlation length of the monopole-antimonopole pairs and the penetration length of the chromo-electric flux. This suggests that the monopole condensation could induce two physical states, one 0^{++} and one 1^{++} vacuum fluctuation modes [3, 4].

However, the confinement in QCD is not exactly dual to the superconductivity in QED. First of all, in QED the magnetic field to be confined is generated by the electric current, not by the magnetic charge. But in QCD the colored flux to be confined comes from the color charge, not the chromo-magnetic current. Second, in superconductor the Cooper pair has electric charge but the monopole-antimonopole pair in QCD obviously has no chromo-magnetic charge.

Third, in the superconductor the magnetic field is actually screened by the supercurrent, not confined by the Cooper pair. But in QCD the chromo-electric field is confined by the monopole-antimonopole pair. In other words, it is not the monopole supercurrent which provides the confinement. QCD has no monopole supercurrent. Fourth, the chormo-electric flux is described by the Coulomb (i.e., scalar) potential, not by the vector potential, in QCD. But in superconductor the magnetic field is described by the vector potential.

Finally, in the superconductor the Higgs mechanism takes place. The Landau–Ginzburg theory of superconductivity is a classic example of Higgs mechanism, where the spontaneous symmetry breaking generates the massive vector field which screens the magnetic field. But in QCD there is no spontaneous symmetry breaking. It is the dynamical symmetry breaking which generates the confinement. This tells that the confinement mechanism in QCD is not exactly dual to the Meissner effect. They are different.

In particular, this implies that the penetration length in QCD could be represented by a scalar field, not by a spin-one field. This is because the chromo-electric field is described by the Coulomb (i.e., scalar) potential. This strongly suggests that both the correlation length and the penetration length in QCD must be represented by the scalar mode. In other words, there might be no 1^{++} vacuum fluctuation mode in QCD.

The remaining question is if the two scalar modes are different or not. In principle they could be different, but as we have shown in (2.30) the monopole condensation generates only one mass scale. This, together with the fact that QCD has only one scale Λ_{QCD}, strongly suggests that they are the same.

From this we may conclude that the monopole condensation could have only one 0^{++} vacuum fluctuation mode which could naturally be called the magnetic glueball or simply the monoball. Clearly this fluctuation mode must be different from the glueballs made of the chromons because this characterizes the monopole condensation.

The importance of the monoball is that this represents the monopole condensation, so that the experimental verification of this monoball can be interpreted as the confirmation of the monopole condensation. This makes the experimental identification of the monoball a most urgent issue in QCD.

Ultimately, however, the nature of the monopole condensation (and the number the vacuum fluctuation modes) should be determined by experiment, and it could well be that the monopole condensation has no vacuum fluctuation mode at all. To understand this possibility consider the Dirac sea, the vacuum of Dirac's theory of electron. It has vacuum bubbles made of electron-positron pairs, but is not the electron-positron pair condensation and apparently has no fluctuation mode.

So, if the QCD vacuum is like the Dirac sea, there will be no vacuum fluctuation and thus no monoball. At the moment it is not clear if the QCD vacuum is similar to Dirac sea, and only experiments can tell whether the nature of the QCD vacuum is different from the Dirac sea or not. This makes the experimental confirmation of the monoball more interesting.

One might ask if there is any candidate of the monoball. Actually PDG has several isoscalar 0^{++} states, in particular $f_0(500)$ and $f_0(980)$, which do not fit well in the quark model. It would be very interesting to find which of them (if at all) could be interpreted as the monoball.

2.7 Vacuum decomposition

The Abelian decomposition has another important application. Refining the A-belian decomposition we can have the vacuum decomposition which separates the vacuum part gauge independently from the gauge potential. An advantage of this vacuum decomposition is that we can express the most general vacuum potential in a compact form. This allows us to study the topological structure of the non-Abelian gauge theory very easily.

Unlike the Abelian decomposition we can have the vacuum decomposition in Abelian gauge theory. So we start from the vacuum decomposition of U(1) gauge theory first. Let $U = \exp(i\theta)$ be an element of U(1) where θ is an arbitrary U(1) phase angle. With this we can construct a most general vacuum and obtain the gauge independent vacuum decomposition of the gauge potential as follows [41, 42]

$$\Omega_\mu = -i\, U^{-1}\partial_\mu U = \partial_\mu\theta,$$
$$A_\mu = \Omega_\mu + Z_\mu. \tag{2.33}$$

Clearly the decomposition is gauge independent because U is arbitrary. Moreover, Ω_μ describes the most general vacuum potential which has the full gauge degrees of freedom but Z_μ becomes gauge invariant physical potential which describes the physical photon.

To see this let α be the (infinitesimal) gauge parameter and consider the gauge transformation

$$\delta A_\mu = \frac{1}{e}\partial_\mu\alpha. \tag{2.34}$$

Under the gauge transformation the U(1) phase changes by α, so that we have

$$\delta\theta = \alpha, \quad \delta\Omega_\mu = \frac{1}{e}\partial_\mu\alpha, \quad \delta Z_\mu = 0. \tag{2.35}$$

This tells that the gauge transformation affects only Ω_μ, which has the full gauge degrees of freedom. Moreover, Z_μ becomes a gauge invariant Lorentz four-vector. Again this is because the connection space forms an affine space.

In this decomposition Z_μ should describe the photon, so that we have to remove the unphysical (longitudinal) degree with a gauge condition to keep it massless. But (2.35) tells that Z_μ no longer has a gauge freedom. To understand this apparent contradiction notice that the decomposition (2.33) put the theory to the background field formalism in which Ω_μ and Z_μ become the slow-varying classical field and the fluctuating quantum field. In this case the gauge symmetry is automatically enlarged to two gauge symmetries, the classical and quantum gauge symmetries [17, 16].

So, in addition to the classical (slow) gauge transformation given by (2.35), we have the quantum (fast) gauge transformation given by

$$\delta\theta = 0, \quad \delta\Omega_\mu = 0, \quad \delta Z_\mu = \frac{1}{e}\partial_\mu\alpha. \tag{2.36}$$

Because of this Z_μ has an extra (quantum) gauge freedom, and we have to fix this freedom to quantize it. This allows us to impose the (Lorentz invariant) transversality condition to Z_μ

$$\partial_\mu Z_\mu = 0, \tag{2.37}$$

to keep it massless. This is how the photon field remains massless after the vacuum decomposition.

We can generalize the above vacuum decomposition to non-Abelian gauge theory [43]. To do that we first have to identify the most general vacuum potential. Let G be an n-dimensional non-Abelian gauge group, U be an arbitrary matrix element of G determined by n non-Abelian phase angles, and t_i ($i = 1, 2, ..., n$) be the generators. With this we can obtain the most general vacuum potential by

$$\hat{\Omega}_\mu = \frac{1}{2g}Tr \ (\vec{t}\,U^{-1}\partial_\mu U), \tag{2.38}$$

where $\vec{t} = (t_1, t_2, ..., t_n)$. Clearly $\hat{\Omega}_\mu$ represents the most general vacuum, because here U is an arbitrary element of G. Moreover, under the gauge transformation (2.38) transforms by itself, so that $\hat{\Omega}_\mu$ forms its own connection space which has the full gauge degrees of freedom.

With (2.38), we can have the vacuum decomposition of the gauge potential

$$\vec{A}_\mu = \hat{\Omega}_\mu + \vec{Z}_\mu.$$

(2.39)

Notice that the decomposition is gauge independent, because this applies in any gauge. Moreover, $\vec{Z}_\mu$ transforms covariantly. This again is because the connection space made of $\vec{A}_\mu$ forms an affine space. Obviously (2.39) with (2.38) is a straightforward generalization of (2.33) to non-Abelian gauge theory.

There is a more convenient way to obtain the vacuum potential which is very useful for our purpose. Consider SU(2) QCD for simplicity, and let $\hat{n}_i$ ($i = 1, 2, 3$) be an arbitrary gauge covariant right-handed orthonormal local SU(2) basis. To obtain the pure vacuum we impose the vacuum isometry to the potential

$$\forall_i \quad D_\mu \hat{n}_i = 0, \qquad (\hat{n}_i^2 = 1).$$

(2.40)

Clearly this assures the vacuum because

$$\forall_i \, [D_\mu, \, D_\nu] = g\vec{F}_{\mu\nu} \times \hat{n}_i = 0 \Rightarrow \vec{F}_{\mu\nu} = 0.$$

(2.41)

Solving (2.40) we obtain the most general vacuum potential $\hat{\Omega}$ [43]

$$\hat{\Omega}_\mu = \frac{1}{2g} \epsilon_{ijk}(\hat{n}_i \cdot \partial_\mu \hat{n}_j) \, \hat{n}_k = C_\mu^{\ k} \hat{n}_k,$$

$$C_\mu^{\ k} = \frac{1}{2g} \epsilon_{ij}^{\ \ k}(\hat{n}_i \cdot \partial_\mu \hat{n}_j).$$

(2.42)

Notice that, just like U in (2.38), $\hat{n}_i$ (and thus $\hat{\Omega}$) here is completely fixed by three SU(2) phase angles. So (2.42) has exactly the same information as (2.38).

Moreover, this vacuum is able to describe all topologically distinct vacua classified by integer n,

$$n = -\frac{g^3}{96\pi^2} \int \epsilon_{0\alpha\beta\gamma} \epsilon_{ijk} C_\alpha^{\ i} C_\beta^{\ j} C_\gamma^{\ k} d^3 x.$$

(2.43)

which describes the topology $\Pi_3(S^3) \simeq \Pi_3(S^2)$ determined by $\hat{n} = \hat{n}_3$ which defines the mapping from the compactified space S^3 to the group space S^3. This confirms that (2.42) describes the most general vacuum.

To understand the physical meaning of this, notice that the above vacuum comes from the trivial vacuum $\vec{A}_\mu = 0$ by a gauge transformation. To see this let $(\hat{e}_1, \hat{e}_2, \hat{e}_3)$ be the trivial (i.e., space-time independent) SU(2) basis, and consider the gauge transformation

$$(\hat{e}_1, \hat{e}_2, \hat{e}_3) \rightarrow (\hat{n}_1, \hat{n}_2, \hat{n}_3), \quad \hat{n}_i = U_{ij} \, \hat{e}_j.$$

(2.44)

Under this we have

$$\vec{A}_\mu = 0 \rightarrow \hat{\Omega}_\mu = \frac{1}{2g} Tr\left[t^k \, U\partial_\mu U^{-1}\right] \hat{n}_k$$

$$= \frac{1}{2g} \epsilon_{ij}^{\ \ k}(\hat{n}_i \cdot \partial_\mu \hat{n}_j) \, \hat{n}_k.$$

(2.45)

This shows that the non-trivial vacuum (2.42) is related to the trivial vacuum $\vec{A}_\mu = 0$ by a gauge transformation.

We can have another expression of the vacuum potential in the trivial basis. Transforming (2.45) back to the trivial basis, we have

$$U^{-1}\hat{\Omega}_\mu U + \frac{1}{2g}Tr\left[\vec{t}\, U^{-1}\partial_\mu U\right] = 0,$$
$$\hat{e}_i = (U^{-1})_{ij}\,\hat{n}_j. \tag{2.46}$$

So we have another vacuum expressed by

$$\hat{\Omega}'_\mu \equiv \frac{1}{2g}Tr\left[\vec{t}\, U^{-1}\partial_\mu U\right] = -U^{-1}\hat{\Omega}_\mu U = -\Omega_\mu{}^k\hat{e}_k. \tag{2.47}$$

This means that the same vacuum potential (2.42), with the change of the signature, describes the physically equivalent vacuum in the trivial basis. The only difference is that $\hat{\Omega}_\mu$ is obtained from the trivial vacuum with the gauge transformation U, but $\hat{\Omega}'_\mu$ with the inverse gauge transformation U^{-1}. This is remarkable.

Under the infinitesimal gauge transformation we have

$$\delta\hat{n}_i = -\vec{\alpha}\times\hat{n}_i, \tag{2.48}$$

so that

$$\delta\hat{\Omega}_\mu = \frac{1}{g}\bar{D}_\mu\vec{\alpha}, \qquad \delta\vec{Z}_\mu = -\vec{\alpha}\times\vec{Z}_\mu,$$
$$\bar{D}_\mu = \partial_\mu + g\,\hat{\Omega}\times, \tag{2.49}$$

where $\vec{\alpha}$ is the (infinitesimal) gauge parameter. This confirms that $\hat{\Omega}_\mu$ has the full $SU(2)$ gauge freedom and that $\vec{Z}_\mu$ transforms covariantly and thus forms a genuine Lorentz four-vector. But what is really remarkable is that the decomposition is gauge independent. Once $\hat{n}_i$ is chosen, it follows independent of the choice of a gauge.

Just as in QED, the decomposition (2.39) put the theory to the background field formalism, and $\vec{Z}_\mu$ acquires the quantum gauge degrees. We can fix this quantum gauge freedom choosing the generalized Lorentz gauge

$$\bar{D}_\mu\vec{Z}_\mu = 0. \tag{2.50}$$

Again we emphasize that this is not the gauge condition of the classical gauge degrees, but the self-consistent quantum gauge condition which keeps $\vec{Z}_\mu$ massless.

Obviously (2.39) and (2.50) are the straightforward generalization of (2.33) and (2.37) to $SU(2)$ which provides the gauge (and Lorentz) independent decomposition of the non-Abelian potential to the vacuum and physical parts.

2.8   Einstein's theory: gauge theory of Lorentz group

The Abelian decomposition has an important application in general relativity. Applying the Abelian decomposition to general relativity we can simplify Einstein's theory and obtain the restricted gravity (RG) which has the full general invariance but is simpler than Einstein's theory [44, 45, 46].

This is because the general invariance can be viewed as a particular type of gauge symmetry [47, 48]. For example it can be viewed as a gauge theory of 4-dimensional translation group, because the general coordinate transformation can be identified as the local 4-dimensional translation [48, 49]. Or, it can be viewed as a gauge theory of Lorentz group, because the tetrad has the Lorentz gauge symmetry.

In fact, when we include a spinor field in Einstein's theory we have to deal with the γ-matrices which are defined on the local orthonormal basis (the tetrad). In this case the consistency requires us to make the theory local Lorentz invariant. This necessitates a gauge theory of Lorentz group, where the tetrad (not the metric) plays the fundamental role. Constructing a gauge theory of Lorentz group this way we can obtain the Einstein's theory, or more precisely the Einstein–Cartan theory, in which the spinor creates the torsion [50, 51]. Indeed this is a natural way to rediscover Einstein's theory.

To see this we introduce a coordinate basis ∂_μ $(\mu = t, x, y, z)$ and an orthonormal tetrad basis e_a $(a = 0, 1, 2, 3)$

$$[\partial_\mu, \ \partial_\nu] = 0, \quad [e_a, \ e_b] = f_{ab}{}^c \, e_c,$$
$$\partial_\mu = e_\mu{}^a \, e_a, \quad e_a = e_a{}^\mu \partial_\mu,$$
$$f_{ab}{}^c = (e_a{}^\mu \partial_\mu e_b{}^\nu - e_b{}^\mu \partial_\mu e_a{}^\nu) e_\nu{}^c. \tag{2.51}$$

where $e_\mu{}^a$ and $e_a{}^\mu$ are the tetrad and inverse tetrad. Let $J_{ab} = -J_{ba}$ be the generators of Lorentz group which act on the tetrad,

$$[J_{ab}, \ J_{cd}] = \eta_{ac} J_{bd} - \eta_{bc} J_{ad} + \eta_{bd} J_{ac} - \eta_{ad} J_{bc}$$
$$= f_{ab,cd}{}^{mn} \, J_{mn},$$
$$f_{ab,cd}{}^{mn} = \eta_{ac} \delta_b{}^{[m} \delta_d{}^{n]} - \eta_{bc} \delta_a{}^{[m} \delta_d{}^{n]}$$
$$+ \eta_{bd} \delta_a{}^{[m} \delta_c{}^{n]} - \eta_{ad} \delta_b{}^{[m} \delta_c{}^{n]}, \tag{2.52}$$

where $\eta_{ab} = diag\,(-1, 1, 1, 1)$ is the Lorentz invariant Minkowski metric.

Now we introduce the anti-symmetric unit tensor $\mathbf{I}_{ab}$ which forms an adjoint representation of Lorentz group (which we denote by six component vector) defined by

$$I_{ab}{}^{mn} = \left(\delta_a{}^m \delta_b{}^n - \delta_a{}^n \delta_b{}^m\right) = -(J_{ab})^{mn}. \tag{2.53}$$

With this we can define the Lorentz covariant 4-index anti-symmetric metric $\mathbf{g}_{\mu\nu}$

$$\mathbf{g}_{\mu\nu} = e_\mu{}^a e_\nu{}^b \, \mathbf{I}_{ab} = \frac{1}{2} g_{\mu\nu}{}^{ab} \, \mathbf{I}_{ab},$$
$$g_{\mu\nu}{}^{ab} = (e_\mu{}^a e_\nu{}^b - e_\nu{}^a e_\mu{}^b) = e_\mu{}^c e_\nu{}^d I_{cd}{}^{ab}. \tag{2.54}$$

Clearly $\mathbf{I}_{ab}$ is nothing but the anti-symmetric metric $\mathbf{g}_{ab}$ written in the tetrad basis. But notice that the algebraic structure of this metric is identical to that of the curvature tensor. Because of this $\mathbf{g}_{\mu\nu}$, not the popular Einstein metric $g_{\mu\nu} = \eta_{ab}\, e_\mu{}^a e_\nu{}^b$, plays a crucial role in the gauge formalism of Einstein's theory.

To proceed we express the connection (potential) $\Gamma_\mu{}^{ab}$ and the curvature (field strength) $R_{\mu\nu}{}^{ab}$ by Lorentz sextets $\mathbf{\Gamma}_\mu$ and $\mathbf{R}_{\mu\nu}$. In this notation we have

$$\mathbf{R}_{\mu\nu} = \partial_\mu \mathbf{\Gamma}_\nu - \partial_\nu \mathbf{\Gamma}_\mu + \mathbf{\Gamma}_\mu \times \mathbf{\Gamma}_\nu. \tag{2.55}$$

Now, since $\mathbf{g}_{\mu\nu}$ and $\mathbf{R}_{\mu\nu}$ have the same algebraic structure, $\mathbf{g}_{\mu\nu} \cdot \mathbf{R}^{\mu\nu}$ becomes the simplest Lorentz invariant scalar. So we have the simplest Lorentz invariant action [50]

$$S[e_\mu{}^a, \mathbf{\Gamma}_\mu] = \frac{1}{16\pi G_N} \int e\, (\mathbf{g}_{\mu\nu} \cdot \mathbf{R}^{\mu\nu})\, d^4x,$$
$$e = \mathrm{Det}\, e_\mu{}^a, \tag{2.56}$$

which is nothing but the Einstein–Hilbert action. This is the gauge formalism of Einstein's theory. We emphasize that it is $\mathbf{g}_{\mu\nu}$ which allows the Einstein–Hilbert action linear in $\mathbf{R}_{\mu\nu}$. This makes the dynamics of Einstein's theory different from the non-Abelian dynamics of the Yang–Mills theory.

In this first order formalism we have two equations of motion

$$\delta e_\mu{}^a : \quad \mathbf{g}_{\mu\nu} \cdot \mathbf{R}^{\nu a} = 0,$$
$$\delta \mathbf{\Gamma}_\mu : \quad \mathscr{D}_\mu{}^{\mu\nu} = 0 \quad (\mathscr{D}_\mu = \nabla_\mu + {}_\mu\times), \tag{2.57}$$

where ∇_μ is the covariant derivative defined by the Levi–Civita connection. The first equation assures that, in the absence of matter fields, the Ricci tensor must vanish. The second equation is nothing but the metric compatibility of the connection [44, 45, 46],

$$\mathscr{D}_\mu{}^{\mu\nu} = 0 \quad \Longleftrightarrow \quad \nabla_\alpha g_{\mu\nu} = 0. \tag{2.58}$$

This confirms that (2.57) is identical to Einstein's equation.

To understand this notice that componentwise the second equation of (2.57) is expressed by

$$\nabla_\mu g^{\mu\nu ab} + \Gamma_{\mu c}{}^a\, g^{\mu\nu cb} + \Gamma_{\mu c}{}^b\, g^{\mu\nu ac} = 0. \tag{2.59}$$

From this we have

$$e^{b\nu} \mathscr{D}_\mu e^{a\mu} - e^{a\nu} \mathscr{D}_\mu e^{b\mu} + e^{a\mu} \mathscr{D}_\mu e^{b\nu} - e^{b\mu} \mathscr{D}_\mu e^{a\nu} = 0, \tag{2.60}$$

so that

$$\mathscr{D}_\mu e^{a\mu} = -\frac{1}{2} e^{a\mu} e_{b\nu} (\partial_\mu e^{b\nu} + \Gamma_{\mu\sigma}{}^\nu e^{b\sigma})$$
$$= -\frac{1}{2} e^{a\mu} e_{b\nu} \nabla_\mu e^{b\nu} = \frac{1}{2} e^{a\mu} e_{b\nu} \nabla_\mu e^{b\nu}, \tag{2.61}$$

where the last equality follows from $\nabla_\alpha g_{\mu\nu} = 0$.

2.9   Abelian decomposition of Einstein's theory: restricted gravity

With (2.56) we can make the Abelian decomposition of Einstein's theory. Of course, Einstein's theory (2.56) is different from the Yang–Mills Lagrangian. In gauge theory the fundamental field is the gauge potential, but here the fundamental field is the metric. And in the gauge formulation the gauge potential of Lorentz group corresponds to the gravitational connection, not the metric. Also, in gauge theory the Lagrangian is quadratic in field strength. But the Einstein–Hilbert Lagrangian is linear in field strength [50]. So the dynamics of Einstein's theory is different from QCD.

Nevertheless we can still make the Abelian decomposition of the gravitational connection, and express the Einstein–Hilbert Lagrangian in terms of the restricted connection and the valence connection. With this we can show that the theory can be interpreted as a restricted theory of gravity which has the valence connection as the gravitational source. Moreover, we can show that the restricted gravity describes the core dynamics of Einstein's theory which describes the gravitational plane wave, and express it as a Maxwell-type gauge theory.

This tells the restricted gravity retains the essential features of Einstein's theory. It describes the free graviton, and inherits the topological properties of Einstein's theory. As importantly, this implies that the graviton could also be described by a massless spin-one gauge potential, since the equation of motion of the free gravitational wave can be put in to Maxwell-type wave equation. This implies the Abelian dominance in Einstein's theory.

To obtain the Abelian decomposition of Einstein's theory, it is important to keep in mind that Lorentz group has two 2-dimensional maximal Abelian subgroups. This means that there are two different Abelian decompositions. Moreover, each Abelian subgroup has two Abelian directions. Let $\mathbf{p}$ (or p^{ab}) be a gauge covariant sextet which forms an adjoint representation of Lorentz group, which we can choose to be an Abelian direction. Then $\tilde{\mathbf{p}}$ defined by $\tilde{p}^{ab} = \epsilon_{cd}{}^{ab} p^{cd}$ automatically describes the other Abelian direction. This is because ϵ_{abcd} is an invariant tensor of Lorentz group.

Now, we can make the Abelian projection imposing the isometry to the connection,

$$D_\mu \mathbf{p} = \left(\partial_\mu + \boldsymbol{\Gamma}_\mu \times\right) \mathbf{p} = 0, \tag{2.62}$$

Notice that the above condition automatically assures

$$D_\mu \tilde{\mathbf{p}} = \left(\partial_\mu + \boldsymbol{\Gamma}_\mu \times\right) \tilde{\mathbf{p}} = 0. \tag{2.63}$$

This tells that when $\mathbf{p}$ is an isometry, $\tilde{\mathbf{p}}$ also becomes an isometry. Solving this isometry equation we can obtain the restricted connection $\hat{\boldsymbol{\Gamma}}_\mu$.

With this we can obtain the full connection of Lorentz group adding the Lorentz covariant valence connection $\mathbf{Z}_\mu$,

$$\boldsymbol{\Gamma}_\mu = \hat{\boldsymbol{\Gamma}}_\mu + \mathbf{Z}_\mu. \tag{2.64}$$

The corresponding field strength $\mathbf{R}_{\mu\nu}$ which describes the curvature tensor is decomposed to

$$\mathbf{R}_{\mu\nu} = \hat{\mathbf{R}}_{\mu\nu} + \mathbf{Z}_{\mu\nu},$$
$$\mathbf{Z}_{\mu\nu} = \hat{D}_\mu \mathbf{Z}_\nu - \hat{D}_\nu \mathbf{Z}_\mu + \mathbf{Z}_\mu \times \mathbf{Z}_\nu, \tag{2.65}$$

where $\hat{D}_\mu = \partial_\mu + \hat{\boldsymbol{\Gamma}}_\mu \times$ and $\mathbf{Z}_{\mu\nu}$ is the valence part of the curvature tensor.

Now, decomposing the metric $\mathbf{g}_{\mu\nu}$ to the restricted and valence parts [45, 46]

$$\mathbf{g}_{\mu\nu} = \hat{\mathbf{g}}_{\mu\nu} + \mathbf{G}_{\mu\nu}, \tag{2.66}$$

we can express the Einstein–Hilbert Lagrangian as

$$\begin{aligned}
\mathcal{L} &= \frac{e}{16\pi G_N} \Big[(\hat{\mathbf{g}}_{\mu\nu} + \mathbf{G}_{\mu\nu}) \cdot (\hat{\mathbf{R}}^{\mu\nu} + \mathbf{Z}^{\mu\nu}) \Big] \\
&= \frac{e}{16\pi G_N} \Big[\hat{\mathbf{g}}_{\mu\nu} \cdot \hat{\mathbf{R}}^{\mu\nu} + \hat{\mathbf{g}}_{\mu\nu} \cdot \mathbf{Z}^{\mu\nu} + \mathbf{G}_{\mu\nu} \cdot \mathbf{Z}^{\mu\nu} \Big].
\end{aligned} \tag{2.67}$$

This is the Abelian decomposition of Einstein's theory. With $\mathbf{Z}_\mu = 0$, we have the restricted Lagrangian

$$\mathcal{L} = \frac{e}{16\pi G_N} \hat{\mathbf{g}}_{\mu\nu} \cdot \hat{\mathbf{R}}^{\mu\nu}, \tag{2.68}$$

which describes the restricted gravity.

As we have pointed out, there are two different Abelian decompositions because the Lorentz group has two maximal Abelian subgroups. So, unlike QCD, Einstein's theory allows two Abelian decompositions, the light-like (null) decomposition and the non light-like (rotation/boost) decomposition [44, 45, 46]. And the explicit form of the Abelian decomposition depends on which subgroup we choose.

Independent of the details, however, the decomposition of Einstein's theory has deep implications. First of all, this tells that we can construct two restricted theory of gravitation simpler than Einstein's theory. In other words, we can separate the Abelian part of gravity which describes the core dynamics of Einstein's theory without compromising the general invariance.

Moreover, the decomposition makes the topology of Einstein's theory more transparent, because the topological characteristics are imprinted in the isometry. For example, the rotation-boost decomposition makes it clear that the topology of Einstein's theory is closely related to the topology of $SU(2)$ gauge theory. This is because $SU(2)$ forms the rotation subgroup of Lorentz group. This strongly implies that Einstein's theory may have the multiple vacua similar to what we find in $SU(2)$ gauge theory.

In fact we can show that Einstein's theory has exactly the same multiple vacua that we have in $SU(2)$ gauge theory [44, 45, 46]. This tells that the vacuum space-time can be classified by the knot topology $\pi_3(S^3) \simeq \pi_3(S^2)$. This could have a far reaching consequence. Just as in$SU(2)$ gauge theory, the multiple vacua in Einstein's theory can be unstable against quantum fluctuation. And there is a real possibility that Einstein's theory may admit the gravito-instantons which can connect topologically distinct vacua with the vacuum tunneling.

Finally, this leads us to the Abelian dominance in Einstein's theory, that the restricted gravity defined by the Abelian projection of Einstein's theory plays the dominant role in Einstein's theory. It describes the graviton and thus the core dynamics of Einstein's theory, and inherits basic topological properties of Einstein's theory. The Abelian dominance has first been proposed in QCD, which asserts that the Abelian part determines the essential features of QCD (in particular the color confinement). Of course, in Einstein's theory we have no confinement. Nevertheless our analysis implies that, just as in QCD, the essential dynamical and topological features of Einstein's theory is inherited by the restricted gravity.

2.10 Canonical momentum versus kinematic momentum

A fundamental problem in theoretical physics is to find if it is possible to decompose the momentum and spin of a composite particle to those of the constituents, and if so how to do so [41, 42]. Intuitively this must be possible, but in gauge theory the answer is not clear for two reasons. First, in gauge theory it is not simple to obtain a gauge invariant momentum operator of a charged particle which is not contaminated by the gauge bosons. Second, it is not easy to tell what are the constituents of a given composite particle. Moreover, in non-Abelian gauge theory there are two types gluons, so that it is not clear which one we should treat as the constituent [2, 3]. The vacuum decomposition plays a crucial role to resolve this problem.

In gauge theory a charged particle has two momentum operators, the canonical momentum $\mathbf{p}_\mu$ given by $-i\partial_\mu$ and the kinematic momentum $\mathbf{\Pi}_\mu$ given by $-iD_\mu$. And it has generally been believed that only the kinematic momentum is gauge invariant and thus measurable. If this is true, the momentum decomposition in strict sense becomes impossible.

In fact it has been thought that the electron momentum generated by the Lorentz force is the kinematic momentum, and the conserved momentum of a charged particle moving in an electromagnetic field is the sum of the electromagnetic field momentum (the Poynting vector) and the kinematic momentum of the charged particle [52].

In quantum theory, however, this looks strange. Here the momentum operator must satisfy the canonical commutation relation, but the kinematic momentum does not satisfy this criterion. The canonical momentum (as the generator of the space-time translation) satisfies the canonical commutation relation required for the momentum operator,

$$[\mathbf{p}_\mu, \ \mathbf{p}_\nu] = 0, \tag{2.69}$$

but obviously is not gauge covariant. On the other hand, the kinematic momentum is gauge covariant but does not satisfy the canonical commutation relation.

Clearly the vacuum decomposition allows us to define the canonical momentum operator made of the covariant derivative which does not contain the physical photon,

$$\bar{\mathbf{p}}_\mu = -i\bar{D}_\mu, \qquad \bar{D}_\mu = \partial_\mu + ie\Omega_\mu, \tag{2.70}$$

Obviously this canonical momentum operator satisfies the canonical commutation relation,

$$[\bar{\mathbf{p}}_\mu, \ \bar{\mathbf{p}}_\nu] = 0. \tag{2.71}$$

And it does not include the photon. So with this operator we can obtain the gauge invariant canonical momentum for a charged particle which does not contain the contribution of the photon field.

In non-Abelian gauge theory we have

$$\begin{aligned}
[\mathbf{\Pi}_\mu, \ \mathbf{\Pi}_\nu] &= g\vec{F}_{\mu\nu}\times \ \neq 0, \\
\mathbf{\Pi}_\mu &= \partial_\mu + g\vec{A}_\mu \times \ .
\end{aligned} \tag{2.72}$$

As a result the angular momentum operator made of $\mathbf{\Pi}_\mu$ does not satisfy the angular momentum canonical commutation relation. So in quantum theory there seems no reason why kinematic momentum must be more fundamental.

Now, with(2.39) we can define the desired canonical momentum operator in QCD by

$$\bar{\mathbf{p}}_\mu = -i\bar{D}_\mu, \quad [\bar{\mathbf{p}}_\mu, \ \bar{\mathbf{p}}_\nu] = 0, \quad \bar{D}_\mu = \partial_\mu + g\,\hat{\Omega}_\mu \times \ . \tag{2.73}$$

This is the generalization of (2.70) to SU(2), which confirms that we can construct the gauge covariant canonical momentum operator which satisfies the canonical commutation relation and does not involve gluon. Clearly this momentum operator allows us to define the gauge independent quark canonical momentum which is not contaminated by the gluons. So we can use this quark canonical momentum to decompose the nucleon momentum to the momentums of quarks and gluon field [41, 42]. This has been thought to be impossible.

2.11 Proton spin crisis problem: present status and a new resolution

An outstanding problem in nuclear physics is the so-called proton spin crisis problem. The problem originated from the European Muon Collaboration (EMC) experiment on muon scattering of polarized proton, which indicated that the sum of quark spin was much smaller than proton spin [53]. To settle this problem we need more precise measurements of the spin and angular momentum of quark in nucleons.

But theoretically this puts us back to the fundamental issue: Can we decompose the momentum and spin of a composite particle to those of the constituents? This becomes a central issue when we try to decompose momentum and spin of atoms to those of electron and electromagnetic field in QED [41,42]. Here we need to define the electron momentum independent of photon. But we can not use the canonical momentum because it is not gauge invariant. Neither can we use the kinematic momentum because this momentum contains the photon field.

We face a similar problem in QCD when we try to decompose the momentum and spin of nucleons to those of quarks and gluon field [41,42]. Here again an outstanding issue is if it is possible to define the gauge invariant quark canonical momentum or not. Jaffe and Manohar first showed that the conserved total angular momentum in QCD can be decomposed to four terms, quark spin and orbital angular momentum and gluon spin and orbital angular momentum [55]. But the decomposition was not gauge invariant, so that it was thought to be unacceptable. This has generated controversies and confusions in the literature [56,57,58].

One school has tried to obtain the desired decomposition decomposing the gauge potential to the "physical" and "unphysical" parts and defining the gauge covariant canonical momentum operator with only the "unphysical" potential [56]. But their decomposition of the gauge potential was *ad hoc*, and has been criticized to have its own problems [57,58]. Another school claimed that the momentum and spin decomposition is impossible, asserting that the gauge covariant quark momentum operator is the kinematic momentum operator which contains the gluon field [57, 58]. If this is true, the nucleon momentum and spin decompositions should be impossible in the strict sense. The problem with this school is that they do not seem to understand the existence of the gauge covariant canonical momentum operator which does not contain gluons.

On the other hand we have shown that it is possible to have an explicitly gauge invariant momentum and spin decomposition, if we use the gauge covariant canonical momentum operator defined by the vacuum decomposition [41,42]. To show how the gauge covariant canonical momentum allows us to obtain a gauge invariant momentum and spin decompositions of composite particles to those of the constituents and the binding fields, we first consider the hydrogen atom made of electron and photons in QED.

The QED Lagrangian has the conserved momentum given by Noether theorem,

$$
\begin{aligned}
P_\mu^{(qed)} &= P_\mu^{(e)} + P_\mu^{(\gamma)} \\
&= i \int \bar{\psi}\gamma^0 \partial_\mu \psi d^3x + \int [(\partial_\mu A_\alpha)F^{\alpha 0} + \frac{1}{4}\delta_\mu^0 F_{\alpha\beta}^2]d^3x.
\end{aligned}
\tag{2.74}
$$

Obviously this decomposition is not gauge invariant. To cure this defect one might like to change it to the following gauge invariant expression adding a surface term

$$
P_\mu^{(qed)} = i \int \bar{\psi}\gamma^0 D_\mu \psi d^3x + \int \left(F_{\mu\alpha}F^{\alpha 0} + \frac{1}{4}\delta_\mu^0 F_{\alpha\beta}^2 \right) d^3x,
\tag{2.75}
$$

where $D_\mu = \partial_\mu - ieA_\mu$. This, however, is not the desired decomposition because the first term contains both electron and photon, so that strictly speaking it does not represent the electron momentum. Moreover, it does not satisfy the canonical momentum commutation relation.

With (2.70), however, we can easily change (2.74) to

$$
P_\mu^{(qed)} = i \int \bar{\psi}\gamma^0 \bar{D}_\mu \psi d^3x + \int [(\partial_\mu Z_\alpha)F^{\alpha 0} + \frac{1}{4}\delta_\mu^0 F_{\alpha\beta}^2]d^3x,
\tag{2.76}
$$

adding a surface term. Unlike (2.74) or (2.75), each term now is gauge invariant, contains only electron or photon, and at the same time satisfies the canonical commutation relation. This confirms that we can indeed have a gauge invariant decomposition of the total momentum to those of the electron and the photon field in QED. Clearly this would have been impossible without (2.70).

For the spin the QED Lagrangian provides us the following decomposition

$$
\begin{aligned}
J_{\mu\nu}^{(qed)} &= S_{\mu\nu}^{(e)} + L_{\mu\nu}^{(e)} + S_{\mu\nu}^{(\gamma)} + L_{\mu\nu}^{(\gamma)} \\
&= \int \bar{\psi}\gamma^0 \frac{\Sigma_{\mu\nu}}{2}\psi d^3x - i \int \bar{\psi}\gamma^0 x_{[\mu}\partial_{\nu]}\psi d^3x \\
&\quad - \int A_{[\mu}F_{\nu]0}d^3x - \int F_{0\alpha}x_{[\mu}\partial_{\nu]}A_\alpha d^3x.
\end{aligned}
\tag{2.77}
$$

But again this is not gauge invariant. Because of this the following gauge invariant decomposition has become popular

$$
\begin{aligned}
J_{\mu\nu}^{(qed)} &= S_{\mu\nu}^{(e)} + L_{\mu\nu}^{(e)} + J_{\mu\nu}^{(\gamma)} \\
&= \int \bar{\psi}\gamma^0 \frac{\Sigma_{\mu\nu}}{2}\psi d^3x - i \int \bar{\psi}\gamma^0 x_{[\mu}D_{\nu]}\psi d^3x - \int F_{0\alpha}x_{[\mu}F_{\nu]\alpha}d^3x.
\end{aligned}
\tag{2.78}
$$

But this is also unsatisfactory. The second term involves electron and photon, and does not satisfy the angular momentum commutation relation. Moreover, the third term does not decompose the photon angular momentum to intrinsic and orbital parts.

Now, with (2.70) we can modify (2.77) to (adding a surface term)

$$J_{\mu\nu}^{(qed)} = \int \bar{\psi}\gamma^0 \frac{\Sigma_{\mu\nu}}{2}\psi d^3x - i\int \bar{\psi}\gamma^0 x_{[\mu}\bar{D}_{\nu]}\psi d^3x$$
$$- \int Z_{[\mu}F_{\nu]0}d^3x - \int F_{0\alpha}x_{[\mu}\partial_{\nu]}Z_\alpha d^3x. \tag{2.79}$$

Here each term involves only electron or photon, is gauge invariant, and independently satisfies the angular momentum algebra. This shows that one can also decompose the atomic spin to those of electron and photon field in a self-consistent and gauge independent way.

Notice that the photon contribution in the decompositions (2.76) and (2.79) are the contribution of the electromagnetic field, not the contribution of the photon as the constituent of the atom. No finite number of photons can provide this contribution.

This result is remarkable because this is against the common belief that such decomposition is impossible, which is based on the assertion that the gauge invariant canonical momentum does not exist. Our analysis shows that this belief has no foundation. So the assertion that the canonical momentum is not measurable because it is not gauge invariant is simply not true.

Similarly, we can have the gauge invariant momentum and spin decomposition in QCD. In QCD the conserved momentum obtained by Noether's theorem is given by

$$P_\mu^{(qcd)} = i\int \bar{\psi}\gamma^0\partial_\mu\psi d^3x + \int \left[(\partial_\mu\vec{A}_\alpha)\cdot\vec{F}^{\alpha 0} + \frac{1}{4}\delta_\mu^0\vec{F}_{\alpha\beta}^2\right]d^3x. \tag{2.80}$$

But obviously this is not gauge invariant, so that the following decomposition has become popular [57]

$$P_\mu^{(qcd)} = i\int \bar{\psi}\gamma^0 D_\mu\psi d^3x + \int \left[\vec{F}_{\mu\alpha}\cdot\vec{F}^{\alpha 0} + \frac{1}{4}\delta_\mu^0\vec{F}_{\alpha\beta}^2\right]d^3x. \tag{2.81}$$

However, this also has critical defects [56]. For example, the first term contains quarks and gluons, and does not satisfy the momentum commutation relation.

Now, with the help of the gauge invariant canonical momentum (2.73) we can change (2.81) to

$$P_\mu^{(qcd)} = i\int \bar{\psi}\gamma^0\bar{D}_\mu\psi d^3x + \int \left[(\bar{D}_\mu\vec{Z}_\alpha)\cdot\vec{F}^{\alpha 0} + \frac{1}{4}\delta_\mu^0\vec{F}_{\alpha\beta}^2\right]d^3x, \tag{2.82}$$

adding the surface term

$$- \int \partial_\alpha(\hat{\Omega}_\mu\cdot\vec{F}^{\alpha 0})d^3x. \tag{2.83}$$

Clearly each term contains only one constituent, is explicitly gauge invariant, and independently satisfies the momentum commutation relation. So this becomes a gauge invariant momentum decomposition in QCD.

Exactly the same argument applies to the spin decomposition in QCD. Consider the canonical spin decomposition obtained from QCD Lagrangian by Noether's theorem [55],

$$
\begin{aligned}
J_{\mu\nu}^{(qcd)} &= S_{\mu\nu}^q + L_{\mu\nu}^q + S_{\mu\nu}^g + L_{\mu\nu}^g \\
&= \int \bar{\psi}\gamma^0 \frac{\Sigma_{\mu\nu}}{2}\psi d^3x - i\int \bar{\psi}\gamma^0 x_{[\mu}\partial_{\nu]}\psi d^3x \\
&\quad - \int \vec{A}_{[\mu}\cdot\vec{F}_{\nu]0}d^3x - \int \vec{F}_{0\alpha}\cdot x_{[\mu}\partial_{\nu]}\vec{A}_\alpha d^3x.
\end{aligned}
\tag{2.84}
$$

Clearly this decomposition is not acceptable because it is not gauge invariant. Because of this the following spin decomposition has become popular [57]

$$
\begin{aligned}
J_{\mu\nu}^{(qcd)} &= S_{\mu\nu}^{(q)} + L_{\mu\nu}^{(q)} + J_{\mu\nu}^{(g)} = \int \bar{\psi}\gamma^0\frac{\Sigma_{\mu\nu}}{2}\psi d^3x \\
&\quad - i\int \bar{\psi}\gamma^0 x_{[\mu}D_{\nu]}\psi d^3x - \int \vec{F}_{0\alpha}\cdot x_{[\mu}\vec{F}_{\nu]\alpha}d^3x,
\end{aligned}
\tag{2.85}
$$

where $D_\mu = \partial_\mu + g\vec{A}_\mu\times$. This is similar to (2.78) in QED, and has similar critical defects. For example, the second term involves both quark and gluon. Worse, the second and third term can not be interpreted as angular momentum since they do not satisfy the expected angular momentum commutation relation [56].

With (2.73), however, we can express it to (again adding a surface term) [41,42]

$$
\begin{aligned}
J_{\mu\nu}^{(qcd)} &= S_{\mu\nu}^{(q)} + L_{\mu\nu}^{(q)} + S_{\mu\nu}^{(g)} + L_{\mu\nu}^{(g)} \\
&= \int \bar{\psi}\gamma^0\frac{\Sigma_{\mu\nu}}{2}\psi d^3x - i\int \bar{\psi}\gamma^0 x_{[\mu}\bar{D}_{\nu]}\psi d^3x \\
&\quad - \int \vec{Z}_{[\mu}\cdot\vec{F}_{\nu]0}d^3x - \int \vec{F}_{0\alpha}\cdot x_{[\mu}\bar{D}_{\nu]}\vec{Z}_\alpha d^3x.
\end{aligned}
\tag{2.86}
$$

Now each term involves only one constituent, and satisfies the angular momentum commutation relation independently. Clearly this is made possible with (2.73).

Notice that, on the surface the decompositions (2.82) and (2.86) look very much like the decomposition proposed with the "physical" and "unphysical" potentials [56]. The difference is that the separation of the "physical" and "unphysical" potentials in this proposal is *ad hoc*, while the vacuum decomposition in our proposal is unambiguous and well-defined.

We emphasize, however, that this does not necessarily mean that (2.82) and (2.86) describe the desired nucleon momentum and spin decompositions. This is because there two types of gluons, neurons and chromons in QCD [2,3]. So we have to know which gluons are in the nucleons to bind the quarks.

The quark model assures that the nucleons are not hybrid hadrons made of quarks and chromons. It tells that the low-lying nucleons are made of three valence quarks [38]. This strongly implies that there are no chromons in nucleons.

This, of course, does not mean that there is no gluon at all in nucleons. Clearly there should be the neurons to bind the quarks. Of course, just as the quarks in nucleons are not the bare quarks but the fully dressed valence quarks which include virtual quark pairs, there should be the sea component of the chromons (the chromon pairs) in nucleons. But they are the vacuum bubbles coming from higher order corrections, which can not be justified as the valence constituents.

This tells that chromons have no place in nucleon momentum and spin decompositions. Only quarks and neurons should contribute to the decompositions. But this important point has completely been ignored in (2.82) and (2.86), because they include the chromons [41, 42].

It is straightforward to exclude the chromons from (2.82) and (2.86). All we have to do is to replace $\vec{A}_\mu$ by $\hat{A}_\mu$, or equivalently $\vec{X}_\mu$ by $\vec{B}_\mu$ and $\vec{F}_{\mu\nu}$ by $\hat{F}_{\mu\nu}$, in (2.82) and (2.86). Replacing $\vec{Z}_\mu$ by $\vec{B}_\mu$ and $\vec{F}_{\mu\nu}$ to $\hat{F}_{\mu\nu}$ from (2.82) we have the desired nucleon momentum decomposition

$$\hat{P}_\mu = i\int \bar{\psi}\gamma^0 \bar{D}_\mu \psi d^3x + \int \left[(\bar{D}_\mu \vec{B}_\alpha)\cdot \hat{F}^{\alpha 0} + \frac{1}{4}\delta_\mu^0 \hat{F}_{\alpha\beta}^2\right] d^3x, \qquad (2.87)$$

where $\vec{B}_\mu$ is the binding gluon. Just like (2.82) each term involves one constituent, is explicitly gauge invariant, and satisfies the momentum commutation relation separately, except that here we have no valence gluons.

Exactly the same argument applies to the nucleon spin decomposition. Replacing $\vec{Z}_\mu$ by $\vec{B}_\mu$ and $\vec{F}_{\mu\nu}$ to $\hat{F}_{\mu\nu}$ from (2.86) we can change it to the desired spin decomposition (again adding a surface term)

$$\hat{J}_{\mu\nu} = \int \bar{\psi}\gamma^0 \frac{\Sigma_{\mu\nu}}{2}\psi d^3x - i\int \bar{\psi}\gamma^0 x_{[\mu}\bar{D}_{\nu]}\psi d^3x$$
$$- \int \vec{B}_{[\mu}\cdot \hat{F}_{\nu]0}d^3x - \int \hat{F}_{0\alpha}\cdot x_{[\mu}\bar{D}_{\nu]}\vec{B}_\alpha d^3x. \qquad (2.88)$$

Again, each term involves only one constituent, and satisfies the angular momentum commutation relation independently. But unlike (2.85), only the binding gluons contribute in the nucleon spin here.

This tells that there exist two logically acceptable gauge invariant nucleon momentum and spin decompositions, (2.82) and (2.86) or (2.87) and (2.88). This is because there are two types of gluons in QCD, the binding gluons and the valence gluons. So depending on which gluons we decide to be in nucleons, we can have two different momentum and spin decompositions of nucleons. Although we have pointed out that the quark model prefers (2.87) and (2.88), we emphasize that ultimately experiments should determine which decomposition is correct.

At this point one might ask whether there is any theory which could justify (2.87) and (2.88). Of course, there is. Naturally, we can derive (2.87) and (2.88) from RCD. From RCD Lagrangian we have the following conserved momentum

(with Noether's theorem)

$$P_\mu^{(rcd)} = i \int \bar\psi \gamma^0 \partial_\mu \psi d^3x + \int \left[(\partial_\mu \hat A_\alpha) \cdot \hat F^{\alpha 0} + \frac{1}{4} \delta_\mu^0 \hat F_{\alpha\beta}^2 \right] d^3x. \qquad (2.89)$$

But just like (2.80) this decomposition is gauge dependent. To cure this defect one could change it to (adding a surface term)

$$P_\mu^{(rcd)} = i \int \bar\psi \gamma^0 \hat D_\mu \psi d^3x + \int \left(\hat F_{\mu\alpha} \cdot \hat F^{\alpha 0} + \frac{1}{4} \delta_\mu^0 \hat F_{\alpha\beta}^2 \right) d^3x. \qquad (2.90)$$

But again this is unacceptable because this has the same defect as (2.81).

On the other hand, notice that with (2.73) we can change (2.89) to (2.87) adding a surface term

$$\int \partial_\alpha (\vec F^{\alpha 0} \cdot \hat\Omega_\mu) d^3x. \qquad (2.91)$$

This confirms that (2.87) can be viewed as the momentum decomposition of RCD. In this sense $\hat P_\mu$ in (2.87) should really be written as $P_\mu^{(rcd)}$.

As for the spin, we have the conserved spin from RCD Lagrangian

$$\begin{aligned}
J_{\mu\nu}^{(rcd)} &= \int \bar\psi \gamma^0 \frac{\Sigma_{\mu\nu}}{2} \psi d^3x - i \int \bar\psi \gamma^0 x_{[\mu} \partial_{\nu]} \psi d^3x \\
&\quad - \int \hat A_{[\mu} \cdot \hat F_{\nu]0} d^3x - \int \hat F_{0\alpha} \cdot x_{[\mu} \partial_{\nu]} \hat A_\alpha d^3x.
\end{aligned} \qquad (2.92)$$

From this one could have the gauge invariant decomposition similar to (2.84)

$$\begin{aligned}
J_{\mu\nu}^{(qcd)} &= S_{\mu\nu}^{(q)} + L_{\mu\nu}^{(q)} + J_{\mu\nu}^{(g)} = \int \bar\psi \gamma^0 \frac{\Sigma_{\mu\nu}}{2} \psi d^3x \\
&\quad - i \int \bar\psi \gamma^0 x_{[\mu} \vec D_{\nu]} \psi d^3x - \int \hat F_{0\alpha} \cdot x_{[\mu} \hat F_{\nu]\alpha} d^3x.
\end{aligned} \qquad (2.93)$$

But again this has the the same defect that (2.84) suffers. Fortunately, with (2.73) we can modify it to (2.88), adding a surface term

$$- \int \partial_\alpha (\hat F^{\alpha 0} \cdot x_{[\mu} \hat\Omega_{\nu]}) d^3x. \qquad (2.94)$$

This confirms that (2.88) are indeed from RCD. In this sense $\hat J_{\mu\nu}$ in (2.88) should also be viewed as $J_{\mu\nu}^{(rcd)}$.

Before we leave this section we make a few comments. First, (unlike QED) QCD has two kinematic momentums, $-i\hat D_\mu$ made of the restricted potential and $-iD_\mu$ made of the full (restricted and valence) potential. So, even if we accept the view that only the kinematic momentum is measurable, we actually have two momentum decompositions (2.80) and (2.89), depending on whether we include the valence gluons or not. Similarly, we have two spin decompositions (2.84) and (2.92). In comparison, we emphasize that there is only one nucleon momentum decomposition which we can make with the gauge invariant canonical momentum.

Second, (2.87) and (2.88) are very similar to the QED decompositions (2.76) and (2.79). In fact we can make them almost identical to (2.76) and (2.79) by Abelianizing them. Indeed in the gauge independent Abelianization, (2.82) and (2.86) become formally identical to (2.76) and (2.79), except that $G_{\mu\nu}$ and B_μ plays the role of $F_{\mu\nu}$ and X_μ. This observation should play important role when we try to compare (2.82) and (2.86) with experiments, because this tells that the momentum and spin decompositions in QCD are actually not so different from decompositions in QED.

Third, so far we have discussed how to decompose the momentum and spin operators. To predict how much of each term in the decomposition actually contributes to the total momentum and spin of the proton, we have to evaluate the expectation values of these operators with the proton wave function.

Finally, here we have discussed the decomposition of the momentum and spin operators theoretically. How to measure the contribution of each term experimentally is a totally different question. This raises a deep experimental question: How can we measure the canonical momentum of electron moving in an electromagnetic field, and how can we distinguish it from the kinematic momentum? And we have the same question on the definition on the electron orbital angular momentum. In particular, we have to clarify the meaning of the electron orbital angular momentum and ask which momentum is used in the measurement of the electron orbital angular momentum in the experiments.

2.12   Closing remarks

In this review article we have discussed the importance of the Abelian decomposition in QCD and Professor Duan's contribution in the decomposition. The Abelian decomposition simplifies QCD dynamics greatly. It tells that QCD has two gluons which play different roles. Perturbatively (in terms of the Feynman diagrams) the neurons play the role of the photon and the chromons play the role of (massless) charged vector fields in QED. Non-perturbatively, however, (the monopole part of) the neurons become the confining agents. In contrast, the chromons become the confined prisoners. Without the Abelian decomposition we can not tell this difference because all gluons are treated on equal footing.

The Abelian decomposition tells that it is the monopole which confines the color. In particular, it allows us to demonstrate the monopole condensation in QCD. Moreover, it generalize the quark model to the quark and chromon model, and allows us to identify the glueballs experimentally. Furthermore, the Abelian decomposition allows us to have the vacuum decomposition and helps us to resolve the proton spin crisis problem. Indeed I personally believe that the Abelian decomposition probably is the most important development in QCD after the discovery of Yang–Mills theory.

The Abelian decomposition allows us to simplify Einstein's theory and construct the restricted gravity which describes the core dynamics of Einstein's theory. This allows us to prove the Abelian dominance in Einstein's theory. This is because Einstein's theory can be viewed as a gauge theory of Lorentz group.

What is really striking is that Professor Duan was able to discover this Abelian decomposition during the chaotic cultural revolution in China. Doing this he has demonstrated us how the true creativity works in hardship.

Of course, he did not pursue the applications of the Abelian decomposition in QCD much, but this was probably because his main interest was in gravitation. To me this is not a drawback. The fact that he could discover the Abelian decomposition as a general relativist tells that he was truly a great physicist. Unfortunately, it appears that this fact has not been properly appreciated in China. Chinese physics community would do well paying the due appreciation to this truly great physicist and gentleman, and the sooner the better.

Acknowledgement

I thank Mo-Lin Ge suggesting me to write this article, Pengming Zhang for the proof reading, and Rong-Gen Cai for the editing. The work is supported in part by the Korean National Research Foundation funded by the Ministry of Education (Grant 2015-R1D1A1A0-1057578).

Bibliography

[1] Y. S. Duan and Mo-Lin Ge, Sci. Sinica **11**, 1072 (1979).

[2] Y. M. Cho, Phys. Rev. **D21**, 1080 (1980).

[3] Y. M. Cho, Phys. Rev. Lett. **46**, 302 (1981).

[4] Y. M. Cho, Phys. Rev. **D23**, 2415 (1981).

[5] L. Faddeev and A. Niemi, Phys. Rev. Lett. **82**, 1624 (1999).

[6] S. Shabanov, Phys. Lett. **B458**, 322 (1999); **B463**, 263 (1999); H. Gies, Phys. Rev. **D63**, 125023 (2001).

[7] R. Zucchini, Int. J. Geom. Meth. Mod. Phys. **1**, 813 (2004).

[8] W. S. Bae, Y. M. Cho, and S. W. Kimm, Phys. Rev. **D65**, 025005 (2002).

[9] Y. M. Cho, Phys. Rev. Lett. **44**, 1115 (1980).

[10] Y. M. Cho, Franklin H. Cho, and J. H. Yoon, Phys. Rev. **D87**, 085025 (2013).

[11] Y. M. Cho, Int. J. Mod. Phys. **A29**, 1450013 (2014).

[12] G. 't Hooft, Nucl. Phys. **B190**, 455 (1981).

[13] Y. M. Cho, Phys. Rev. **D62**, 074009 (2000).

[14] S. Kato, K. Kondo, T. Murakami, A. Shibata, T. Shinohara, and S. Ito, Phys. Lett. **B632**, 326 (2006); S. Ito, S. Kato, K. Kondo, T. Murakami, A. Shibata, and T. Shinohara, Phys. Lett. **B645**, 67 (2007).

[15] N. Cundy, Y. M. Cho, W. Lee, and J. Leem, Phys. Lett. **B729**, 192 (2014); N. Cundy, Y. M. Cho, W. Lee, and J. Leem, Nucl. Phys. **B**, in press.

[16] W. S. Bae, Y. M. Cho, and S. W. Kim, Phys. Rev. **D65**, 025005 (2001).

[17] B. de Witt, Phys. Rev. **162**, 1195 (1967); 1239 (1967).

[18] See for example, C. Itzikson and J. Zuber, *Quantum Field Theory* (McGraw-Hill) 1985; M. Peskin and D. Schroeder, *An Introduction to Quantum Field Theory* (Addison-Wesley) 1995; S. Weinberg, *Quantum Theory of Fields* (Cambridge University Press) 1996.

[19] G. K. Savvidy, Phys. Lett. **B71**, 133 (1977).

[20] N. Nielsen and P. Olesen, Nucl. Phys. **B144**, 485 (1978); N. Nielsen and P. Olesen, Nucl. Phys. **B160**, 380 (1979); C. Rajiadakos, Phys. Lett. **B100**, 471 (1981).

[21] A. Yildiz and P. Cox, Phys. Rev. **D21**, 1095 (1980); M. Claudson, A. Yilditz, and P. Cox, Phys. Rev. **D22**, 2022 (1980); J. Ambjorn and R. Hughes, Phys. Lett. **B113**, 305 (1982).

[22] W. Dittrich and M. Reuter, Phys. Lett. **B128**, 321, (1983); **B144**, 99 (1984); C. Flory, Phys. Rev. **D28**, 1425 (1983); S. K. Blau, M. Visser, and A. Wipf, Int. J.

Mod. Phys. **A6**, 5409 (1991); M. Reuter, M. G. Schmidt, and C. Schubert, Ann. Phys. **259**, 313 (1997).

[23] Y. M. Cho and D. G. Pak, Phys. Rev. **D65**, 074027 (2002); Y. M. Cho, H. W. Lee, and D. G. Pak, Phys. Lett. **B525**, 347 (2002).

[24] Y. M. Cho, M. L. Walker, and D. G. Pak, JHEP **05**, 073 (2004); Y. M. Cho and M. L. Walker, Mod. Phys. Lett **A19**, 2707 (2004).

[25] F. Gliozzi, J. Scherk, and D. Olive, Nucl. Phys. **B122**, 253 (1977).

[26] See, e. g., M. Green, J. Schwarz, and E. Witten, *Superstring Theory* Vol. I (Cambridge University Press) 1987; M. Kaku, *Introduction to Superstrings* (Springer-Verlag) 1988.

[27] D. Gross and F. Wilczek, Phys. Rev. Lett. **30**, 1343 (1973); H. Politzer, Phys. Rev. Lett. **30**, 1346 (1973).

[28] Y. M. Cho and D. G. Pak, Phys. Rev. Lett. **86**, 1947 (2001); **91**, 039151; W. S. Bae, Y. M. Cho, and D. G. Pak, Phys. Rev. **D64**, 017303 (2001).

[29] V. Schanbacher, Phys. Rev. **D26**, 489 (1982).

[30] J. Schwinger, Phys. Rev. **82**, 664 (1951).

[31] H. Fritzsch and P. Minkowski, Nuovo Cimento **30A**, 393 (1975).

[32] P. G. O. Freund and Y. Nambu, Phys. Rev. Lett. **34**, 1645 (1975); J. Kogut, D. Sinclair, and L. Susskind, Nucl. Phys. **B114**, 199 (1976).

[33] R. L. Jaffe and K. Johnson, Phys. Lett. **B60**, 201(1976); P. Roy and T. Walsh, Phys. Lett. **B78**, 62 (1978).

[34] J. Coyne, P. Fishbane, and S. Meshkov, Phys. Lett **B91**, 259 (1980); M. Chanowitz, Phys. Rev. Lett. **46**, 981 (1981).

[35] C. Amsler and N. Tornqvist, Phys. Rep. **389**, 61 (2004).

[36] V. Mathieu, N. Kochelev, and V. Vento, Int. J. Mod. Phys. **E18**, 1 (2009); W. Ochs, J. Phys. **G40**, 043001 (2013).

[37] G. Bali *et al.*, Phys. Lett. **B309**, 378 (1993); C. Morningstar and M. Peardon, Phys. Rev. **D60**, 034509 (1999); W. Lee and D. Weingarten, Phys. Rev. **D61**, 014015 (2000); Y. Chen *et al.*, Phys. Rev. **D73**, 014516 (2006).

[38] K. Olive *et al.*, [Particle Data Group], Review of Particle Physics, Chin. Phys. **38**, 090001 (2014).

[39] Y. M. Cho, X. Y. Pham, Pengming Zhang, Ju-Jun Xie, and Li-Ping Zou, Phys. Rev. **D91**, 114020 (2015).

[40] Pengming Zhang, Ju-Jun Xie, Li-Ping Zou, and Y. M. Cho, to be published.

[41] Y. M. Cho, AIP Conf. Proc. **1418**, 29 (2011).

[42] Y. M. Cho, Mo-Lin Ge, and Pengming Zhang, Mod. Phys. Lett. **A27**, 1230032 (2012).

[43] Y. M. Cho, Phys. Lett. **B644**, 208 (2007).

[44] Y. M. Cho, Int. J. Mod. Phys. **A24**, 3327 (2009); Int. J. Mod. Phys. **CS7**, 116 (2012).

[45] Y. M. Cho, S. H. Oh, and Sangwoo Kim, Class. Quant. Grav. **29**, 205007 (2012).

[46] Y. M. Cho, Franklin H. Cho, and J. H. Yoon, Class. Quant. Grav. **30**, 055003 (2013).

[47] R. Utiyama, Phys. Rev. **101**, 1597 (1956); R. Utiyama and T. Fukuyama, Prog. Theor. Phys. **45**, 612 (1971).

[48] T. W. B. Kibble, J. Math. Phys. **2**, 212 (1961); D. Sciama, Rev. Mod. Phys. **36**, 463 (1964); M. Carmelli, J. Math. Phys. **11**, 2728 (1970).

[49] Y. M. Cho, Phys. Rev. **D14**, 2521 (1976); See also *100 Years of Gravity and Accelerated Frames—The Deepest Insights of Einstein and Yang–Mills*, edited by J. P. Hsu, World Scientific, 2006.

[50] Y. M. Cho, Phys. Rev. **D14**, 3335 (1976).

[51] F. Hehl, P. von der Heide, G. Kerlick, J. Nester, Rev. Mod. Phys. **48**, 393 (1976).

[52] R. Feynman, R. Leighton, and M. Sands, *The Feynman Lectures on Physics Vol. III* (Addison-Weseley) 1965; J. Sakurai, *Modern Quantum Mechanics* (Addison-Weseley) 1995.

[53] J. Ashman et al. (EMC), Nucl. Phys. **B328**, 1 (1989); B. Adeva et al. (SMC), Phys. Rev.**D60**, 072004 (1999).

[54] V. Alexakhin et al. (COMPASS), Phys. Lett. **B647**, 8 (2007); A. Airapetian et al. (HERMES), Phys. Rev. **D76**, 39901 (2007); A. Adare et al. (PHENIX), Phys. Rev. **D76**, 051106 (2007); B. Abelev et al. (STAR), Phys. Rev. Lett. **100**, 232003 (2008).

[55] R. Jaffe and A. Manohar, Nucl. Phys. **B337**, 509 (1990); R. Jaffe, Phys. Lett. **B365**, 359 (1996).

[56] X. S. Chen, X. F. Lu, W. M. Sun, F. Wang, and T. Goldman, Phys. Rev. Lett. **100**, 232002 (2008); X. S. Chen, X. F. Lu, W. M. Sun, F. Wang, and T. Goldman, Phys. Rev. Lett. **103**, 062001 (2009).

[57] X. Ji, Phys. Rev. Lett. **78**, 610 (1997); X. Ji, Phys. Rev. Lett. **104**, 039101 (2010).

[58] M. Wakamatsu, Phys. Rev. **D81**, 114010 (2010); **D83**, 014012 (2011); Y. Hatta, hep-ph/1101.5989; E. Leader, hep-ph/1101.5956.

Chapter 3

How Can We Understand Quark Confinement in Quantum Yang–Mills Theory?

Kei-Ichi Kondo [1]

Department of Physics, Graduate School of Science,
Chiba University, Chiba 263-8522, Japan

Abstract: We give a short review on the recent progress in understanding quark confinement within the framework of the quantum Yang–Mills theory. We focus on the achievements obtained along the line proposed by Yi-Shi Duan and Mo-Lin Ge and independently by Yong-Min Cho.

Keywords: Quark confinement, dual superconductor, magnetic monopole, mass gap, Yang–Mills theory

3.1   Introduction

The Yang–Mills theory [1] proposed by Chen Ning Yang and Robert L. Mills in 1954 is now an indispensable field theoretical framework for describing the fundamental interactions among elementary particles. In the classical level, the Yang–Mills theory is a quite natural and probably unique extension of the gauge field theory with an Abelian gauge group to that with the non-Abelian gauge group G. The classical Yang–Mills theory in four spacetime dimensions is a scale invariant (conformal) theory which does not have the intrinsic scale just as the Maxwell theory, since it is specified by only dimensionless parameters such as a dimensionless gauge coupling constant g and the number of colors N.

However, it is expected that the *quantum* Yang–Mills theory confine quarks and gluons, and exhibits a mass gap. This issue is taken up as one of the Millennium Prize Problems in Clay Mathematics Institute, [2]
Yang–Mills Existence and Mass Gap: Prove that for any compact simple gauge group G, quantum Yang–Mills theory on $\mathbb{R}^4$ exists and has a mass gap $\Delta > 0$.

---

[1] Email: kondok@faculty.chiba-u.jp

59

In fact, the numerical simulations of the Yang–Mills theory on a lattice indicates that the static potential between a pair of quark and antiquark exhibits the linear potential in the long distance of separation. This is an indication of quark confinement. Remarkably, the string tension as a coefficient of proportionality in the linear potential has the mass dimension two. The dimensionful physical observables such as string tension and mass gap must be of quantum origin. These facts force us to investigate the Yang–Mills theory in the non-perturbative way.

The linear potential between color electric charges resulting from the squeezed chromoelectric flux tube can be naively explained based on the *dual superconductor* picture of the vacuum, as proposed by Nambu, 't Hooft and Mandelstam [3] in the beginning of 1970s. The dual superconductivity could be caused by condensation of magnetic monopoles, which is regarded as the electro-magnetic dual of the ordinary superconductor where the squeezed magnetic flux is realized by the condensation of electric charges pairs called the Cooper pairs. Indeed, confinement due to magnetic monopole condensations was exemplified by Polyakov for simpler models, four-dimensional compact $U(1)$ gauge theory and the three-dimensional Yang–Mills–Higgs model [4] in the end of 1970s. However, it is quite non-trivial to define magnetic monopoles in the pure Yang–Mills theory without scalar field, whereas the magnetic monopole was constructed by 't Hooft and Polyakov for the Yang–Mills theory in the presence of scalar fields [5].

In this situation, a procedure called the *Abelian projection* was proposed by 't Hooft [6] in 1981, which enables us to define a magnetic monopole in the Yang–Mills theory without any scalar field. However, the Abelian projection is a partial gauge fixing which explicitly breaks the original gauge symmetry G to the maximal torus subgroup (Cartan subgroup) H, $G \to H$.

It is rather useful to take a specific gauge fixing called the *Maximal Abelian gauge* [11] as a realization of the Abelian projection. In this gauge, indeed, the Abelian dominance [7] was confirmed for the string tension [8] and the correlation function [9, 10] by numerical simulations on the lattice, while the magnetic monopole dominance was also confirmed for the string tension [12]. However, this approach cannot sweep away the criticism that the results obtained under the Abelian projection might be gauge artifact.

In the seminal paper written with Mo-Lin Ge [13], Professor Yi-Shi Duan has proposed the *field decomposition method* already in 1979. This method gives the gauge-independent definition of magnetic monopole and includes the 't Hooft one (Abelian magnetic monopole) as a special case. However, the paper did not draw much attention in the community until relatively recently, at least, as far as I know. One reason is that the paper has been published in Chinese, although the English translation by Peng-Ming Zhang is currently available. [14] It should be noted that the same idea was proposed independently by Yong-Min Cho [15] in 1979. Nevertheless, the issue has been readdressed twenty years later by Faddeev and Niemi [16] and developed by Shabanov [17]. Another reason is that the Abelian projection was immediately implemented to the lattice and extensively used to introduce the Abelian magnetic momopole in the Yang–Mills theory, while the field decomposition is more involved and the lattice formulation has been completed quite recently.

In what follows, we try to show how the idea of Duan and Ge as well as Cho has been elaborated and expanded to understand confinement in recent years. The following description is based on a review by the author and collaborators recently published as a volume of Physics Report [18].

3.2 Decomposition of the Yang–Mills field and extension to SU(N) gauge group

In what follows, we use our notations [18] instead of the original ones appeared in the works of Duan and Ge (and Cho). [2]

(I) The original Yang–Mills field $\mathbf{A}_\mu$ is decomposed into $\mathbf{V}_\mu$ and $\mathbf{X}_\mu$,

$$\mathbf{A}_\mu(x) = \mathbf{V}_\mu(x) + \mathbf{X}_\mu(x), \tag{3.1}$$

where $\mathbf{V}_\mu$ follows the same gauge transformation $U \in G$ as that of $\mathbf{A}_\mu$,

$$\mathbf{V}_\mu(x) \to \mathbf{V}'_\mu(x) := U(x)\mathbf{V}_\mu(x)U(x)^\dagger + ig^{-1}U(x)\partial_\mu U(x)^\dagger, \tag{3.2}$$

while $\mathbf{X}_\mu$ obeys the adjoint gauge transformation,

$$\mathbf{X}_\mu(x) \to \mathbf{X}'_\mu(x) := U(x)\mathbf{X}_\mu(x)U(x)^\dagger. \tag{3.3}$$

We call $\mathbf{V}_\mu(x)$ the *restricted field* and $\mathbf{X}_\mu(x)$ the *remaining field*.

(Ia) The restricted field $\mathbf{V}_\mu$ can be constructed in terms of more fundamental fields c_μ and $\boldsymbol{n}$:

$$\mathbf{V}_\mu(x) = c_\mu(x)\boldsymbol{n}(x) - ig^{-1}[\partial_\mu \boldsymbol{n}(x), \boldsymbol{n}(x)], \quad c_\mu(x) = \mathbf{A}_\mu(x) \cdot \boldsymbol{n}(x), \tag{3.4}$$

We call $\boldsymbol{n}(x)$ the *color (direction) field*. They can describe the magnetic monopole in a gauge-independent way. Indeed, a gauge-invariant magnetic monopole current k is defined by

$$k_\mu(x) = \partial_\nu {}^* f_{\mu\nu}(x), \tag{3.5}$$

using the gauge-invariant field strength $f_{\mu\nu}$ defined by

$$f_{\mu\nu}(x) := \boldsymbol{n}(x) \cdot \mathbf{F}_{\mu\nu}[\mathbf{V}](x), \tag{3.6}$$

---

[2]Translation from our notations to Duan and Ge is as follows.

$$\mathbf{A}_\mu \to B_\mu, \quad \mathbf{V}_\mu \to \Gamma_\mu, \quad \mathbf{X}_\mu \to b_\mu$$

since the field strength $\mathbf{F}_{\mu\nu}[\mathbf{V}](x) := \partial_\mu \mathbf{V}_\nu(x) - \partial_\nu \mathbf{V}_\mu(x) - ig[\mathbf{V}_\mu(x), \mathbf{V}_\nu(x)]$ of the restricted field $\mathbf{V}_\mu$ and the color field $\boldsymbol{n}$ transform according to the adjoint representation under the gauge transformation:

$$\mathbf{F}_{\mu\nu}[\mathbf{V}](x) \to U(x)\mathbf{F}_{\mu\nu}[\mathbf{V}](x)U^\dagger(x), \quad \boldsymbol{n}(x) \to U(x)\boldsymbol{n}(x)U^\dagger(x). \tag{3.7}$$

(Ib) For the remaining field $\mathbf{X}_\mu$, therefore, $\mathrm{tr}(\mathbf{X}_\mu\mathbf{X}_\mu)$ is gauge invariant and

$$M^2\mathrm{tr}(\mathbf{X}_\mu\mathbf{X}_\mu) \tag{3.8}$$

provides a mass term of $\mathbf{X}_\mu$ without introducing the spontaneously symmetry breaking and the Higgs mechanism. Thus the remaining field $\mathbf{X}_\mu$ plays the role of massive vector particles.

Although Duan and Ge discussed only the $SU(2)$ gauge group in the paper [13], the idea of field decomposition was immediately extended to $SU(N)$ $(N \geq 3)$ by Cho [19] and later carefully examined by Fadeev and Niemi [20].

The field decomposition of Duan and Ge (and Cho) is enough to study the classical Yang–Mills theory, e.g., solutions of classical Yang–Mills equation of motion [16]. In order to study the quantum Yang–Mills theory, however, we need to show that the original Yang–Mills theory written in terms of the Yang–Mills field $\mathbf{A}_\mu(x)$ is equivalently rewritten in terms of new field variables, e.g., $\boldsymbol{n}(x), c_\mu(x), \mathbf{X}_\mu(x)$ for $SU(2)$ group. This step was quite non-trivial, but it is achieved first by Cho and his collaborators [21] and subsequently from a different viewpoint by us [22, 23] for $SU(2)$ group. The key point is how to construct the color field n from the original Yang–Mills theory. The procedure is called the *reduction condition*. The color field is obtained by minimizing a functional under the extended gauge transformation. The local minimum condition is written in the differential form. The new theory equipollent to the original Yang–Mills theory is called the *reformulated Yang–Mills theory*.

For $G = SU(N)$ $(N \geq 3)$, however, such a gauge-independent reformulation of the Yang–Mills theory is not unique, although the extension along the line proposed by Cho, Faddeev and Niemi is possible. In fact, it has been found [25] that there are a number of possible options of the reformulations, which are discriminated by the maximal stability group $\tilde{H}$ of G, while for $G = SU(2)$ there is a unique option $\tilde{H} = U(1)$ [22, 23, 24]. The maximal stability group depends on the representation of the gauge group, to that the quark source belongs. For the fundamental quark of $SU(3)$, the maximal stability group is $U(2)$, which is different from the maximal torus group $U(1) \times U(1)$ suggested from the Abelian projection. Therefore, the chromomagnetic monopole inherent in the Wilson loop operator responsible for confinement of quarks in the fundamental representation for $SU(3)$ is the *non-Abelian magnetic monopole*, which is distinct from the Abelian magnetic monopole for the $SU(2)$ case. Therefore, we claim that the mechanism for quark confinement for $SU(N)$ $(N \geq 3)$ is the non-Abelian dual superconductivity caused by condensation of non-Abelian magnetic monopoles. We have given some theoretical considerations and numerical results supporting this picture.

3.3   Non-Abelian Stokes theorem for the Wilson loop operator and gauge-invariant magnetic monopole

We can use a version of the non-Abelian Stokes theorem for the Wilson loop operator for the following purposes. First, the Wilson loop operator $W_C[\mathbf{A}]$ defined in terms of $\mathbf{A}$ is completely rewritten in terms of the restricted field $\mathbf{V}$, $W_C[\mathbf{A}] = W_C[\mathbf{V}]$. This is used to exhibit the infrared restricted-field dominance or infrared Abelian dominance in quark confinement. Second, a magnetic monopole current k in the Yang–Mills theory can be constructed in a manifestly gauge-invariant way starting from the gauge-invariant Wilson loop operator. Therefore, the contribution of the magnetic monopoles to the Wilson loop average $\langle W_C[\mathbf{A}] \rangle$ and hence the static quark potential $V(R)$ can be extracted from the Wilson loop operator in a gauge-independent way. This is used to exhibit magnetic monopole dominance in quark confinement. This type of the non-Abelian Stokes theorem was originally found by Diakonov and Petrov [26, 27] for $SU(2)$. Later, it was rederived in a unified way based on the path integral method through the coherent state for $SU(2)$ group [28], for $SU(3)$ group [29, 30], and for $SU(N)$ group [31, 32, 33].

Thus the Wilson loop average $\langle W_C[\mathbf{A}] \rangle$ is calculated according to the new reformulation written in terms of new field variables obtained from the original Yang–Mills field based on change of variables. The Maximally Abelian gauge in the original Yang–Mills theory is also reproduced by taking a specific gauge fixing with a uniform color field $\boldsymbol{n}(x) = \boldsymbol{n}_0$ in the reformulated Yang–Mills theory. This observation justifies the preceding results obtained in the maximal Abelian gauge at least for gauge-invariant quantities for $SU(2)$ gauge group, which eliminates the criticism of gauge artifact raised for the Abelian projection. The claim has been confirmed based on the numerical simulations as exemplified below.

3.4   Lattice formulations and numerical simulations

The lattice gauge theory for the Yang–Mills theory was reformulated using the new variables first for $SU(2)$ group [34] and subsequently generalized to $SU(N)$ group [35, 36], which enables one to perform the numerical simulations on the lattice. In fact, some numerical evidences were obtained to support the hypothesis of the dual superconductivity for quark confinement as follows.

First of all, the numerical simulations have shown the existence of the linear potential for *quark-antiquark potential* $V(R)$, i.e., non-vanishing string tension σ. The "infrared Abelian dominance" (or "restricted field $\mathbf{V}$ dominance") and magnetic monopole dominance in the string tension have been confirmed first for $SU(2)$ group [34, 37] and subsequently for $SU(3)$ group [40, 41]. The "infrared Abelian dominance" has been also shown by measuring the correlation functions for $SU(2)$ group [38, 39] and subsequently for $SU(3)$ group [40, 41] in the sense that the restricted-field correlator $\langle \mathbf{X}_\mu(x)\mathbf{X}_\mu(y)\rangle$ decreases quickly just like the massive propagator.

The followings are the results of numerical simulations for the Yang–Mills theory [40, 41] performed on the lattice according to the lattice reformulation explained above [35, 36]. The static quark-antiquark potential $V(R)$ is obtained by extrapolating the Wilson loop average $\langle W_C[U]\rangle$ for a rectangular loop $C = R \times T$ to the limit $T \to \infty$. In order to see the mechanism of quark confinement, the three potentials which are gauge invariant by construction were calculated:

(i) the *full potential* $V_{\text{full}}(R)$ from the standard Wilson loop average $\langle W_C[U]\rangle$:

$$V_{\text{full}}(R) = - \lim_{T\to\infty} \frac{1}{T} \ln\langle W_C[U]\rangle, \tag{3.9}$$

(ii) the *restricted potential* $V_{\text{rest}}(R)$ from the restricted Wilson loop average $\langle W_C[V]\rangle$ defined by the restricted field $\mathbf{V}$ alone:

$$V_{\text{rest}}(R) = - \lim_{T\to\infty} \frac{1}{T} \ln\langle W_C[V]\rangle, \tag{3.10}$$

(iii) the *magnetic-monopole potential* $V_{\text{mono}}(R)$ from the magnetic monopole part $\langle W_C[k]\rangle = \langle e^{i(k,\omega_\Sigma)}\rangle$ through the magnetic monopole current k:

$$V_{\text{mono}}(R) = - \lim_{T\to\infty} \frac{1}{T} \ln\langle W_C[k]\rangle, \tag{3.11}$$

where k is extracted using the non-Abelian Stokes theorem for the Wilson loop operator [31].

Fig. 3.1 show the comparison of the three quark-antiquark potentials [40] (i), (ii) and (iii). The curve is fitted to the potential of the Cornell type:

$$V_{q\bar{q}}(R) = \sigma R + b + c/R. \tag{3.12}$$

Here the coefficient σ of the linear part of the potential $V(R)$ is the string tension which equals the slope of the curve for large R.

The results exhibit the *infrared restricted field* $\mathbf{V}$ *dominance* in the string tension, e.g., for $SU(3)$,

$$\frac{\sigma_{\text{rest}}}{\sigma_{\text{full}}} = \frac{0.0380}{0.0413} \simeq 0.92, \tag{3.13}$$

and the *non-Abelian magnetic monopole dominance* in the string tension, e.g.,

$$\frac{\sigma_{\text{mono}}}{\sigma_{\text{full}}} = \frac{0.0352}{0.0413} \simeq 0.85. \tag{3.14}$$

Thus, we have obtained the *infrared restricted variable* $\mathbf{V}$ *dominance* in the string tension and the *non-Abelian magnetic monopole dominance* in the string tension. Both dominance are obtained in the gauge independent way. [3] These results should

---

[3] The similar attempts including numerical simulations on the lattice were given by Cho and his collaborators [43].

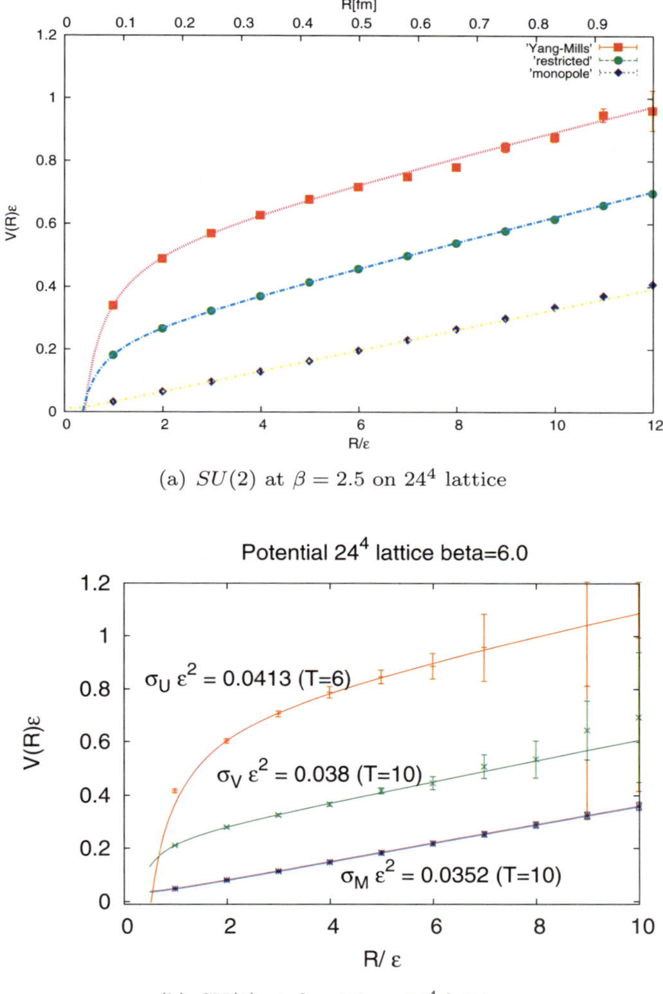

(a) $SU(2)$ at $\beta = 2.5$ on 24^4 lattice

(b) $SU(3)$ at $\beta = 6.0$ on 24^4 lattice

Fig. 3.1 The static quark-antiquark potentials as functions of the quark-antiquark distance R: (from above to below) (i) full potential $V_{\text{full}}(R)$, (ii) restricted part $V_{\text{rest}}(R)$ and (iii) magnetic–monopole part $V_{\text{mono}}(R)$. (a) $SU(2)$ at $\beta = 2.5$ on 24^4 lattice, (b) $SU(3)$ at $\beta = 6.0$ on 24^4 lattice.

be compared with the counterparts obtained in the Maximal Abelian gauge, e.g., see the result [42] for $SU(3)$.

Moreover, some numerical evidences supporting the dual superconductor picture are obtained by the numerical simulations on the lattice for $SU(2)$ group [39] and for $SU(3)$ group [40, 41]. The dual Meissner effect is examined by the simultaneous formation of the chromoelectric flux tube and the associated magnetic-monopole

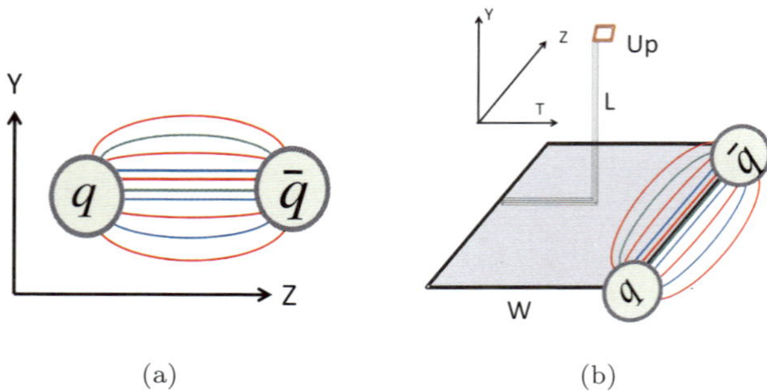

Fig. 3.2   (a) The setup of measuring the chromo-flux produced by a quark–antiquark pair. (b) The gauge-invariant connected correlator $(U_p LWL^\dagger)$ between a plaquette U_p and the Wilson loop W.

current induced around it.

First, we observe the formation of the chromoelectric flux tube. Fig.3.2 shows that the chromoelectric and chromomagnetic fields created by a quark-antiquark pair are extracted by using the gauge-invariant connected correlator between a plaquette variable U_P and the Wilson loop variable W in $SU(N)$ Yang–Mills theory according to Di Giacomo et al [44]:

$$\rho_{U_P} := \frac{\langle \mathrm{tr}\,(U_P L^\dagger WL)\rangle}{\langle \mathrm{tr}\,(W)\rangle} - \frac{1}{N}\frac{\langle \mathrm{tr}\,(U_P)\,\mathrm{tr}\,(W)\rangle}{\langle \mathrm{tr}\,(W)\rangle}, \tag{3.15}$$

where L is the Wilson line connecting the source W and the probe U_P, which is necessary to guarantee the gauge invariance of the correlator ρ_{U_P}. In the continuum limit, ρ_{U_P} reduces to the field strength in presence of $q\bar{q}$ source:

$$\rho_{U_P} \overset{\varepsilon\to 0}{\simeq} g\epsilon^2\,\langle \mathbf{F}_{\mu\nu}\rangle_{q\bar{q}} := \frac{\langle \mathrm{tr}\,(ig\epsilon^2 \mathbf{F}_{\mu\nu}L^\dagger WL)\rangle}{\langle \mathrm{tr}\,(W)\rangle} + O(\epsilon^4). \tag{3.16}$$

Thus, the *gauge-invariant chromo-field strength* $F_{\mu\nu}[U]$ produced by a $q\bar{q}$ pair is given by $F_{\mu\nu}[U] := \epsilon^{-2}\sqrt{\frac{\beta}{2N}}\rho_{U_P}$.

Fig.3.3 is the result of measurement for the chromo-field strength $F_{\mu\nu}[U]$. We find

(1) only the component E_z of the *chromoelectric field* $(E_x, E_y, E_z) = (F_{10}, F_{20}, F_{30})$ along the axis connecting q and $\bar{q}$ has non-zero value. The magnitude E_z quickly decreases in the distance y from the axis connecting q and $\bar{q}$.

(2) The other components are zero within the numerical errors. The chromomagnetic field $(B_x, B_y, B_z) = (F_{23}, F_{31}, F_{12})$ connecting q and $\bar{q}$ does not exist. (3) These results hold for the original Yang–Mills field U and the restricted field V.

Fig. 3.4 shows the distribution of E_z component of the chromoelectric field. The position of a quark and an antiquark is marked by the solid (blue) box. The

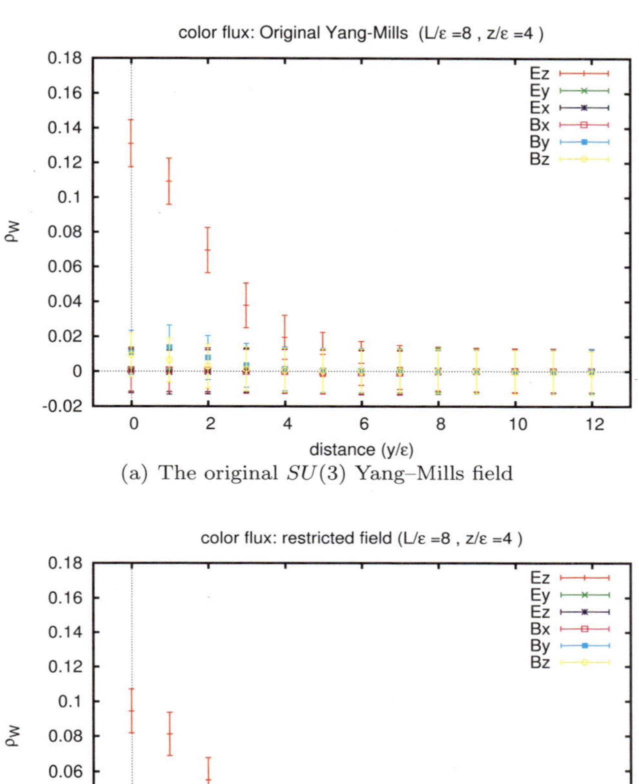

(a) The original $SU(3)$ Yang–Mills field

(b) The restricted field

Fig. 3.3   Measurement of components of the chromoelectric field $\boldsymbol{E}$ and chromomagnetic field $\boldsymbol{B}$ as functions of the distance y from the z axis. (a) The original $SU(3)$ Yang–Mills field. (b) The restricted field.

magnitude of E_z is shown by the height of the 3D plot and also the contour plot in the bottom plane. The left panel of Fig. 3.4 shows the plot of E_z for the $SU(3)$ Yang–Mills field U, and the right panel of Fig. 3.4 for the restricted field V. We find that the magnitude E_z is quite uniform for the restricted part V, while it is almost uniform for the original part U except for the neighborhoods of the locations of q, $\bar{q}$ source. This difference will be due to the contributions from the remaining part X which affects only the short distance.

Next, the magnetic-monopole current k induced around the chromoelectric flux

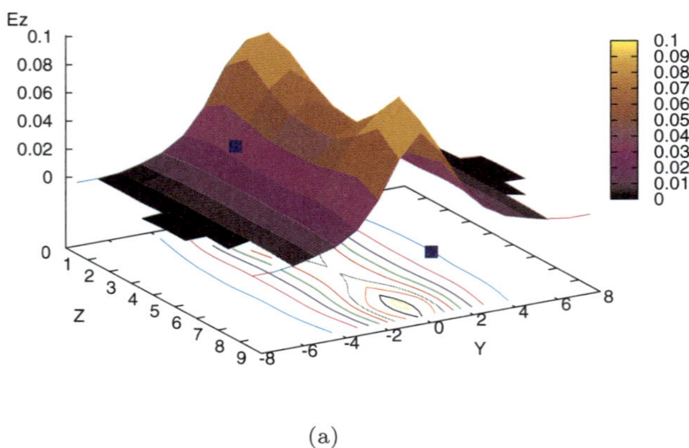

(a)

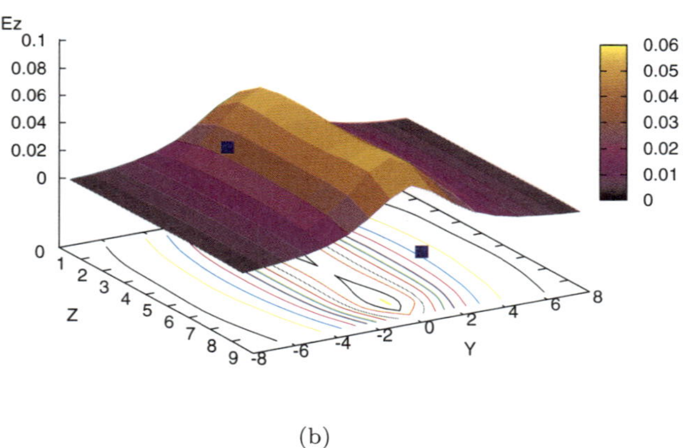

(b)

Fig. 3.4   The distribution in Y-Z plane of the chromoelectric field E_z connecting a pair of quark and antiquark: (a) chromoelectric field produced from the original Yang–Mills field, (b) chromoelectric field produced from the restricted field.

tube can be calculated as $k = \delta^* F[V] = {}^* dF[V]$. Fig. 3.5 shows the magnetic current measured in X-Y plane at the midpoint of quark and antiquark pair in the Z-direction. The left panel of Fig. 3.5 shows the positional relationship between chromoelectric flux and magnetic current. The right panel of Fig. 3.5 shows the magnitude of the chromoelectric field E_z (left scale) and the magnetic current k (right scale). The existence of non-vanishing *magnetic current* k around the *chromoelectric field* E_z supports the *dual superconductor* which is the dual of the ordinary superconductor exhibiting the electric current J around the magnetic field

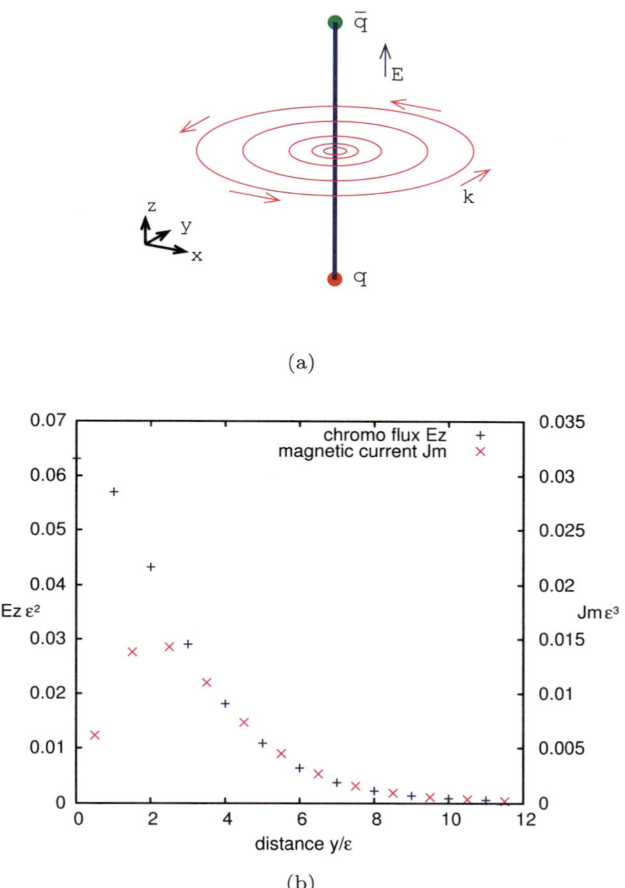

Fig. 3.5   The magnetic-monopole current $\mathbf{k}$ induced around the flux along the z axis connecting a quark-antiquark pair. (a) The positional relationship between the chromoelectric field E_z and the magnetic current $\mathbf{k}$. (b) The magnitude of the chromo-electronic current E_z and the magnetic current $J_m = |\mathbf{k}|$ as functions of the distance y from the z axis.

B.

These are the numerical evidences which support *"non-Abelian" dual superconductivity due to non-Abelian magnetic monopoles as a mechanism for confinement in $SU(3)$ Yang–Mills theory*.

Finally, the type of the dual superconductor is examined. Fig. 3.6 shows the fitting of the numerical simulation data to the *Ginzburg–Landau (GL) theory*: the penetration length $\lambda = 0.1207(17)$fm and the coherence length $\xi = 0.2707(86)$fm is obtained in units of the string tension $\sigma_{\mathrm{phys}} = (440\mathrm{MeV})^2$. Then the *dual*

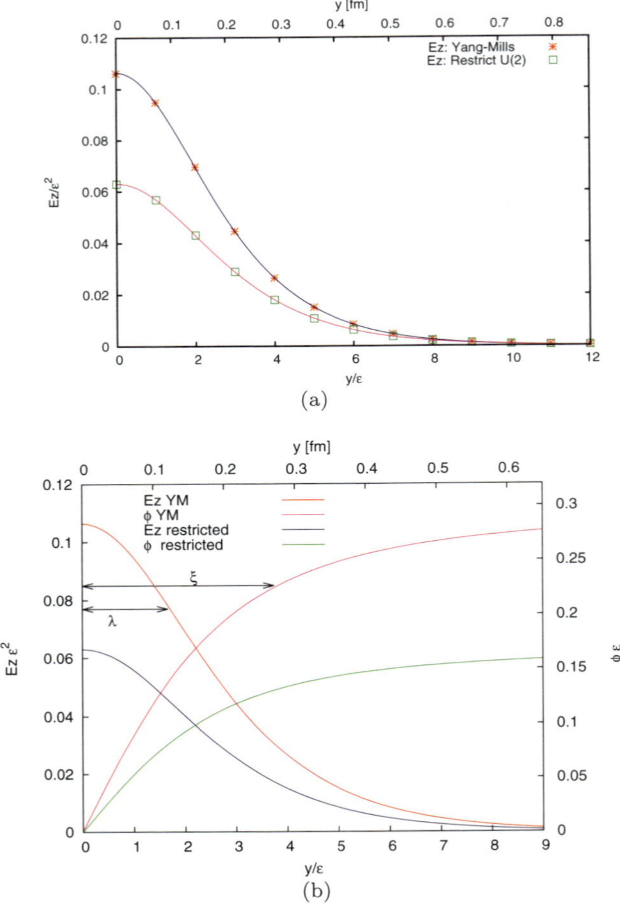

Fig. 3.6 Type of the dual superconductor. (a) The plot of the chromoelectric field E_z versus the distance y in units of the lattice spacing ϵ and the fitting as a function $E_z(y)$ of y according to the Clem fit. The red cross for the original $SU(3)$ field and the green square symbol for the restricted field. (b) The order parameter ϕ reproduced as a function $\phi(y)$ of y.

superconductor of $SU(3)$ Yang–Mills theory is type I, since the Ginzburg–Landau (GL) parameter is given by

$$\kappa := \frac{\lambda}{\xi} = 0.45 \pm 0.01 < \kappa_c := 1/\sqrt{2} \simeq 0.707. \tag{3.17}$$

The GL parameter measures the coherence of the magnetic monopole condensate (the dual version of the Cooper pair condensate). This conclusion is against the preceding studies, since the dual superconductivity was believed to be of type II for a long time. However, this result is consistent with that of the other group [45]. This fact influences the possible interactions between chromoelectric flux tubes and needs further studies. Moreover, the restricted field dominance is observed also for the type:

$$\lambda = 0.132(3)\text{fm}, \quad \xi = 0.277(14)\text{fm} \quad \Longrightarrow \kappa = 0.48 \pm 0.02. \tag{3.18}$$

This is a novel feature overlooked in the preceding studies. Thus the *restricted-field dominance* can be seen also in the determination of the type of dual superconductivity where the discrepancy is just the normalization of the chromoelectric field at the core $y = 0$, coming from the difference of the total flux Φ. These are gauge-invariant results.

In addition, we have given the analytical and numerical investigations on a direct connection between the topological configuration of the Yang–Mills field such as instantons/merons and the magnetic monopole. See the review [18] for more details.

3.5   Summary

We have given a brief review on the recent progress in understanding quark confinement. The emphasis of this review is placed on how to obtain a manifestly gauge-independent picture for quark confinement motivated by the dual superconductivity in the Yang–Mills theory, which should be compared with the Abelian projection proposed by 't Hooft. The basic tools are novel reformulations of the Yang–Mills theory based on change of variables extending the decomposition of the $SU(N)$ Yang–Mills field due to Duan-Ge, Cho, and Faddeev-Niemi, together with the combined use of extended versions of the Diakonov-Petrov version of the non-Abelian Stokes theorem for the $SU(N)$ Wilson loop operator.

In this way, we have obtained the infrared restricted variable $\mathbf{V}$ dominance in the string tension and the non-Abelian magnetic monopole dominance in the string tension. Both dominance are obtained in the gauge independent way.

Moreover, some numerical evidences supporting the dual superconductor picture are obtained by the numerical simulations on the lattice for $SU(2)$ group [39] and for $SU(3)$ group [40, 41]. The dual Meissner effect has been confirmed by measuring

(a) the chromoelectric flux tube squeezed between quark-antiquark pair,

(b) the magnetic-monopole current induced around the flux tube,

However, the dual superconductivity is of type I, rather than type II which was believed for a long time. This fact influences the possible interactions between chromoelectric flux tubes and needs further studies.

Bibliography

[1] C.N. Yang and R.L. Mills, Phys. Rev. **96**, 191–195 (1954).
R. Utiyama, Phys. Rev. **101**, 1597–1607 (1956).

[2] http : //www.claymath.org/millennium − problems

[3] Y. Nambu, Phys. Rev. **D10**, 4262–4268 (1974).
G. 't Hooft, in: High Energy Physics, edited by A. Zichichi (Editorice Compositori, Bologna, 1975).
S. Mandelstam, Phys. Report **23**, 245–249 (1976).

[4] A.M. Polyakov, Phys. Lett. **B59**, 82–84 (1975). Nucl. Phys. **B120**, 429–458 (1977).

[5] G. 't Hooft, Magnetic Monopoles in Unified Gauge Theories, Nucl. Phys. **B79**, 276–284 (1974).
A. M. Polyakov, Particle Spectrum in the Quantum Field Theory, JETP Lett. **20**, 194–195 (1974). Pisma Zh. Eksp. Teor. Fiz. **20**, 430–433 (1974).

[6] G. 't Hooft, Nucl.Phys. **B190** [FS3], 455–478 (1981).

[7] Z.F. Ezawa and A. Iwazaki, Phys. Rev. **D25**, 2681–2689 (1982).

[8] T. Suzuki and I. Yotsuyanagi, Phys. Rev. **D42**, 4257–4260 (1990).

[9] K. Amemiya and H. Suganuma, Phys. Rev. **D60**, 114509 (1999). [hep-lat/9811035]

[10] V.G. Bornyakov, M.N. Chernodub, F.V. Gubarev, S.M. Morozov and M.I. Polikarpov, Phys. Lett. **B559**, 214–222 (2003). [hep-lat/0302002]

[11] A. Kronfeld, M. Laursen, G. Schierholz and U.-J. Wiese, Phys.Lett. **B198**, 516–520 (1987).

[12] J.D. Stack, S.D. Neiman and R.J. Wensley, Phys. Rev. **D50**, 3399–3405 (1994). [hep-lat/9404014]
H. Shiba and T. Suzuki, Phys. Lett. **B333**, 461–466 (1994). [hep-lat/9404015]

[13] Y.S. Duan and M.L. Ge, Sinica Sci., **11**, 1072 (1979).

[14] Peng-Ming Zhang, private communications.

[15] Y.M. Cho, Phys. Rev. **D21**, 1080–1088 (1980). Y.M. Cho, Phys. Rev. **D23**, 2415–2426 (1981).

[16] L. Faddeev and A.J. Niemi, Phys. Rev. Lett. **82**, 1624–1627 (1999). [hep-th/9807069]

[17] S.V. Shabanov, Phys. Lett. **B458**, 322–330 (1999). [hep-th/9903223]
S.V. Shabanov, Phys. Lett. **B463**, 263–272 (1999). [hep-th/9907182]

[18] K.-I. Kondo, S. Kato, A. Shibata and T. Shinohara, Phys. Rept. **579**, 1 (2015). arXiv:1409.1599 [hep-th]

[19] Y.M. Cho, Unpublished preprint, MPI-PAE/PTh 14/80 (1980).

Y.M. Cho, Phys. Rev. Lett. **44**, 1115–1118 (1980).

[20] L. Faddeev and A.J. Niemi, Phys. Lett. **B449**, 214–218 (1999). [hep-th/9812090]
L. Faddeev and A.J. Niemi, Phys. Lett. **B464**, 90–93 (1999). [hep-th/9907180]
T.A. Bolokhov and L.D. Faddeev, Theoretical and Mathematical Physics, **139**, 679–692 (2004).

[21] W.S. Bae, Y.M. Cho and S.W. Kimm, Phys. Rev. **D65**, 025005 (2002). [hep-th/0105163]

[22] K.-I. Kondo, T. Murakami and T. Shinohara, Prog. Theor. Phys. **115**, 201–216 (2006). [hep-th/0504107]

[23] K.-I. Kondo, T. Murakami and T. Shinohara, Eur. Phys. J. **C42**, 475–481 (2005). [hep-th/0504198]

[24] K.-I. Kondo, Phys. Rev. **D74**, 125003 (2006). [hep-th/0609166]

[25] K.-I. Kondo, T. Shinohara, and T. Murakami, Prog. Theor. Phys. **120**, 1-50 (2008). arXiv:0803.0176 [hep-th]

[26] D.I. Diakonov and V.Yu. Petrov, Phys. Lett. **B224**, 131–135 (1989).

[27] D. Diakonov and V. Petrov, [hep-th/9606104]
D. Diakonov and V. Petrov, [hep-lat/0008004]
D. Diakonov and V. Petrov, [hep-th/0008035]

[28] K.-I. Kondo, Phys. Rev. **D58**, 105016 (1998). [hep-th/9805153]

[29] K.-I. Kondo and Y. Taira, Mod. Phys. Lett. **A15**, 367–377 (2000); [hep-th/9906129]

[30] K.-I. Kondo and Y. Taira, Prog. Theor. Phys. **104**, 1189–1265 (2000). [hep-th/9911242]

[31] K.-I. Kondo, Phys. Rev. **D77**, 085029 (2008). arXiv:0801.1274 [hep-th]

[32] K.-I. Kondo, J. Phys. G: Nucl. Part. Phys. **35**, 085001 (2008). arXiv:0802.3829 [hep-th]

[33] R. Matsudo and K.-I. Kondo, Phys. Rev. **D92**, 125038 (2015). arXiv:1509.04891 [hep-th]

[34] S. Kato, K.-I. Kondo, T. Murakami, A. Shibata, T. Shinohara and S. Ito, Phys. Lett. **B632**, 326–332 (2006). [hep-lat/0509069]

[35] K.-I. Kondo, A. Shibata, T. Shinohara, T. Murakami, S. Kato, S. Ito, Phys. Lett. **B669**, 107–118 (2008). arXiv:0803.2451 [hep-lat]

[36] A. Shibata, K.-I. Kondo and T. Shinohara, Phys. Lett. **B691**, 91–98 (2010). arXiv:0911.5294 [hep-lat]

[37] S. Ito, S. Kato, K.-I. Kondo, T. Murakami, A. Shibata and T. Shinohara, Phys. Lett. **B645**, 67–74 (2007). [hep-lat/0604016]

[38] A. Shibata, S. Kato, K.-I. Kondo, T. Murakami, T. Shinohara and S. Ito, Phys. Lett. **B653**, 101–108 (2007). arXiv:0706.2529 [hep-lat]

[39] S. Kato, K.-I. Kondo, and A. Shibata, Phys. Rev. **D91**, 034506 (2015).arXiv:1407.2808 [hep-lat]

[40] K.-I. Kondo, A. Shibata, T. Shinohara, and S. Kato, Phys. Rev. **D83**, 114016 (2011). arXiv:1007.2696 [hep-th]

[41] A. Shibata, K.-I. Kondo, S. Kato and T. Shinohara, Phys. Rev. **D87**, 054011 (2013). arXiv:1212.6512 [hep-lat]

[42] N. Sakumichi and H. Suganuma, Phys.Rev. **D90**, 111501 (2014), e-Print: arXiv:1406.2215 [hep-lat]

[43] Nigel Cundy, Y.M. Cho, Weonjong Lee, Jaehoon Leem, Nucl. Phys. **B895**, 64–131

(2015). arXiv:1503.07033 [hep-lat]

[44] A. Di Giacomo, M. Maggiore and S. Olejnik, Nucl. Phys. **B347**, 441–460 (1990).
A. Di Giacomo, M. Maggiore and S. Olejnik, Phys. Lett. **B236**, 199–202 (1990).

[45] P. Cea, L. Cosmai and A. Papa, Phys. Rev. **D86**, 054501 (2012). arXiv:1208.1362 [hep-lat]
P. Cea, L. Cosmai, F. Cuteri, and A. Papa, Phys. Rev. **D89**, 094505 (2014). arXiv:1404.1172 [hep-lat]

Chapter 4

Asymmetrical Input-output Control in Cavity Quantum Electrodynamics

Yifu Zhu

Department of Physics, Florida International University, Miami, Florida 33199, USA

Abstract: I show that in a composite cavity and two-level absorber system coupled by two coherent input fields, the output light fields from the cavity are asymmetrical and can be controlled by the relative phase between the two input fields. Under appropriate conditions, when the two input fields have a particular phase difference, one of the cavity output field increases linearly with the input field while the other output field vanishes. Thus, the CQED system functions like an optical transistor and the two output fields can be switched on or off alternatively by varying the relative phase between the input fields.

4.1 Introduction

Cavity quantum electrodynamics (CQED) studies interactions of atoms and photons in a controlled structure of electromagnetic modes and has a variety of applications in quantum physics and quantum electronics [1,2,3]. The fundamental cavity QED system consists of a single two-level atom coupled to a single cavity mode [4,5,6]. The composite atom-cavity system exhibits a double-peaked transmission spectrum representing the two eigenstates of the coupled atomic and photonic states commonly referred to as the normal modes or the polaritons. When the CQED system is resonantly coupled, the two normal modes are separated in frequency by $2g$ (g is the atom-cavity coupling coefficient), which is often called the vacuum Rabi splitting [6,7]. If N two-level atoms collectively interact with the cavity mode, the coupling coefficient becomes $G = \sqrt{N}g$ and the vacuum Rabi splitting of the normal modes for the collectively coupled atom-cavity system becomes $2G$ [8,9]. Such collective enhancement of the atom and photon coupling in CQED was essential for earlier studies of CQED and plays an important role in recent studies of quantum measurements with spin squeezing and optical nonlinearities induced by the Rydberg blockade effects [10, 11, 12, 13, 14, 15, 16].

In a standard configuration for CQED studies, a light field is coupled into the cavity only from one end and the output light field from the other end is collected and analyzed. However, if two input fields are coupled into the cavity from opposite ends of the cavity mirrors, the interference manifested by the intra-cavity atomic medium is induced by the two input fields and may give rise to new physical phenomena. For example, it has been shown recently that when the two input fields are identical, the CQED system behaves as a perfect photon absorber, the two input fields are completely absorbed coherently and no light can leak out from the cavity mirrors [17, 18, 19]. It is also shown that in the nonlinear excitation regime of the CQED, the optical bistability can be controlled by the relative phase of the two input fields, which may be useful for the CQED system to function as a phase dependent logical device [20]. As a light-control-light, optical logical device, the CQED system can be used to study all-optical switching, optical transistor, and optical modulator, and may have applications in optical communications and quantum information network.

Here I propose an application of a CQED system with the two-sided inputs as an asymmetrical input-output device. I show that a CQED system containing an absorbing medium and coupled by the two-sided input fields can be used to realize all optical switching and provide the output characteristics of an optical transistor. In particular, when the two input fields have the same amplitude, one can selectively turn on or off one of the two output channels by varying the relative phase of the two input fields and realize the flip-flop optical logical function.

4.2   Theoretical model and analysis

Fig. 4.1 shows the schematic diagram for a coupled cavity and two-level absorber system, in which the absorber is modeled as N two-level atoms confined in the cavity mode volume. The cavity mode is excited by two input light fields a_{in}^r and a_{in}^l from the opposite ends of the cavity as shown in Fig. 4.1 (the drawing is for a standing-wave cavity, but a similar schematic diagram can be drawn for a running-wave cavity). The intra-cavity field couples the atomic transition $|g > -|e >$ with frequency detuning $\Delta_c = \nu_c - \nu_{ge}$. The two input fields have the same frequency ν_p and are tuned away from the atomic transition by $\Delta_p = \nu_p - \nu_{ge}$. We define the collective atomic operators $S^z = \frac{1}{2}\sum_{i=1}^N (\sigma_{ee}^i - \sigma_{gg}^i) = \sum_{i=1}^N \sigma_{eg}^i (\hat{\sigma}_{lm}^{(i)} = |l><m|^{(i)} (l, m = e$ or $g))$ is the atomic operator for the ith atom). The Hamiltonian for the coupled cavity and atom system is

$$H = (\Delta_c - \Delta_p)a^\dagger a + \Delta_p S_z + \left\{ gaS^\dagger + ia^\dagger \left(\frac{\sqrt{\kappa_r}}{\tau} a_{in}^r + \frac{\sqrt{\kappa_l}}{\tau} a_{in}^l \right) + H.C. \right\}. (4.1)$$

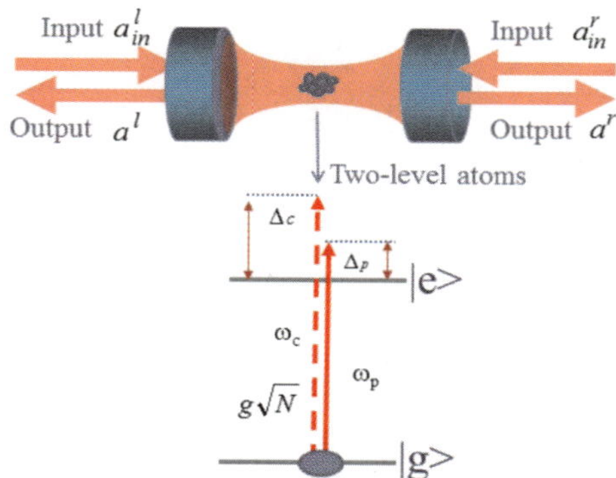

Fig. 4.1 A CQED system with N two-level atoms confined in a single cavity mode and coupled by two input light fields from opposite ends of the cavity. The dashed arrow denotes the cavity mode and the solid arrow denotes the coupling of the input light fields.

Here $\hat{a}(\hat{a}^\dagger)$ is the annihilation (creation) operator of the cavity photons, a_{in}^l and a_{in}^r are two input fields to the cavity, $\kappa_i = \frac{T_i}{\tau}$ ($i = r$ or l) is the loss rate of the cavity field on the mirror i (T_i is the mirror transmission and τ is the photon round trip time inside the cavity), and $g = \mu_{eg}\sqrt{\omega_c/2\hbar\epsilon_0 V}$ is the cavity-atom coupling coefficient and is assumed to be uniform for the N identical atoms inside the cavity.

The equations of motion for the coupled cavity system are given by $\frac{d\hat{\sigma}}{dt} = -i[H, \hat{\sigma}] + \hat{L}\hat{\sigma}$. After dropping the quantum fluctuation terms, the system equations of motion for the expectation values of S^+, S^-, S^z, and a are given by

$$\dot{S}^z = -2\Gamma(S_z + N/2) - i(gaS^+ - ga^\dagger S^-), \tag{4.2a}$$

$$\dot{S}^+ = -(\Gamma - i\Delta_p)S^+ - 2iga^\dagger S^z, \tag{4.2b}$$

$$\dot{a} = -((\kappa_l + \kappa_r)/2 + i(\Delta_c - \Delta_p))a - igS^- + \sqrt{\kappa_1/\tau}a_{in}^r + \sqrt{\kappa_2/\tau}a_{in}^l. \tag{4.2c}$$

Here 2Γ is the decay rate of the excited state $|e>$. Assume a symmetric cavity for the subsequent analysis, $T_l = T_r = T$, then $\kappa_l = \kappa_r = \kappa$. In the weak excitation limit ($S^z \approx -N/2$), the steady-state solution of the intra-cavity light field is then given by

$$a = \frac{\sqrt{\kappa/\tau}(a_{in}^r + a_{in}^l)}{\kappa - i(\Delta_p - \Delta_c) + \dfrac{g^2 N}{(\Gamma - i\Delta_p)(1 + \frac{2g^2|a|^2}{\Gamma^2 + \Delta_p^2})}}. \tag{4.3}$$

The denominator in the right side of Eq. (5.88) contains the Intra-cavity intensity term $|a|^2$ and indicates the nonlinear dependence of the intra-cavity field on the input light fields.

For studies of the optical logical functions, the CQED system can be viewed as a photon input-output device with four channels: two input channels and two output

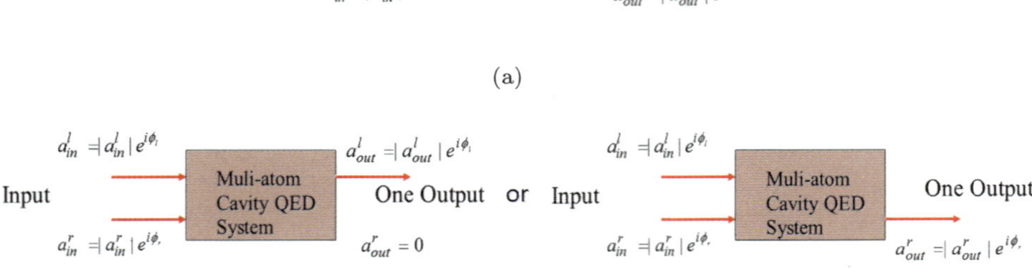

Fig. 4.2   (a) The CQED system in Fig. 4.1 can be viewed as a 4-channel input-output network. (b) Seek asymmetrical input-output in the CQED system in which there is only one nonzero output, the other output is zero. By varying the relative phase of the input fields, the two outputs can be alternated.

channels as shown in Fig. 4.2. The output fields from the left mirror and right mirror of the cavity are related to the intra-cavity field and the input fields, and are given by

$$a^l = \sqrt{T}a - a^l_{in}, \tag{4.4}$$

$$a^r = \sqrt{T}a - a^r_{in}. \tag{4.5}$$

Assume $a^l_{in} = a_{in}$ and $a^r_{in} = \alpha|a^l_{in}|e^{i\Phi}$ (α is a real number), the nonlinear equation relates the input intensity $|a^l_{in}|^2 = |a_{in}|^2$ ($|a^r_{in}|^2 = \alpha^2|a^r_{in}|^2 = \alpha^2|a_{in}|^2$) and the intra-cavity intensity $|a|^2$ is given by

$$|a^l_{in}|^2 = \frac{T|a|^2}{\kappa^2(1 + \alpha^2 + 2\alpha\cos(\varphi))}\left\{\kappa^2 + (\Delta_p - \Delta_c)^2 \right.$$
$$\left. + \frac{g^2 N[\kappa\Gamma - 2\Delta_p(\Delta_p - \Delta_c)]}{\Gamma^2 + \Delta_p^2 + 2g^2|a|^2} + \frac{g^4 N^2(\Gamma^2 + \Delta_p^2)}{(\Gamma^2 + \Delta_p^2 + 2g^2|a|^2)^2}\right\}. \tag{4.6}$$

Here the light intensity is equal to the expectation value of the number of photons, i.e., $|a|^2 = n$, $|a_{in}|^2 = n_{in}$, $|a^l|^2 = n^l$, and $|a^r|^2 = n^r$. Then the output intensity from the left mirror is

$$|a^r|^2 = \left|\frac{(1 + \alpha e^{i\Phi})\kappa}{\kappa - i(\Delta_p - \Delta_c) + \frac{g^2 N}{(\Gamma - i\Delta_p)(1 + \frac{2g^2|a|^2}{\Gamma^2 + \Delta_p^2})}} - \alpha e^{i\Phi}\right|^2 |a_{in}|^2 \tag{4.7}$$

and the output light intensity from the right mirror is

$$|a^l|^2 = \left|\frac{(1 + \alpha e^{i\Phi})\kappa}{\kappa - i(\Delta_p - \Delta_c) + \frac{g^2 N}{(\Gamma - i\Delta_p)(1 + \frac{2g^2|a|^2}{\Gamma^2 + \Delta_p^2})}} - 1\right|^2 |a_{in}|^2. \tag{4.8}$$

Eq. (4.7) and Eq. (4.8) give the steady-state solutions of the two output fields of the CQED system coupled by two input fields and show explicitly the phase dependence. Here we seek the asymmetrical output fields from the CQED system such that one output field vanishes but the other output field is finite as shown in Fig. 4.2(b). For example, one may choose $|a^l|^2 = 0$, then the right output becomes $|a^r|^2 = (1 + \alpha^2 - 2\alpha \cos\varphi)|a_{in}|^2$. Such asymmetrical outputs of the CQED system is valid provided the following condition is satisfied

$$\alpha\kappa e^{i\varphi} = i(\Delta_p - \Delta_c) + \frac{g^2 N}{(\Gamma - i\Delta_p)(1 + \frac{2g^2|a|^2}{\Gamma^2 + \Delta_p^2})}. \tag{4.9}$$

Eq. (4.9) can be separated into two equations relating the relative phase φ and the other system parameters:

$$\alpha\kappa \cos\varphi = \frac{g^2 N \Gamma}{\kappa\alpha(\Gamma^2 + \Delta_p^2 + 2g^2|a|^2)}$$

and

$$\sin\varphi = \frac{g^2 N \Delta_p - (\Delta_p - \Delta_c)(\Gamma^2 + \Delta_p^2 + 2g^2|a|^2)}{\kappa\alpha(\Gamma^2 + \Delta_p^2 + 2g^2|a|^2)}.$$

On the other hand, one may also set $|a^r|^2 = 0$, then the left output becomes $|a^l|^2 = (1 + \alpha^2 - 2\alpha \cos\varphi)|a_{in}|^2$. The required phase relations become

$$\cos\varphi = \frac{\alpha g^2 N \Gamma}{\kappa(\Gamma^2 + \Delta_p^2 + 2g^2|a|^2)}$$

and

$$\sin\varphi = -\frac{\alpha(g^2 N \Delta_p - (\Delta_p - \Delta_c)(\Gamma^2 + \Delta_p^2 + 2g^2|a|^2))}{\kappa(\Gamma^2 + \Delta_p^2 + 2g^2|a|^2)}.$$

In particular, when $\alpha = 1$, the two output fields are flip-flopped when the phase is switched from φ to $-\varphi$. Therefore, the CQED system can perform the optical switching operation by changing the phase φ to $-\varphi$. Also note that when $\alpha > 2\cos\varphi$, $|a^l|^2 > |a_{in}|^2$, the left output field is amplified relative to the left input field; when $\alpha < 2\cos\varphi$, $|a^l|^2 > |a_{in}|^2$, the left output field is de-amplified relative to the left input field. Similar amplification or de-amplification condition is also obtained for the CQED system when the condition is met for $|a^l|^2 = 0$ and $|a^r| \neq 0$. That is, the cavity system can be made to behave as an optical amplifier or 扌Cde-amplifier by adjusting the parameter α.

Eq. (4.7) and (4.8) show that a variety of combination of system parameters can be chosen to meet the conditions of the asymmetrical cavity outputs in which the left (right) output field is linearly amplified (before the nonlinearity becomes dominant) while the right (left) output field is suppressed. We discuss three specific cases below.

1. $\Delta_p = \Delta_c$, the input light frequency matches the empty cavity resonant frequency. In order to obtain the asymmetrical cavity output with only one nonzero

output field, the input light has to be detuned from the atomic resonant frequency by an amount $\Delta_p = \pm\sqrt{(g^2 N/\alpha\kappa)^2 - \Gamma^2}$, the relative phase φ between the two input fields satisfies $\cos\varphi = \frac{\Gamma}{\sqrt{\Gamma^2 + \Delta_p^2}}$. Under such conditions, the required collective coupling coefficient $g\sqrt{N} \geq \sqrt{\alpha\kappa\Gamma}$, which can be viewed the strong coupling regime.

2. $\Delta_p = 0$, the input light frequency matches the atomic transition frequency. Then the required detuning of the empty cavity resonant frequency from the atomic transition frequency is $\Delta_c = \pm\sqrt{(\alpha\kappa)^2 - (\frac{g^2 N}{\Gamma^2 + 2g^2|a|^2})^2}$ and the relative phase φ between the two input fields is decided by $\cos\varphi = \frac{g^2 N}{\alpha\kappa\Gamma}$. Such operating conditions require the collective coupling coefficient $g\sqrt{N} \leq \sqrt{\alpha\kappa(\Gamma + 2g^2|a|^2/\Gamma)}$. Therefore, the CQED system can be viewed as in the weakly coupling regime.

3. $\Delta_c = 0$, the resonant frequency of the empty cavity matches the atomic resonant frequency. The corresponding value for Δ_p can be derived from the e-quation $(g^2 N\Gamma)^2 + \Delta_p^2(g^2 N - \Gamma^2 - \Delta_p^2 - 2g^2|a|^2)^2 = (\alpha\kappa)^2(\Gamma^2 + \Delta_p^2 + 2g^2|a|^2)^2$, which leads to the specific phase equation $\cos\varphi = \frac{g^2 N\Gamma}{\kappa\alpha(\Gamma^2 + \Delta_p^2 + 2g^2|a|^2)}$ and $\sin\varphi = \frac{g^2 N\Delta_p + \Delta_p(\Gamma^2 + \Delta_p^2 + 2g^2|a|^2)}{\kappa\alpha(\Gamma^2 + \Delta_p^2 + 2g^2|a|^2)}$. Then the collective coupling coefficient must satisfies $g\sqrt{N} \leq \sqrt{\alpha(\kappa\Delta_p^2/\Gamma + \kappa\Gamma + 2\kappa g^2|a|^2/\Gamma)}$.

4.3   Numerical results

The above analysis shows that one can vary the parameters of a CQED system coupled by two input fields and enable the CQED system to function as an optical logical device, in which one output field is nonzero while the other output field is suppressed. Practically, such input-output states of the CQED system can be achieved by varying the relative phase of the two input fields. It is desirable to find the range of the input light intensities in which such asymmetrical output properties of the CQED can follow the input intensity change. To find answers to this question, I plot in Fig. 4.3 and 4.4 the intensities of the both output fields versus the input light intensity under a variety of conditions discussed above under which there only one nonzero output field from the CQED system is coupled by two input fields.

Fig. 4.3 plots the two output intensities of the CQED system versus the input light intensity $|a_{in}|^2$ with the matching input light frequency and the empty cavity frequency, $\Delta_c = \Delta_p$ when the CQED system parameters are taken under the condition (9) for the asymmetrical cavity output in the linear excitation regime (a weak input intensity). It shows that there is a large dynamic range of the input light intensities in which the asymmetrical output of the CQED system, particularly when the matching frequency of the cavity and the input light fields ($\Delta_c = \Delta_p$) are tuned sufficiently away from the atomic resonance. One of the output field increases linearly with the input light field while the other output field is essentially zero.

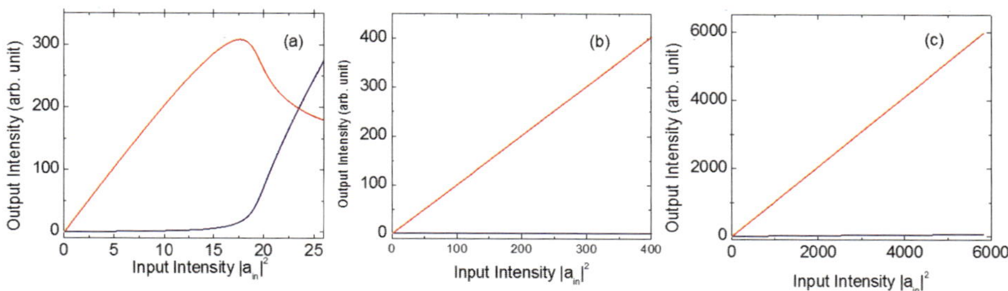

Fig. 4.3   The output intensities $|a^l|^2$ (blue lines) an $|a^r|^2$ (red lines) versus the input intensity $|a_{in}|^2$. (a) $\Delta_c = \Delta_p = 1.37\Gamma$, $\alpha = 5$ and $\varphi = 1.318$ rad. (b) $\Delta_c = \Delta_p = 9.95\Gamma$, $\alpha = 1$, and $\varphi = 1.32$ rad. (c) $\Delta_c = \Delta_p = 100\Gamma$, $\alpha = 0.1$, and $\varphi = 1.56$ rad. The other parameters are $\kappa = 2\Gamma$, $g\sqrt{N} = 4.47\Gamma$, $g = 0.02\Gamma$, $T = 0.01$.

This behavior indicates that as a 4 channel input-output device, the CQED system behaves like an optical transistor. If the detuning $\Delta_c = \Delta_p$ is small (Fig. 4.3(a)), the operating range of the logical CQED system is narrower and the leakage of the zero output channel becomes noticeable at large input intensities.

Fig. 4.4 plots the two output intensities of the CQED system versus the phase difference φ with the matching input light frequency and the empty cavity frequency, $\Delta_c = \Delta_p$ when the CQED system parameters are taken under the condition (4.9) for the asymmetrical cavity output. The two output intensities varies sinusoidally versus the phase φ. When $\alpha \neq 1$, only one output becomes zero at a specific φ value given by Eq. (4.9). When the two input light fields have the same amplitude ($\alpha = 1$), one output intensity vanishes while the other output is at the maximum. That is, the two outputs can be flip-flopped if one switches the phase difference from φ to $\varphi + \pi$. Thus, the CQED system can be made to function as a phase controlled, pure optical logical device.

Fig. 4.5 plots the two output intensities of the CQED system versus the input light intensity $|a_{in}|^2$ when the input light frequency does not match the empty cavity frequency, $\Delta_c \neq \Delta_p$. With the input light frequency tuned to be resonant with the atomic transition ($\Delta_p = 0$), the one-sided light output in the CQED system occurs when the cavity frequency matches Eq. (9). When the CQED system parameters are taken under the condition (9) for the asymmetrical cavity output in the linear excitation regime (a weak input intensity), it shows again that the larger the cavity detuning Δ_c, the larger the dynamic range of the input light intensities in which the asymmetrical output of the CQED system can maintain only one nonzero output channel while the other output channel is essentially zero. Again this behavior indicates that as a 4 channel input-output device, the CQED system behaves like an optical transistor.

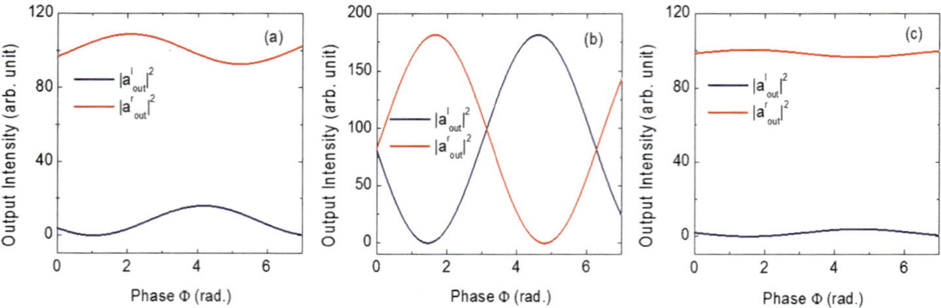

Fig. 4.4 $|a^l|^2$ (blue lines) an $|a^r|^2$ (red lines) versus the relative phase difference φ of the two input fields. (a) $\Delta_p = \Delta_c = 1.73\Gamma$, $\alpha = 5$ and $|a_{in}|^2 = 5$, $\varphi = 1.318$ rad. (b) $\Delta_p = \Delta_c = 9.95\Gamma$, $\alpha = 1$, and $|a_{in}|^2 = 100$ and $\varphi = 1.32$ rad. (c) $\Delta_p = \Delta_c = 100\Gamma$, $\alpha = 0.1$, $|a_{in}|^2 = 100$, and $\varphi = 1.56$ rad. The other parameters are $\kappa = 2\Gamma$, $g\sqrt{N} = \Gamma$, $g = 0.02\Gamma$, $T = 0.01$.

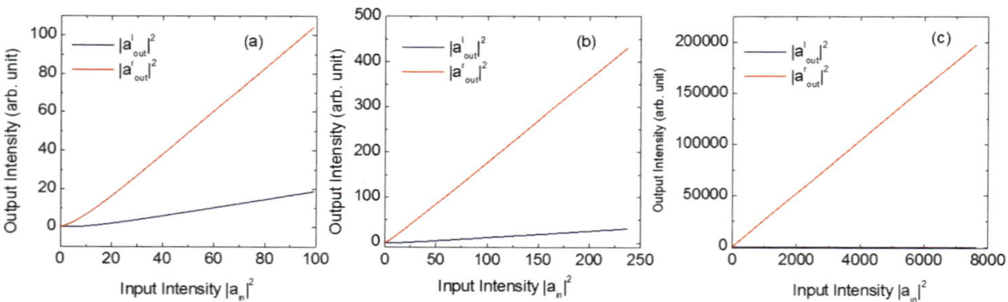

Fig. 4.5 $|a^l|^2$ (blue lines) an $|a^r|^2$ (red lines) versus the input intensity $|a_{in}|^2$. (a) $\Delta_p = 0$, $\Delta_c = 0.66\Gamma$, $\alpha = 0.6$ and $\varphi = 0.585$ rad. (b) $\Delta_p = 0$, $\Delta_c = 1.73\Gamma$, $\alpha = 1$, and $\varphi = 1.05$ rad. (c) $\Delta_p = 0$, $\Delta_c = 9.95\Gamma$, $\alpha = 5$, and $\varphi = 1.47$ rad. The other parameters are $\kappa = 2\Gamma$, $g\sqrt{N} = \Gamma$, $g = 0.02\Gamma$, $T = 0.01$.

Fig. 4.6 plots the two output intensities of the CQED system versus the phase difference φ when the input light frequency does not match the empty cavity frequency, $\Delta_c \neq \Delta_p$. Again, when the two input light fields have the same amplitude ($\alpha = 1$), the two output intensities varies sinusoidally versus the phase φ. At a specific φ value given by Eq. (9), one output intensity vanishes while the other output is at the maximum. The two outputs are flip-flopped when the phase difference from φ to $\varphi + \pi$. The CQED system therefore can be made to function as a phase controlled, pure optical logical device.

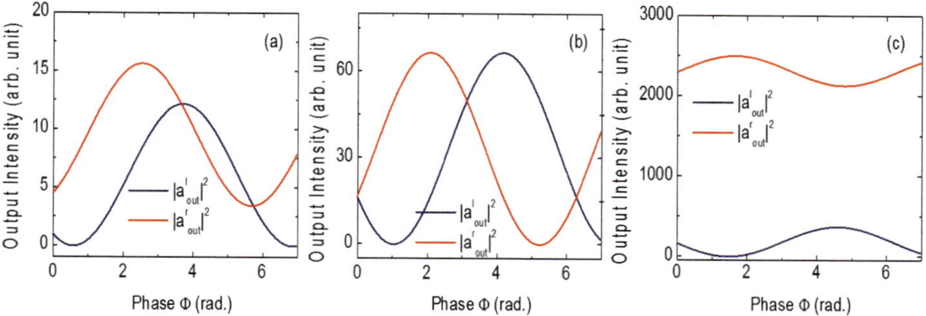

Fig. 4.6　$|a^l|^2$ (blue lines) and $|a^r|^2$ (red lines) versus the relative phase difference φ of the two input fields. (a) $\Delta_p = 0$, $\Delta_c = 0.66\Gamma$, $\alpha = 0.6$ and $|a_{in}|^2 = 20$, $\varphi = 0.585$ rad. (b) $\Delta_p = 0$, $\Delta_c = 1.73\Gamma$, $\alpha = 1$, and $|a_{in}|^2 = 50$, $\varphi = 1.05$ rad. (c) $\Delta_p = 0$, $\Delta_c = 9.95\Gamma$, $\alpha = 5$, $|a_{in}|^2 = 100$, and $\varphi = 1.47$ rad. The other parameters are $\kappa = 2\Gamma$, $g\sqrt{N} = \Gamma$, $g = 0.02\Gamma$, $T = 0.01$.

4.4　Conclusion

In recent years, chiral quantum optics has emerged as an active research field, in which the directional coupling of the photons and atoms leads to asymmetrical input-output characteristics for photonic devices [21]. Here I show that while the photon-atom interaction does not possess the chiral asymmetry in a symmetrical CQED system, it is possible to realize the asymmetrical input-output behavior in a CQED system. Based on a semiclassical analysis, I have derived the light input-output relationship in a multiatom CQED system with two-sided input fields and the results show that the CQED system can be configured to exhibit asymmetrical output behavior in which one output field is suppressed while the other output field is preserved. Such chiral input-output characteristics is induced by the interference of the two input fields manifested by the intra-cavity two-level absorber and therefore is controllable by varying the phase difference between the two input fields. When the two input fields have the same amplitude, the two output fields can be flip-flopped by a π phase ramp and therefore realizing the phase-controlled all-optical switching. With unequal amplitudes of the two input fields, the optical transistor behavior can be observed for the CQED system. Such a CQED system coupled by two laser fields from two output ends of the cavity can be treated as a four-channel input-output optical system and may be useful for practical applications in optics communication and information processing network.

Bibliography

[1] *Cavity Quantum Electrodynamics*, edited by P. R. Berman (Academic, San Diego, 1994).

[2] S. J. van Enk, H. J. Kimble, and H. Mabuchi, Quantum information processing 3, 75(2004).

[3] S. Haroche, Reviews of modern physics, 85, 1083 (2013).

[4] E. T. Jaynes and F. W. Cummings, Proc. IEEE 51, 89 (1963).

[5] A. Boca, R. Miller, K. M. Birnbaum, A. D. Boozer, J. McKeever, and H. J. Kimble, Phys. Rev. Lett. **93**, 233603(2004).

[6] J. J. Sanchez-Mondragon, N. B. Narozhny, and J. H. Eberly, Phys. Rev. Lett. **51**, 550(1983).

[7] G. S. Agarwal, Phys. Rev. Lett. **53**, 1732 (1984).

[8] M. G. Raizen, R. J. Thompson, R. J. Brecha, H. J. Kimble, and H. J. Carmichael, Phys. Rev. Lett. **63**, 240 (1989).

[9] Y. Zhu, D. J. Gauthier, S. E. Morin, Q. Wu, H. J. Carmichael, and T. W. Mossberg, Phys. Rev. Lett. **64**, 2499(1990).

[10] Ian D. Leroux, Monika Schleier-Smith, Hao Zhang, and Vladan Vuletić, Phys. Rev. **A85**, 013803 (2012).

[11] Monika Schleier-Smith, Ian D. Leroux, and Vladan Vuletić, Phys. Rev. **A81**, 021804(R) (2010).

[12] K. C. Cox, G. P. Greve, J. M. Weiner, J. K. Thompson, Phy. Rev. Lett. **116**, 093602 (2016).

[13] Z. Chen, J. G. Bohnet, S. R. Sankar, J. Dai, J. K. Thompson, Phy. Rev. Lett. **106**, 133601 (2011).

[14] O. Hosten, N. J. Engelsen, R. Krishnakumar, and M. A. Kasevich, Nature, 529, 505 (2016).

[15] A. Grankin, E. Brion, R. Boddeda, S. Ćuk, I. Usmani, A. Ourjoumtsev, and P. Grangier, Phys. Rev. Lett. **117**, 253602 (2016).

[16] V. Parigi, E. Bimbard, J. Stanojevic, A. J. Hilliard, F. Nogrette, R. Tualle-Brouri, A. Ourjoumtsev, and P. Grangier, Phys. Rev. Lett. **109**, 233602 (2012).

[17] W. Wan, Y. Chong, L. Ge, H. Noh, A. D. Stone, and H. Cao, Science **331**, 889 (2011).

[18] X. Fang, M. L. Tseng, J. Ou, K. F. MacDonald, D. P. Tsai, and N. I. Zheludev, Appl. Phys. Lett. **104**, 141102 (2014).

[19] G. S. Agarwal and Y. Zhu, Phys. Rev. **A92**, 023824 (2015).

[20] G. S. Agarwal, K. Di, L. Wang, and Y. Zhu, Phys. Rev. **A93**, 063805 (2016).

[21] T. Ramos, B. Vermersch, P. Hauke, H. Pichler, and P. Zoller, Phys. Rev. **A93**, 062104 (2016).

Chapter 5

Energy and Angular Momentum in Gravity Theories

Rong-Gen Cai[a,1] and Li-Ming Cao[b,2]

[a]CAS Key Laboratory of Theoretical Physics, Institute of Theoretical Physics, Chinese Academy of Sciences, P.O. Box 2735, Beijing 100190, China

[b]Interdisciplinary Center for Theoretical Study, University of Science and Technology of China, Hefei, Anhui 230026, China

Abstract: We give a brief review on the basic ideas to get the energy and angular momentums in gravity theories. This includes the method from the Noether currents based on the Lagrangian formalism of the gravity theories and the method based on the Hamilton formulation of the theories. Based on the Noether currents, we provide a simple way to define the canonical energy-momentum tensors and canonical spin tensors. This is helpful to define the pseudotensor and superpotential in the study of the energy-momentum tensor of gravitational field. For the global defined conserved quantities, the conserved quantities at the spacelike infinity of a spacetime are briefly summarized in the usual coordinate dependent form as well as in the coordinate free form by Ashtekar and Hansen. We give a brief introduction to the covariant phase space method by Wald and Zoupas to define the conserved quantities dual to BMS supertranslations at null infinity. For the conserved quantities defined for a quasi-local region, we mainly focus on the Hawking mass, Hayward mass, Brown–York energy-momentum tensor and their later development. In the case with a cosmological constant, mimicking the Hayward energy, we propose a new mass which includes the contribution of gravitational radiation and approaches Ashtekar–Magnon–Das conformal mass at the infinity of an asymptotical AdS spacetime. For the asymptotically Schwarzschild-de Sitter spacetimes, this mass approaches to the definition by Ashtekar, Bonga, and Kesavan.

---

[1]Email: cairg@itp.ac.cn
[2]Email: caolm@ustc.edu.cn

5.1 Introduction

The terminology "energy" received its modern meaning during the development of mechanics and thermodynamics in the 17th century and the 18th century. In the early time of the 19th century, it was realized that the energy is conserved. However, until the 20th century, the mathematical foundation of the law of the conservation of energy was not established based on the celebrated work by Noether, i.e., the Noether theorem [1]. This reason for the conservation of energy can be attributed to the consequence of the time translation invariant of the system under considering. Similarly, the conservation laws of the linear momentum and angular momentum of a system are the consequences of the spatial translation and rotation invariant of the system, respectively.

In Newton mechanics, the transformations which preserve the structure of the background spacetime (for example, a flat Newton–Cartan spacetime) form the so-called Galileo group. Once the Lagrangian of the system does not depend on time explicitly, the invariant of the action of the system under the time translation of the Galileo group ensures that the total energy of the system is a constant. In special relativity, the isometries of Minkowski spacetime form Poincaré group. The symmetry of this background spacetime also implies the conservations of energy, linear momentum, and angular momentum of the matter system living on this flat background spacetime. In both cases—Newton mechanics and special relativity—these conservations are the direct results of the relativistic principle which reflects the symmetries of the underline spacetimes. So the conservations of energy, linear momentums, and angular momentums are related to the uniformity and homogeneity of the spacetimes, and the Noether theorem is the precisely mathematic formalization for this deep connection between the conservation laws and the symmetries. For each infinitesimal isometry, i.e., the Killing vector field of the Minkowski spacetime, we have a Noether current. Although the current might be nonunique (unique up to a curl), one can remedy it by some natural selection and get the so-called canonical current. In this sense, we can obtain an unique Noether current for each isometry.

The spacetime, usually being modeled as a Lorentzian manifold, in general relativity or more general diffeomorphsim invariant gravity theories, however, has no isometry in general. The way to construct the Noether currents from the Killing fields in principal does not work any more. However, as diffeomorphsim invariant theories, they have a natural "symmetry" coming from the covariance under coordinate transformations. Of course, the covariance under the coordinate transformations can not be understood as the true symmetry of a covariant theory because any theory which can be expressed by the langue of differential geometry is automatically covariant[3]. Nevertheless, for any one-parameter diffeomorphsim (or

---

[3]A typical example is the Newton–Cartan theory for the Newton mechanics and Newton gravity.

infinitesimal coordinate transformation), we can construct a Noether current. For example, in general relativity, the divergence free of the Noether current (for the one-parameter diffeomorphsim invariant) corresponds to the Bianchi identity or the covariant divergence free of the total energy momentum tensor of matter fields. However, this does not imply the conservation of energy or linear momentums because the origin of this Noether current is not the truly symmetry of the spacetime (unless the one-parameter diffeomorphsim is a true symmetry along some direction of the spacetime, i.e., isometry). Furthermore, the Noether current is not unique (up to a curl) as in the case of Minkowski, and we have no canonical way to choose one of them and say it is better than others.

Background independence is the distinguished feature of general relativity. Actually, any general diffeomorphsim invariant gravity theory also has such a feature. There is no nondynamical field in these theories, and the metric of the spacetime itself is also dynamical. As mentioned in the above, our knowledge on the energy or linear momentums is based on some fixed background spacetime, for example, Minkowski spacetime, and our knowledge on the conservation is also based on the symmetries of the spacetime. Once fix the metric of the spacetime to be some special form, we can only get some part of the whole system, and the Noether current for the matter fields is generally not conserved unless the spacetime we have fixed admits enough isometries. The reason is that matter fields alone do not form a closed system. Nevertheless, based on the construction of the Noether current from the Lagrangian of the theory, we can get the canonical energy-momentum tensor and the symmetric energy-momentum tensor for the matter fields, and conservation law can be established when the isometries of the spacetime are present. The situation becomes worse when we consider the gravitational field itself. The problem is sharpened in the case where matter fields are absent, i.e., a pure gravity system. We can not fix a metric on the event manifold because this departures from the spirit of general relativity—background independence—and introduces a nondynamical obvolute object to the theory. This means we have no idea how to construct the canonical energy-momentum tensor for gravitational field unless some background structure is allowed. Once this background structure has been introduced, we can get a background dependent energy-momentum tensor for gravitational field, i.e, pseudotensor, and the corresponding conservation law is not covariant. Furthermore, this energy-momentum tensor for gravitational field is quite nonunique, and a lot of pseudotensors have been developed in the history of general relativity. For example, if the background structure is a preferred coordinate system, the energy-momentum tensor of the gravitational field is coordinate dependent.

The difficulty to construct the locally well defined energy-momentum tensor for gravitational field probably stems from some fundamental principle of general relativity, for example, equivalence principle. Regardless this principle is fundamental

---

The most important difference between this theory and general relativity is the presence of some nondyanmical objects or some background structure.

or not, locally, on the world line of a free falling observer, the pseudotensor might be vanishing in one coordinate system and nonvanishing in another coordinate system. This is a very strange behavior and cannot be accepted. So, nowadays, most of people think that the local energy-momentum tensor of gravitational field is meaningless. Of course, this also means that the energy density or momentum density of gravitational field is meaningless. It is argued that the energy or momentum of a gravitational system can only be defined globally or at least quasi-locally. If we hope to globally define the energy or momentum for a gravitational system, firstly, we have to face a question: for what kind of system, we can define the conserved quantities globally. Actually, in general relativity, there are two kinds of system, we can define their conserved quantities globally. The first one is the isolated system which does not interact with its environment, and the second is not isolated, but has some radiation modes. The spacetime which is asymptotically flat at space-like infinity and null infinity satisfies these conditions. Since no matter fields can arrive at spacelike infinity, the definitions of conserved quantities at the spacelike infinity is for the whole spacetime and does not change with time. This quantity at the spacelike infinity are ADM energy or momentums. Others are the quantities defined at the null infinity. When gravitational radiation is absent, for example, a stationary spacetime, one can also get truly conserved quantities at the null infinity. However, when some radiation is present, the quantities defined at the null infinity is not conserved, and one can only get some balance equations for those quantities and some radiation flux. These kinds of conserved quantities are realized as integrals on some codimension-2 surfaces at the infinities of the spacetimes. There are a lot of ways to construct these global quantities. The most interesting way is the construction based on the Hamilton formulation of the gravity theories. The usual Hamilton formulation of gravity based on the space and time decomposition is important to understand the traditional definition of the ADM energy or momentums. We will introduce the ADM energy from this analysis and give a brief introduction to the properties of ADM energy and momentums. The intriguing formalism is the covariant phase space analysis proposed by Wald *et. al.* in general diffeomorphsim invariant gravity theories. This formalism is not limited to general relativity but also to more general diffeomorphsim invariant gravity theories. In this paper, we give a brief review on this method and the algorithm to get ADM energy at the spacelike infinity and the conserved quantities at null infinity dual to BMS super-translations. The coordinate-free form of ADM 4-momentum has been established by Ashtekar *et. al.*, we also give a brief summary on their definition.

Gravitational field has energy. Nearly all physicists believe this point. The energy-momentum tensor for gravitational field can not be locally defined in a co-variant way. Most of physicists think this is right. However, for a given gravitational system, can the energy only be defined at the infinities? This of course is not reasonable. For the spacetime we are living on, we even can not get any information from some kind of infinity. Without the asymptotical behavior of the fields near the infinity, how do we study the energy of our spacetime? On the other hand, the asymptotical conditions on the fields in asymptotical flat spacetime can not be truly realized in practice. Those conditions are used to define an ideal system which can not happen in reality. For these reasons and probably others, we have to consider the conserved quantities for a finite region or quasi-local region of the spacetime. This is the so-called quasi-local conserved quantities. However, similar to the pseudotensor method, there are a lot of quasi-local definition on the energy for a given quasi-local region, and we do not know which one is better. Here, we only list two of them, i.e., Hawking mass, Brown–York energy-momentum tensor. We also briefly summarize some development of these two quasi-local definitions.

This paper is organized as follows: in Sec.2, we briefly review the Noether currents constructed from the Lagrangian of the theory, and list some conserved quantities based on these Noether currents. Sec.3 is devoted to the introduction to the globally defined conserved quantities. In Sec.4, we give the definition of Hawking mass and give a short summary on the Brown–York energy-momentum tensor. Sec.5 is for the conclusion and discussion. In the appendix, we provide a brief review on the definitions of the conserved quantities in asymptotical (A)dS spacetimes.

At the end of this introduction, we would like to say something about this paper: Firstly, this paper is far from a thorough review on this topic. The readers who are interesting to know the details about the conserved quantities in general relativity can find a nice review by Sazbados [2] and other papers in this book. Secondly, although this paper is mainly a review on the conserved charges in gravity theories, there are also some new findings: We provide a simple way to get the canonical energy-momentum tensor and spin tensor, and get correct divergence equations or relations between them. This can be found in subsubsection (5.2.5.2). We also propose a mass definition when cosmological constant is presented. This mass, in some sense, is the generalization of the Hayward mass. See eq.(5.277) and following discussions.

In this paper, we use the notations in the standard text book [3]. For example, for a vector field v^a, the Riemann tensor is defined as $(\nabla_a \nabla_b - \nabla_b \nabla_a)v_c = R_{abcd}v^d$, where ∇_a is the covariant derivative which is compatible with the metric.

5.2 Noether currents and conserved quantities

In this section, we give a brief review on the Neother currents constructed from the Lagrangian of a theory. Based on these currents, one can define some conserved quantities which will used to define conserved quantities of gravity theories.

5.2.1   *Noether currents*

Generally, on an n-dimensional manifold M, the Lagrangian of a field system can be put into an n-form

$$\mathbf{L} = \mathbf{L}\left(\Phi, \overline{\nabla}_a \Phi, \cdots \overline{\nabla}_{(a_1} \cdots \overline{\nabla}_{a_k)}\Phi\,; \Psi, \overline{\nabla}_a \Psi, \cdots \overline{\nabla}_{(a_1} \cdots \overline{\nabla}_{a_l)}\Psi\right)\bar{\epsilon}, \qquad (5.1)$$

where Φ is the collection of all possible dynamical fields, and Ψ denotes all of the background fields. Here, we have omitted all possible indices on the fields. $\mathbf{L}$ is a function of these fields and depends on the selection of volume element $\bar{\epsilon}$, and then is a scalar density with an unit weight. So $\mathbf{L}$ can be viewed as an usual tensor. $\overline{\nabla}_a$ is the derivative operator compatible with the given volume element $\bar{\epsilon}$ on M, i.e., we have $\overline{\nabla}_a \bar{\epsilon} = 0$. For example, if we choose $\bar{\epsilon}$ to be usual coordinates volume element $dx^1 \wedge \cdots \wedge dx^n$, and $\overline{\nabla}_a$ is just the usual partial derivative ∂_a associated with the coordinates.

We assume the theory is covariant or diffeomorphsim invariant, i.e., the action

$$I[\Phi, \Psi] = \int_D \mathbf{L} \qquad (5.2)$$

is invariant under any diffeomorphsim. Here D is a given region of the manifold M. For more details, assume $f : M \to M$ is a diffeomorphsim, we have

$$I[f^*\Phi, \Psi] = I[\Phi, \Psi]\,, \qquad (5.3)$$

where f^* denots the pull back or push forward on the fields on the manifold M. It should be noted that the background field Ψ does not change under the transformation. Obviously, if the Lagrangian is diffeomorphism invariant, i.e.,

$$(f^*\mathbf{L})(\Phi, \Psi) = \mathbf{L}(f^*\Phi, \Psi)\,, \qquad (5.4)$$

then the action is diffeomorphsim invariant. Generally, the action I is diffeomorphism invariant when the Lagrangian n-form $\mathbf{L}$ is diffeomporphsim invariant up to a total derivative term.

Considering a curve Φ_λ through a point $\Phi = \Phi_0$ with parameter λ in the configuration space comprised by kinetically allowed fields on the manifold of events [4,5], then the variation of the fields can be viewed as the tangent vector of the curve, i.e., we have

$$\delta\Phi = \left.\frac{d\Phi_\lambda}{d\lambda}\right|_{\lambda=0}. \qquad (5.5)$$

Other curves passing through the point Φ_0 give corresponding variations of the fields at the point. With this general variation of the fields, the variation of the action can be written as

$$\delta I[\Phi, \Psi] = \left.\frac{dI[\Phi_\lambda, \Psi]}{d\lambda}\right|_{\lambda=0} = \int_D \delta\mathbf{L}\,, \qquad (5.6)$$

where

$$\delta\mathbf{L} = \sum_{i=0}^{k} \mathbf{U}^{a_1\cdots a_i}\overline{\nabla}_{(a_1}\cdots\overline{\nabla}_{a_i)}\delta\Phi\,, \tag{5.7}$$

with

$$\mathbf{U}^{a_1\cdots a_k} = \frac{\partial\mathbf{L}}{\partial\big(\overline{\nabla}_{(a_1}\cdots\overline{\nabla}_{a_i)}\Phi\big)}\,. \tag{5.8}$$

Here, $\overline{\nabla}_a$ has no direct relation to dynamical fields, so it does not change under the variation. After integration by parts, one finds

$$\delta\mathbf{L} = \mathbf{E}\delta\Phi + \mathrm{d}\Theta\,, \tag{5.9}$$

where

$$\mathbf{E} = \sum_{i=0}^{k}(-1)^i\big(\overline{\nabla}_{a_1}\cdots\overline{\nabla}_{a_i}\mathbf{U}^{a_1\cdots a_i}\big)\,, \tag{5.10}$$

and $\mathbf{E} = 0$ is the Euler-Lagrangian equation of the system. The $(n-1)-$form $\Theta = \Theta(\Phi, \Psi, \delta\Phi)$ can be expressed as

$$\Theta_{a_1\cdots a_{n-1}} = \sum_{i=1}^{k}\sum_{j=1}^{i}(-1)^{j+1}\big(\overline{\nabla}_{b_2}\cdots\overline{\nabla}_{b_j}\mathbf{U}^{ab_2\cdots b_j b_{j+1}\cdots b_i}{}_{aa_1\cdots a_{n-1}}\big)$$
$$\times\big(\overline{\nabla}_{b_{j+1}}\cdots\overline{\nabla}_{b_i}\delta\Phi\big)\,. \tag{5.11}$$

Now, let us consider some symmetric transformation of the theory, and the associated variation of the field Φ can be denoted by $\delta_s\Phi$. Under the symmetric transformation, the action is invariant, so the Lagrangian is transformed as $\delta_s\mathbf{L} = \mathrm{d}\mathbf{B}$. From eq.(5.9) we have

$$\mathrm{d}\mathbf{B} = \mathbf{E}\delta_s\Phi + \mathrm{d}\Theta\,, \tag{5.12}$$

This suggests for on-shell field configurations, the current

$$\mathbf{J} = \Theta(\Phi, \Psi, \delta_s\Phi) - \mathbf{B} \tag{5.13}$$

is closed, i.e., $\mathrm{d}\mathbf{J} = 0$. Actually, we have

$$\mathrm{d}\mathbf{J} = -\mathbf{E}\delta_s\Phi\,. \tag{5.14}$$

The $(n-1)-$form $\mathbf{J}$ is the so-called Noether current. Since the current is closed when the fields are on-shell, one can construct Noether charge $(n-2)-$form as [6]

$$\mathbf{J} = \mathrm{d}\mathbf{Q}\,, \tag{5.15}$$

and the integral of this $(n-1)-$form on some codimension-2 submanifold will give the so-called Noether charge.

5.2.2 *Theories without background fields*

Assume Φ being comprised of a metric g_{ab} and a gauge potential A_a on M (a Lie algebra valued one form), and the Lagrangian has a form

$$\mathbf{L}\left(g_{ab}, \overline{\nabla}_{a_1}g_{ab}, \cdots, \overline{\nabla}_{(a_1}\cdots\overline{\nabla}_{a_k)}g_{ab}; A_a, \overline{\nabla}_{a_1}A_a, \cdots, \overline{\nabla}_{(a_1}\cdots\overline{\nabla}_{a_l)}A_a; \Psi\right).\tag{5.16}$$

By requiring that the Lagrangian is diffeomorphism invariant and gauge invariant, i.e., $\mathbf{L}$ satisfies (5.4) and is also invariant under the gauge transformation of the gauge potential

$$A_a \to \mathrm{g}^{-1}A_a\mathrm{g} - (i/\lambda)\mathrm{g}^{-1}(\mathrm{d}\mathrm{g})_a\,,$$

where g is a local gauge transformation which can be viewed as a mapping from M to the gauge group of the theory, Wald, Iyer, and Hollands have proved two Thomas replacement theorems and shown the Lagrangian can be written as [5, 9]

$$\mathbf{L}\left(g_{ab}, R_{abcd}, \nabla_{a_1}R_{abcd}, \cdots, \nabla_{(a_1}\cdots\nabla_{a_{k-2})}R_{abcd};\right.$$
$$\left.F_{ab}, \mathcal{D}_{a_1}F_{ab}, \cdots, \mathcal{D}_{(a_1}\cdots\mathcal{D}_{a_{l-1})}F_{ab}; \psi, \nabla_a\psi\cdots\right),\tag{5.17}$$

where ∇_a is compatible with g_{ab} and $\mathcal{D}_a$ is the covariant derivative defined as

$$\mathcal{D}_a = \nabla_a + i\lambda A_a\,,\tag{5.18}$$

and λ is the coupling constant of the gauge theory. Of course, the Lagrangian (5.17) might include other dynamical fields ψ and their symmetric derivatives. For simplicity, we do not consider the gauge field or put them into ψ and their symmetric derivatives. Now, the Lagrangian can be written as

$$\mathbf{L} = L\epsilon\,,\tag{5.19}$$

where L is a scalar density (according to the definition by Lee and Wald in [4], this definition is a little bit different from the usual definition of a scalar density but the same in essence) and has a form

$$L = L\left(g_{ab}, R_{abcd}, \nabla_{a_1}R_{abcd}, \cdots, \nabla_{(a_1}\cdots\nabla_{a_m)}R_{abcd};\right.$$
$$\left.\psi, \nabla_a\psi, \cdots, \nabla_{(a_1}\cdots\nabla_{a_l)}\psi\right),\tag{5.20}$$

and ϵ is the natural volume element associated with the metric g_{ab}. Under the general variation of the fields, the variation of Lagrangian has a form

$$\delta\mathbf{L} = (\mathbf{E}_g)^{ab}\delta g_{ab} + \mathbf{E}_\psi\delta\psi + \mathrm{d}\Theta\,.\tag{5.21}$$

where

$$\mathbf{E}_\psi = \epsilon E_\psi\,,\tag{5.22}$$

$$(\mathbf{E}_g)^{ab} = \epsilon\left[\frac{\partial L}{\partial g_{ab}} + \frac{1}{2}g^{ab}L + 2\nabla_d\nabla_c E_R^{dabc} + E_R^{cdea}R_{cde}{}^b + \nabla_c S_R^{cab} + \nabla_c S_\psi^{cab}\right].\tag{5.23}$$

In these equations, we have defined

$$E_\psi = \sum_{i=0}^{l} E_\psi^{(i)}, \tag{5.24}$$

$$E_R^{abcd} = \sum_{i=0}^{m} (-1)^i E_R^{(i)abcd}, \tag{5.25}$$

$$S_\psi^{cab} = \sum_{i=1}^{m}\sum_{j=1}^{i} (-1)^{i-j}\left[U_{(\psi)}^{(i;i-j)(a_1\cdots a_j)} (\Delta f_{(\psi)})_{(a_1\cdots a_j)}{}^{cab}\right], \tag{5.26}$$

$$S_R^{cab} = \sum_{i=1}^{m}\sum_{j=1}^{i} (-1)^{i-j}\left[U_{(R)}^{(i;i-j)b_1b_2b_3b_4(a_1\cdots a_j)} (\Delta f_{(R)})_{b_1b_2b_3b_4(a_1\cdots a_j)}{}^{cab}\right]. \tag{5.27}$$

The quantities in the above equations have following definitions

$$f_{(R)abcd(a_1\cdots a_i)} = \nabla_{(a_1}\cdots\nabla_{a_i)}R_{abcd}, \tag{5.28}$$

$$U_{(R)}^{abcd(a_1\cdots a_i)} = \partial L/\partial f_{(R)abcd(a_1\cdots a_i)}, \tag{5.29}$$

$$E_{(R)}^{(i)abcd} = \nabla_{(a_1}\cdots\nabla_{a_i)}U_{(R)}^{abcd(a_1\cdots a_i)}, \tag{5.30}$$

$$U_{(R)}^{(i;i-j)abcd(a_1\cdots a_j)} = \nabla_{(b_1}\cdots\nabla_{b_{i-j})}U_{(R)}^{abcd(b_1\cdots b_{i-j}a_1\cdots a_j)}, \tag{5.31}$$

and

$$f_{(\psi)(a_1\cdots a_i)} = \nabla_{(a_1}\cdots\nabla_{a_i)}\psi, \tag{5.32}$$

$$U_{(\psi)}^{(a_1\cdots a_i)} = \partial L/\partial f_{(\psi)(a_1\cdots a_i)}, \tag{5.33}$$

$$E_{(\psi)}^{(i)} = \nabla_{(a_1}\cdots\nabla_{a_i)}U_{(\psi)}^{(a_1\cdots a_i)}, \tag{5.34}$$

$$U_{(\psi)}^{(i;i-j)(a_1\cdots a_j)} = \nabla_{(b_1}\cdots\nabla_{b_{i-j})}U_{(\psi)}^{(b_1\cdots b_{i-j}a_1\cdots a_j)}. \tag{5.35}$$

The Δf_R and Δf_ψ terms come from the variation of the covariant derivatives ∇_a, which has forms like

$$(\Delta f_{(R)})_{b_1b_2b_3b_4(a_1\cdots a_i)}{}^{c_1c_2c_3}$$
$$= \sum_{l=1}^{i-1} \Delta_{a_ia_lc}^{c_1c_2c_3} g^{ce} f_{(R)b_1b_2b_3b_4(a_1\cdots e\cdots a_{i-1})}$$
$$+ \sum_{s=1}^{4} \Delta_{a_ib_sc}^{c_1c_2c_3} g^{ce} f_{(R)b_1\cdots e\cdots b_4(a_1\cdots a_{i-1})} \tag{5.36}$$

with

$$-2\Delta_{abc}^{def} = \delta_a^d\delta_b^e\delta_c^f + \delta_b^d\delta_a^e\delta_c^f - \delta_c^d\delta_a^e\delta_b^f. \tag{5.37}$$

Here, it should be noted that we have omitted the possible indices of ψ to simplify the expressions. To get detailed form of Δf_ψ, one has to consider the indices of ψ. In eq.(5.23), S_R^{abc} and S_ψ^{abc} are determined by these variations.

The $(n-1)-$form $\boldsymbol{\Theta}$, i.e., symplectic potential $(n-1)-$form, has a detailed form

$$\boldsymbol{\Theta} = 2\mathbf{E}_R^{bcd}\nabla_d\delta g_{bc} - 2\nabla_d\mathbf{E}_R^{bcd}\delta g_{bc} - \mathbf{S}_R^{ab}\delta g_{ab} - \mathbf{S}_\psi^{ab}\delta g_{ab}$$

$$+ \sum_{i=1}^{l}\sum_{j=1}^{i}(-1)^{i-j}\mathbf{U}_\psi^{(i;i-j)(a_1\cdots a_{j-1})}\delta\nabla_{(a_1}\cdots\nabla_{a_{j-1})}\psi$$

$$+ \sum_{i=1}^{m}\sum_{j=1}^{i}(-1)^{i-j}\mathbf{U}_R^{(i;i-j)abcd(a_1\cdots a_{j-1})}\delta\nabla_{(a_1}\cdots\nabla_{a_{j-1})}R_{abcd}\,. \tag{5.38}$$

The tensors $\mathbf{E}_R^{bcd}$, $\mathbf{S}_R^{ab}$, $\cdots$ in the above expressions are defined as follows

$$(\mathbf{S}_R^{ab})_{a_2\cdots a_n} = S_R^{cab}\epsilon_{ca_2\cdots a_n}\,, \tag{5.39}$$

$$(\mathbf{S}_\psi^{ab})_{a_2\cdots a_n} = S_\psi^{cab}\epsilon_{ca_2\cdots a_n}\,, \tag{5.40}$$

$$(\mathbf{E}_R^{bcd})_{a_2\cdots a_n} = E_R^{abcd}\epsilon_{aa_2\cdots a_n}\,, \tag{5.41}$$

$$\left(\mathbf{U}_\psi^{(i;i-j)(b_1\cdots b_{j-1})}\right)_{a_2\cdots a_n} = U_{(\psi)}^{(i;i-j)(b_1\cdots b_{j-1}e)}\epsilon_{ea_2\cdots a_n}\,, \tag{5.42}$$

$$\left(\mathbf{U}_R^{(i;i-j)abcd(b_1\cdots b_{j-1})}\right)_{a_2\cdots a_n} = U_{(R)}^{(i;i-j)abcd(b_1\cdots b_{j-1}e)}\epsilon_{ea_2\cdots a_n}\,. \tag{5.43}$$

For some symmetric transformations, $\delta_s\mathbf{L} = \mathrm{d}\mathbf{B}$, one can define a Noether current as before

$$\mathbf{J} = \boldsymbol{\Theta}(\Phi, \delta_s\Phi) - \mathbf{B}\,. \tag{5.44}$$

Now, let us consider the variation due to the infinitesimal diffeomorphsim, i.e., the Lie derivative the fields along the integral curve of some vector field ξ. The theory is diffeomporphsim invariant, so the Lie derivative is the variation of a symmetric transformation, and $\mathcal{L}_\xi\mathbf{L}$ an exact form. Actually, from the well known Cartan identity, it is easy to find

$$\mathcal{L}_\xi\mathbf{L} = \mathrm{d}(\xi\cdot\mathbf{L})\,, \tag{5.45}$$

and the corresponding Noether current is given by [4,5]

$$\mathbf{J}[\xi] = \boldsymbol{\Theta}(\Phi, \mathcal{L}_\xi\Phi) - \xi\cdot\mathbf{L}\,. \tag{5.46}$$

Of course, one can also obtain the Noether currents associated with the gauge transformation when gauge fields are present. For example, consider the more general Lagrangian (5.17), we can get the current for the gauge transformation. It is easy to find $\mathbf{J}[\xi]$ is closed when all the fields are on-shell. Actually, we have

$$\mathrm{d}\mathbf{J}[\xi] = -(\mathbf{E}_g)^{ab}\mathcal{L}_\xi g_{ab} - \mathbf{E}_\psi\mathcal{L}_\xi\psi\,. \tag{5.47}$$

This is the special case of eq.(5.14).

5.2.3 *Energy-momentum tensor of matter fields*

The general covariant theory has a Lagrangian (5.17), and there are no background fields inside the Lagrangian any more. Assume the Lagrangian can be naturally split into geometric part and matter part, i.e., we have

$$\mathbf{L} = \mathbf{L}_{\mathrm{g}} + \mathbf{L}_{\mathrm{m}} ,$$

where $\mathbf{L}_{\mathrm{g}}$ has nothing to do with the matter field ψ and $\mathbf{L}_{\mathrm{m}}$ has a form

$$\mathbf{L}_m = \mathbf{L}_m \Big(g_{ab} , R_{abcd} , \nabla_{a_1} R_{abcd} , \cdots , \nabla_{(a_1} \cdots \nabla_{a_m)} R_{abcd} ;$$

$$\psi , \nabla_a \psi , \cdots , \nabla_{(a_1} \cdots \nabla_{a_l)} \psi \Big) . \tag{5.48}$$

Then we have equation (5.21) and [8]

$$\delta \mathbf{L}_{\mathrm{m}} = \mathbf{E}_\psi \delta\psi + \frac{1}{2}\epsilon T^{ab}\delta g_{ab} + \mathrm{d}\mathbf{\Theta}_{\mathrm{m}} , \tag{5.49}$$

$$\delta \mathbf{L}_{\mathrm{g}} = (\overline{\mathbf{E}}_g)^{ab}\delta g_{ab} + \mathrm{d}\mathbf{\Theta}_{\mathrm{g}} . \tag{5.50}$$

The symmetric tensor T^{ab} is the energy-momentum tensor of the matter fields. By this definition, if $\mathbf{L}_{\mathrm{m}} = L_{\mathrm{m}}\epsilon$, similar to get eq.(5.23), it is not hard to find

$$T^{ab} = 2\Big[\frac{\partial L_{\mathrm{m}}}{\partial g_{ab}} + \frac{1}{2}g^{ab}L_{\mathrm{m}} + 2\nabla_b\nabla_c E_R^{babc} + E_R^{cdea}R_{cde}{}^b$$

$$+ \nabla_c S_R^{cab} + \nabla_c S_\psi^{cab}\Big] , \tag{5.51}$$

where E_R^{abcd}, S_R^{abc}, and S_ψ^{abc} are defined in equations from (5.25) to (5.37) except that L is replaced by L_{m}. Of course, when ψ is minimally coupled to the metric, the terms including E_R^{abcd} do not exist. This energy-momentum tensor is symmetric, i.e., we always have $T^{[ab]} = 0$.

As the splitting of the Lagrangian, we have $\mathbf{\Theta} = \mathbf{\Theta}_{\mathrm{m}} + \mathbf{\Theta}_{\mathrm{g}}$ and

$$(\mathbf{E}_g)^{ab} = (\overline{\mathbf{E}}_g)^{ab} + \frac{1}{2}\epsilon T^{ab} . \tag{5.52}$$

The Noether current (5.46) now also splits into

$$\mathbf{J}[\xi] = \mathbf{J}_{\mathrm{g}}[\xi] + \mathbf{J}_{\mathrm{m}}[\xi] , \tag{5.53}$$

where we have defined

$$\mathbf{J}_{\mathrm{g}}[\xi] = \mathbf{\Theta}_{\mathrm{g}} - \xi \cdot \mathbf{L}_{\mathrm{g}} , \tag{5.54}$$

and

$$\mathbf{J}_{\mathrm{m}}[\xi] = \mathbf{\Theta}_{\mathrm{m}} - \xi \cdot \mathbf{L}_{\mathrm{m}} . \tag{5.55}$$

However, now we have

$$\mathrm{d}\mathbf{J}_{\mathrm{g}}[\xi] = -(\overline{\mathbf{E}}_g)^{ab}\mathcal{L}_\xi g_{ab} , \tag{5.56}$$

$$\mathrm{d}\mathbf{J}_\mathrm{m}[\xi] = -\mathbf{E}_\psi \mathcal{L}_\xi \psi - \frac{1}{2}\epsilon T^{ab}\mathcal{L}_\xi g_{ab}\,. \tag{5.57}$$

So the currents are generally not conserved even that the matter fields have solved to their equations of motion, i.e., $\mathbf{E}_\psi = 0$. Actually, now we have

$$\begin{aligned}
\mathrm{d}\mathbf{J}_\mathrm{m}[\xi] &= -\frac{1}{2}\epsilon T^{ab}(\nabla_a \xi_b + \nabla_b \xi_a) \\
&= -\epsilon \nabla_a(T^{ab}\xi_b) + \epsilon \nabla_a T^{ab}\xi_b\,.
\end{aligned} \tag{5.58}$$

Considering $\epsilon \nabla_a(T^{ab}\xi_b)$ is an exact form, i.e., $\mathrm{d}(k_\xi \cdot \epsilon)$ with $k_\xi = T^{ab}\xi_b$, then $\epsilon \nabla_a T^{ab}\xi_b$ has to be an exact form too. Since ξ is arbitrary, this is possible only when T^{ab} satisfies

$$\nabla_a T^{ab} = 0\,. \tag{5.59}$$

With this result, we always have

$$\mathrm{d}\mathbf{J}_\mathrm{m}[\xi] + \mathrm{d}(k_\xi \cdot \epsilon) = 0\,. \tag{5.60}$$

This suggests locally one has

$$\mathbf{J}_\mathrm{m}[\xi] + k_\xi \cdot \epsilon = \mathrm{d}\mathbf{K}\,. \tag{5.61}$$

where $\mathbf{K}$ is an $(n-2)-$form constructed from the vector field ξ, g_{ab}, ψ, and their derivatives [5].

It should be mentioned here, from eqs.(5.57) to (5.61), the gravitational equations of motion are not imposed, so these constructions can be thought as a theory of matter fields on an arbitrary fixed spacetime (M, g_{ab}). Another point is: unlike $\mathbf{J}[\xi]$ which is closed when all fields are on-shell (see eq.(5.47)) , $\mathbf{J}_\mathrm{m}[\xi]$ is not closed in general. This reflects the fact that the matter system is not a closed system, and one has to consider the total system including both the matter fields and spacetime geometry in a theory of gravity which is general covariance or diffeomorphsim invariant.

5.2.4 *Noether currents and energy of matter fields*

5.2.4.1 *Conserved quantities for matter fields*

From eq.(5.61), it is easy to find that $k_\xi \cdot \epsilon$ is a closed form when $\mathbf{J}_\mathrm{m}$ is closed. Eq.(5.57) suggests that this will happen when ξ is a Killing vector field. So, when ψ is on-shell and ξ is a Killing vector field of the spacetime, we have

$$\mathrm{d}(k_\xi \cdot \epsilon) = 0\,. \tag{5.62}$$

Let us assume Σ is a spacelike hypersurface embedded in the spacetime (M, g_{ab}), and $D \subset \Sigma$ be closed region inside Σ, then we can define a conserved quantity associated with the Killing vector field ξ as

$$Q_D[\xi] = \int_D (k_\xi \cdot \epsilon) = \int_D T_{ab}\xi^a n^b \epsilon_D\,, \tag{5.63}$$

where ϵ_D is the volume element of D and n^a is the unit normal vector field of Σ. Since $k_\xi \cdot \epsilon$ is closed, by using Stokes theorem, it is not hard to prove $Q_D[\xi]$ does not depend on the slice Σ.

This can be understood as follows. Since $(k_\xi \cdot \epsilon)$ is closed form, locally it can be expressed as

$$(k_\xi \cdot \epsilon) = \mathrm{d}\boldsymbol{U}_\xi\,,$$

where $\boldsymbol{U}_\xi$ is is an $(n-2)$−form which can be viewed as a kind of superpotential. If the $(n-1)$−de Rham cohomology group of D is trivial, this relation is globally defined. Thus we obtain

$$Q_D[\xi] = \int_{\partial D} \boldsymbol{U}_\xi\,. \tag{5.64}$$

This suggests that $Q_D[\xi]$ is actually associated with the codimension-2 surface ∂D but not D and Σ (see Fig.5.1). For a matter field on Minkowski spacetime $(\mathbb{R}^n, \eta_{ab})$,

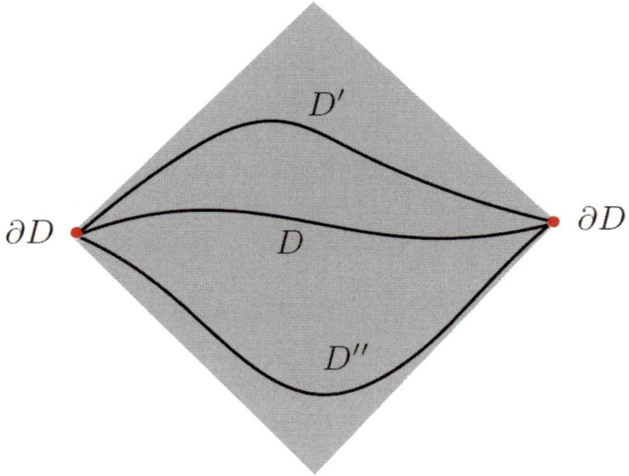

Fig. 5.1 $Q_D[\xi]$ is actually the conserved quantity associated with ∂D or to the domain of dependence of D, see [2]. For D, D', and D'', we have the same value.

assume that ξ is the generator of the Poincaré group, then we get the conserved quantities associated with any element of the Poincaré group. For example, in the usual inertial frame of the Minkowski spacetime, $t^a = (\partial/\partial t)^a$ is the time translation generator of the Poincaré group, and n^a is also given by t^a, then we have

$$E_D = Q_D[t] = \int_D T_{ab}t^a t^b \epsilon_D\,. \tag{5.65}$$

This is just the energy of the matter field inside D. $T_{ab}t^a t^b$ is nothing but the energy density observed by inertial observers. For ten Killing vectors of the Minkowski spacetime, one can define ten conserved quantities. Three spatial infinitesimal translations provide three linear momentums, and Killing vector fields for rotating invariance give us the angular momentums of the system. Furthermore, from the energy and linear momentums one can construct a mass for the system. See [2, 10] for details.

5.2.4.2 *Energy for matter fields*

However, it should be noted here, on a curved spactime (for example, a stationary spactime), that $T_{ab}\xi^a n^b$ for some timelike Killing vector field ξ generally is not the energy density observed by any observer. So $Q_D[\xi]$ is different from the energy of the matter field inside D in principle. On a general curved spacetime, the energy density of the matter field observed by the observer with velocity n^a is given by

$$E_D = \int_D T_{ab} n^a n^b \epsilon_D \,. \tag{5.66}$$

Obviously, it is different from $Q_D[\xi]$ because this energy really depend on the hypersurface Σ where D lays in.

5.2.4.3 *Canonical energy for matter fields*

For the matte system, actually, one can define a canonical energy $\mathcal{E}$ from the Hamiltonian of the system [5] (see also following sec.(5.3.1), and consider the case with fixed metric). For a Cauchy surface Σ_t of an asymptotically flat spacetime, one has

$$\mathcal{E} = \int_\Sigma \mathbf{J}_\mathrm{m} = -\int_\Sigma k_t \cdot \epsilon + \int_\infty (\mathbf{K} - t \cdot \mathbf{B})$$
$$= \int_\Sigma T_{ab} t^a n^b \epsilon_\Sigma + \int_\infty (\mathbf{K} - t \cdot \mathbf{B}) \,, \tag{5.67}$$

where t^a is the time flow vector field and usually different from the normal vector n^a of Σ_t. In the case of hypersurface orthogonal, i.e., $t^a = n^a$, it is easy to find that the difference between $\mathcal{E}$ and E_D (with $D = \Sigma_t$) is just a surface term. Similar to E_D, $\mathcal{E}$ generally is not conserved because it depends on the hypersurface Σ_t.

5.2.5 *Canonical energy-momentum tensors of matter fields*

5.2.5.1 *Matter fields in Minkowski spacetime*

For a general Lagrangian theory, especially a field theory in Minkowski spacetime $(\mathcal{R}^n, \eta_{ab})$, one can define the so-called canonical energy-momentum tensor from the Noether current. The logic is like this: assume ξ^a be an arbitrary Killing vector field in Minkowski spacetime, then the Noether current has to satisfy

$$(\mathbf{J}_\mathrm{m})_{a_1 \cdots a_{n-1}} = -\mathcal{T}^a{}_b \xi^b \epsilon_{a a_1 \cdots a_n} - \mathcal{S}^a{}_{bc} \nabla^b \xi^c \epsilon_{a a_1 \cdots a_n} \,, \tag{5.68}$$

where ∇_a is compatible with η_{ab}, and $\mathcal{T}^a{}_b$, $\mathcal{S}^a{}_{bc}$ are constructed from η_{ab}, ψ and its derivatives. The above expression for the Noether current is based on the fact that any Killing vector fields ξ^a of Minkowski spacetime is determined by $\xi^a|_p$ and $\nabla_a\xi^b|_p$ at any given point p of the spacetime. Since $\mathbf{J}_\mathrm{m}$ is closed when ξ is a Killing vector field, we have

$$\nabla_a\mathcal{T}^a{}_b\xi^b + (\mathcal{T}_{ab} + \nabla_c\mathcal{S}^c{}_{ab})\nabla^a\xi^b = 0. \tag{5.69}$$

Here, we have used the flat condition of the Minkowski spacetime, i.e., we have $R_{abcd} = 0$. Since ξ^a is an arbitrary Killing vector field, we have to impose conditions

$$\nabla_a\mathcal{T}^{ab} = 0, \qquad \mathcal{T}^{[ab]} + \nabla_c\mathcal{S}^{c[ab]} = 0. \tag{5.70}$$

These mean that $\mathcal{T}_{ab}$ is not symmetric in general, and its anti-symmetric part is given by $\nabla_c\mathcal{S}^{c[ab]}$.

With the above investigation on the Noether current, and considering the relation (5.61), we can always find a two form $\alpha_{ab} = \alpha_{[ab]}$ (corresponding to the Hodge dual of $\mathbf{K}$) such that

$$-\mathcal{T}^a{}_b\xi^b - \mathcal{S}^a{}_{bc}\nabla^b\xi^c + T^a{}_b\xi^b = \nabla_b\alpha^{ba}. \tag{5.71}$$

Similar to $\mathbf{K}$, α_{ab} is also constructed from ξ, η_{ab}, ψ, and their derivatives. So α_{ab} has a form

$$\alpha^{ab} = -H^{ab}{}_c\xi^c + \beta^{ab}{}_{cd}\nabla^c\xi^d, \tag{5.72}$$

where $H^{ab}{}_c = H^{[ab]}{}_c$ and $\beta^{ab}{}_{cd} = \beta^{[ab]}{}_{cd}$. Substituting this expansion into eq.(5.71), we have

$$T^{ab} - \mathcal{T}^{ab} = \nabla_c H^{cab}, \tag{5.73}$$

and

$$\mathcal{S}^{a[bc]} = H^{a[bc]} + \nabla_d\beta^{da[bc]} \tag{5.74}$$

or

$$H^{abc} = H^{[ab]c} = (\mathcal{S}^{a[bc]} + \mathcal{S}^{b[ca]} - \mathcal{S}^{c[ab]}) - \nabla_d(\beta^{dabc} + \beta^{dbca} - \beta^{dcab}). \tag{5.75}$$

As a conclusion, in Minkowski spacetime, we can always find a tensor $H^{abc} = H^{[ab]c}$ such that $\mathcal{T}^{ab}$ is symmetric by eq.(5.73). The tensor $\mathcal{T}^{ab}$ comes from Noether current, and is called canonical energy-momentum tensor. The above procedure shows the canonical energy-momentum tensor can be symmetrized by introducing the tensor H^{abc}. This is essentially the so-called Belinfante-Rosenfeld procedure [13,14,15]. In the special case with translation Killing vector field ξ^a, the above discussion can be found in [5].

5.2.5.2   *Matter fields in curved spacetimes*

The above discussion depends on the Killing vector fields in Minkowski spacetime. It seems that we cannot apply this method to discuss the canonical energy-momentum tensor of the matter fields in a general curved spacetime. However, from the expression (5.38) we know, except the term $\mathbf{E}_R^{bcd}\nabla_d\mathcal{L}_\xi g_{bc}$, that $\Theta(\Phi, \mathcal{L}_\xi\Phi)$ can also be expanded by ξ^a and $\nabla^a\xi^b$ as in eq.(5.68) without the requirement that ξ^a is a Killing vector field. Actually the second derivative term of ξ^a inside $\mathbf{E}_R^{bcd}\nabla_d\mathcal{L}_\xi g_{bc}$ can be put into a form $d\mathbf{Q}_m[\xi]$, where

$$(\mathbf{Q}_m[\xi])_{a_1\cdots a_{n-2}} = -E_R^{abcd}\nabla_{[c}\xi_{d]}\epsilon_{aba_1\cdots a_{n-2}} \, . \tag{5.76}$$

It should be noted here that E_R^{abcd} has the same definition as in eq.(5.25) except that L is replaced by L_m. The lower order terms inside $\mathbf{E}_R^{bcd}\nabla_d\mathcal{L}_\xi g_{bc}$ will be put into the form as in the right hand side form of eq.(5.68). Thus we have

$$(\mathbf{J}_m[\xi])_{a_1\cdots a_{n-1}} = -\mathcal{T}^a{}_b\xi^b\epsilon_{aa_1\cdots a_{n-1}} - \mathcal{S}^a{}_{bc}\nabla^b\xi^c\epsilon_{aa_1\cdots a_{n-1}} + (d\mathbf{Q}_m[\xi])_{a_1\cdots a_{n-1}} \, . \tag{5.77}$$

By the expansion in (5.77) and the relation in eq.(5.61), in a general curved spacetime, we have

$$\nabla_a(T^{ab}\xi_b) = \nabla_a\mathcal{T}^a{}_b\xi^b + (\mathcal{T}_{ab} + \nabla_c\mathcal{S}^c{}_{ab})\nabla^a\xi^b + \mathcal{S}_{cab}\nabla^c\nabla^a\xi^b \, , \tag{5.78}$$

or equivalently

$$\begin{aligned}
\nabla_a T^{ab}\xi_b &+ T^{ab}\nabla_{(a}\xi_{b)} \\
&= \nabla_a\mathcal{T}^{ab}\xi_b + (\mathcal{T}^{ab} + \nabla_c\mathcal{S}^{cab})\nabla_{(a}\xi_{b)} \\
&\quad + (\mathcal{T}^{ab} + \nabla_c\mathcal{S}^{cab})\nabla_{[a}\xi_{b]} + \mathcal{S}^{cab}R_{dcab}\xi^d \\
&\quad + \mathcal{S}^{cab}\left[\nabla_c\nabla_{(a}\xi_{b)} + \nabla_a\nabla_{(b}\xi_{c)} - \nabla_b\nabla_{(c}\xi_{a)}\right] . \tag{5.79}
\end{aligned}$$

Comparing two sides in the above equation and considering $\nabla_a T^{ab} = 0$, we have the consistency relations

$$\nabla_a\mathcal{T}^a{}_d = -R_{dcab}\mathcal{S}^{cab} \, , \tag{5.80}$$

$$\mathcal{T}^{[ab]} = -\nabla_c\mathcal{S}^{c[ab]} \, , \tag{5.81}$$

$$T^{ab} - \mathcal{T}^{(ab)} = \nabla_c\mathcal{S}^{c(ab)} \, , \tag{5.82}$$

and

$$\mathcal{S}^{(ab)c} - \mathcal{S}^{(ac)b} - \mathcal{S}^{(bc)a} = f^{abc} \, , \tag{5.83}$$

where $f_{abc} = f_{[abc]}$ and $\nabla_a f^{abc} = 0$, i.e., the 3-form f is co-closed. So locally it can be expressed as

$$f^{abc} = \nabla_e\beta^{eabc} \, ,$$

where β is a 4-form. From eq.(5.83), we have

$$H^{cab} = H^{[ca]b} := \mathcal{S}^{c[ab]} + \mathcal{S}^{a[bc]} - \mathcal{S}^{b[ca]} = \mathcal{S}^{cab} + f^{cab} \, . \tag{5.84}$$

If we still define $\mathcal{T}^{ab}$ to be the canonical energy-momentum tensor, then it has to satisfy the above equation. However, it is not divergence free in general. From the above relations, it is easy to find

$$T^{ab} = \mathcal{T}^{ab} + \nabla_c H^{cab}, \tag{5.85}$$

and

$$
\begin{aligned}
(\mathbf{J}_{\mathrm{m}}[\xi])_{a_1 \cdots a_{n-1}} = &-\mathcal{T}^{ab}\xi_b \epsilon_{aa_1 \cdots a_{n-1}} - H^{abc}\nabla_b \xi_c \epsilon_{aa_1 \cdots a_{n-1}} \\
&+ (\mathrm{d}\mathbf{Q}_{\mathrm{m}}[\xi])_{a_1 \cdots a_{n-1}} + (\mathrm{d}\tilde{\mathbf{Q}}[\xi])_{a_1 \cdots a_{n-1}},
\end{aligned} \tag{5.86}
$$

where $\tilde{\mathbf{Q}}[\xi]$ is defined as

$$(\tilde{\mathbf{Q}}[\xi])_{a_2 \cdots a_n} = \beta^{abcd}\nabla_{[c}\xi_{d]}\epsilon_{aba_1 \cdots a_{n-2}}. \tag{5.87}$$

By considering eq.(5.85), the Noether current can also be expressed as

$$
\begin{aligned}
(\mathbf{J}_{\mathrm{m}}[\xi])_{a_1 \cdots a_{n-1}} = &-T^{ab}\xi_b \epsilon_{aa_1 \cdots a_{n-1}} + (\mathrm{d}\mathbf{S}[\xi])_{a_1 \cdots a_{n-1}} \\
&+ (\mathrm{d}\mathbf{Q}_{\mathrm{m}}[\xi])_{a_1 \cdots a_{n-1}} + (\mathrm{d}\tilde{\mathbf{Q}}[\xi])_{a_1 \cdots a_{n-1}},
\end{aligned} \tag{5.88}
$$

where $\mathbf{S}[\xi]$ is an $(n-2)-$form and satisfies

$$(\mathrm{d}\mathbf{S}[\xi])_{a_1 \cdots a_{n-1}} = -\nabla_b(H^{abc}\xi_c)\epsilon_{aa_1 \cdots a_{n-1}}. \tag{5.89}$$

Obviously, comparing eq.(5.88) and eq.(5.61), we have

$$\mathbf{K} = \mathbf{S}[\xi] + \mathbf{Q}_{\mathrm{m}}[\xi] + \tilde{\mathbf{Q}}[\xi].$$

The tensor $\mathcal{S}^{cab}$ sometimes is called canonical spin tensor (different from the so-called dynamical spin tensor defined by torsion [2, 11]), which has a close relation to the spin of the field and is constructed from the field and its derivatives in the theory. The detailed study of the spin tensor can be found in the nice reviews [2] and [11]. Some discussions are given as follows:

(i). Firstly, it should be pointed out here that the canonical energy-momentum tensor is in general not covariant conserved in a curved spacetime. However, T^{ab} is always covariant conserved.

(ii). Secondly, if $\mathbf{L}_{\mathrm{m}}$ changes as $\mathbf{L}_{\mathrm{m}} \to \mathbf{L}'_{\mathrm{m}} = \mathbf{L}_{\mathrm{m}} + \mathrm{d}\mathbf{\Lambda}$, where $\mathbf{\Lambda}$ is an $(n-1)-$form constructed by fields and their derivatives, then the change of Noether current is a surface term

$$\mathbf{J}_{\mathrm{m}}[\xi] \to \mathbf{J}'_{\mathrm{m}}[\xi] = \mathbf{J}_{\mathrm{m}}[\xi] + \mathrm{d}(\xi \cdot \mathbf{\Lambda}),$$

and $\mathbf{K}$ changes as

$$\mathbf{K} \to \mathbf{K}' = \mathbf{K} + \xi \cdot \mathbf{\Lambda}.$$

This means that T^{ab} does not change under the modification of the Lagrangian. However, the canonical energy-momentum tensor and spin tensor do change under this transformation of the Lagrangian.

(iii). Thirdly, if the spacetime is flat, i.e., the Minkowski spacetime, the spacetime has $n(n+1)/2$ independent Killing vector fields. By using a Lorentz coordinate system $\{x^\mu\} = \{t, x^i\}$, they are given by

$$(\xi_\mu)^a = \left(\frac{\partial}{\partial x^\mu}\right)^a, \qquad (\xi_{\mu\nu})^a = x_\mu\left(\frac{\partial}{\partial x^\nu}\right)^a - x_\nu\left(\frac{\partial}{\partial x^\mu}\right)^a, \qquad (5.90)$$

where $x_\mu = \eta_{\mu\nu}x^\nu$ and μ, ν can be viewed as the name indices of the Killing vector fields. We assume some boundary term $\boldsymbol{\Lambda}$ has been added to the Lagrangian such that the Noether current has a form

$$(\mathbf{J}_m[\xi])_{a_1\cdots a_{n-1}} = -\mathcal{T}^{ab}\xi_b\epsilon_{aa_1\cdots a_{n-1}} - \mathcal{S}^{abc}\partial_b\xi_c\epsilon_{aa_1\cdots a_{n-1}}, \qquad (5.91)$$

where ∂_a is the usual derivative associated with the Lorentz coordinate system. Since the spin tensor and canonical energy-momentum tensor are sensible to $\boldsymbol{\Lambda}$, $\mathcal{T}^{ab}$, $\mathcal{S}^{abc}$, and H^{abc} are different to the original ones. By these considerations and substituting the Killing vector fields, one finds

$$(p_\mu)^a = \mathcal{T}^a{}_b(\xi_\mu)^b = \mathcal{T}^a{}_b\left(\frac{\partial}{\partial x^\mu}\right)^b, \qquad (5.92)$$

$$(j_{\mu\nu})^a = (j_{[\mu\nu]})^a = x_\mu(p_\nu)^a - x_\nu(p_\mu)^a + 2\mathcal{S}^\sigma{}_{[\mu\nu]}\left(\frac{\partial}{\partial x^\sigma}\right)^a, \qquad (5.93)$$

where $\mathcal{S}^\sigma{}_{\mu\nu}$ is the components of $\mathcal{S}^c{}_{ab}$ in the Lorentz coordinate system. $(p_\mu)^a$ is the momentum density of the fields, $(j_{\mu\nu})^a$ corresponds to the total angular-momentum density of the fields. The term with $\mathcal{S}^\sigma{}_{\mu\nu}$ is the contribution of the spin of the fields.

(iv). Finally, we would like to emphasize that the above discussion is based on the fact — the spacetime metric is a fixed background field. It cannot be used for the gravitational field in a diffeomorphsim invariant gravity theory because all of the fields are dynamical there.

5.2.6 *Pseudo tensor and energy-momentum tensor for gravitational field*

It is believed that gravitational field has energy. However, it is not so easy to find a well defined energy density in a general diffeomorphsim invariant gravity theory which does not have any nondynamical background object. For matter fields, T^{ab} is given by the variation of the matter Lagrangian with respect to g_{ab}, and T^{ab} with this definition naturally plays the role of the source of the gravitational field equations. However, for pure gravity (without matter fields), the variation of the Lagrangian with respect to g_{ab} corresponds to the gravitational field equations. So there is no natural definition of T^{ab} for gravitational field. The tensor $\mathcal{T}^{ab}$ for matter fields defined by Noether current, i.e., the canonical energy-momentum tensor, is generally not symmetric, probably can be used to define the energy-momentum tensor of gravitational field. However, in this definition, a nondynamical background field is necessary. To define the canonical energy-momentum tensor as well as canonical spin tensors for gravitational field, one has to introduce some background structure to the general diffeomorphsim invariant gravity theory, for example, Einstein gravity theory or general relativity. This is not the spirit of general relativity, and the resulting energy-momentum tensor for gravitational field is not covariant (so it is called pseudotensor). The background structure might be a preferred coordinate system, a background metric, or a background connection. In any case, the diffeomorphsim invariant is broken.

5.2.6.1 *Background metric*

Here we give a little bit discussion on the energy-momentum tensor of the gravitational field. Let us consider a theory with metric decomposition

$$g_{ab} = \eta_{ab} + h_{ab},\tag{5.94}$$

where η_{ab} is a fixed metric associated with a spacetime and h_{ab} is a dynamical metric. Viewing η_{ab} and h_{ab} as two independent fields, then the dynamical fields now include ψ and h_{ab}. Formally, eqs.(5.52) and (5.57) can be written as

$$(\mathbf{E}_\eta)^{ab} = (\overline{\mathbf{E}}_\eta)^{ab} + \frac{1}{2}\epsilon T^{ab},\tag{5.95}$$

and

$$\mathrm{d}\mathbf{J}_{(\mathrm{m},h)}[\xi] = -\mathbf{E}_\psi \mathcal{L}_\xi \psi - (\mathbf{E}_h)^{ab}\mathcal{L}_\xi h_{ab} - \frac{1}{2}\epsilon T^{ab}\mathcal{L}_\xi \eta_{ab}.\tag{5.96}$$

Since η_{ab} is a background field, $(\overline{\mathbf{E}}_\eta)^{ab}$ is identically vanishing, and eq.(5.95) becomes

$$(\mathbf{E}_\eta)^{ab} = \frac{1}{2}\epsilon T^{ab}.\tag{5.97}$$

From the Noether current $\mathbf{J}_{(\mathrm{m},h)}[\xi]$, we can define the canonical energy-momentum tensor and canonical spin tensor as in the previous subsection. Assume t^{ab} is the canonical energy-momentum tensor, and H^{cab} is partly constructed from the spin tensor, then we have

$$T^{ab} = t^{ab} + \nabla_c H^{cab}.\tag{5.98}$$

However, when h_{ab} satisfies equations of motion, η_{ab} also satisfies equations of motion [5]. So we have $T^{ab} = 0$, and t^{ab} satisfies

$$t^{ab} = -\nabla_c H^{cab}.\tag{5.99}$$

This means the canonical energy-momentum tensor of gravitational field is always the divergence of the tensor H^{cab}. So H^{cab} is called superpotential. Since the definition of t^{ab} depends on the background metric η_{ab}, so it is a pseudotensor in this sense [5]. Relevant discussions can also be found in [16, 17, 18, 19, 20, 21, 22].

5.2.6.2 *Preferred coordinates*

Another way to define the canonical energy-momentum tensor of gravitational field is based on some special coordinate system. The discussion is based on the Lagrangian which has a form of eq.(5.1) without the background field Ψ

$$\mathbf{L} = \mathbf{L}\left(\Phi, \partial_a\Phi, \cdots \partial_{(a_1} \cdots \partial_{a_k)}\Phi\right)\bar{\epsilon}, \tag{5.100}$$

where $\bar{\epsilon}$ is a coordinate volume element, and Φ includes metric g_{ab} and some matter fields. Based on this Lagrangian, one can define the energy-momentum tensor $t^a{}_b$ and canonical spin tensor $\mathcal{S}^{cab}$, which are both pseudotensors. For example, in a second order theory, one has

$$\sqrt{g}t^a{}_b = \mathbf{L}\delta^a{}_b - \frac{\partial\mathbf{L}}{\partial(\partial_a\Phi)}\partial_b\Phi + \partial_c\frac{\partial\mathbf{L}}{\partial(\partial_a\partial_c\Phi)}\partial_b\Phi - \frac{\partial\mathbf{L}}{\partial(\partial_a\partial_c\Phi)}\partial_b\partial_c\Phi, \tag{5.101}$$

where $\sqrt{g}$ is the square root of the determine of the metric in the coordinate system (the symbol of absolute value inside $\sqrt{|g|}$ has been omitted). The divergence of this pseudotensor satisfies

$$\partial_a\left(\sqrt{g}(t^a{}_b + T^a{}_b)\right) = 0, \tag{5.102}$$

where $T^a{}_b$ is the energy-momentum tensor for matter fields. This equation implies there exists a superpotential $\mathcal{H}^{ca}{}_b = \mathcal{H}^{[ca]}{}_b$ such that

$$\sqrt{-g}(t^a{}_b + T^a{}_b) = \frac{1}{2}\partial_c\mathcal{H}^{ca}{}_b. \tag{5.103}$$

Obviously, the superpotential $\mathcal{H}^{ca}{}_b$ is not unique. Actually, if

$$\tilde{\mathcal{H}}^{ca}{}_b = \mathcal{H}^{ca}{}_b + \partial_d\beta^{dca}{}_b, \tag{5.104}$$

where $\beta^{dca}{}_b$ satisfies $\beta^{dca}{}_b = \beta^{[dc]a}{}_b$ and $\beta^{dca}{}_b = \beta^{d[ca]}{}_b$, then $\tilde{\mathcal{H}}^{ca}{}_b$ is also a superpotential. The superpotential has a close relation to the canonical spin tensor as discussed in the previous sections. So the superpotential and canonical energy-momentum tensor for gradational field also depend on some possible boundary term $\mathbf{\Lambda}$ added to the Lagrangian. This boundary term of course does not change the equations of motion but really bring changes to the canonical energy-momentum tensor, canonical spin tensor, and then the superpotential. For these reasons, there are a lot of canonical energy-momentum tensors and superpotentials. For example, Einstein got a first order Lagrangian by introducing a boundary term,

$$\sqrt{g}L_g = \frac{1}{16\pi}\left\{\sqrt{g}R - \partial_a\left[\sqrt{g}\left(\Gamma^a{}_{bc}g^{bc} - g^{ab}\Gamma^b{}_{cb}\right)\right]\right\}$$
$$= \frac{1}{16\pi}\sqrt{g}g^{ab}\left[\Gamma^c{}_{ad}\Gamma^d{}_{bc} - \Gamma^c{}_{cd}\Gamma^d{}_{ab}\right], \tag{5.105}$$

and obtained a pseudotensor and superpotential [23, 24, 25], which have forms as

$$\sqrt{g}t^a{}_b = \frac{1}{16\pi}\sqrt{g}\left[\delta^a{}_b(\Gamma^c{}_{de}\Gamma^d{}_{cf} - \Gamma^c{}_{cd}\Gamma^d{}_{ef})g^{ef}\right.$$
$$+ g^{ac}\Gamma^d{}_{cb}\Gamma^e{}_{ed} - g^{ac}\Gamma^d{}_{bd}\Gamma^e{}_{be} + g^{cd}\Gamma^a{}_{cd}\Gamma^e{}_{be}$$
$$\left. + g^{cd}\Gamma^a{}_{bc}\Gamma^e{}_{ed} - 2g^{cd}\Gamma^a{}_{ec}\Gamma^e{}_{bd}\right], \tag{5.106}$$

and

$$\mathcal{H}^{ab}{}_c = -2\sqrt{g}\left(\Gamma^{[a}{}_{cd}g^{b]d} + \delta^{[a}{}_c\Gamma^{b]}{}_{de}g^{de} - \delta^{[a}{}_c g^{b]d}\Gamma^e{}_{ed}\right). \tag{5.107}$$

See [11, 12] for the details to get the above equations and the review [2] for Papapetrou [26], Bergmann [27], Møller [29], and symmetric Landau-Lifshitz-Goldberg [28, 31] pseudotensors in general relativity. Although most of them are the same near the infinities of spacetime, they are very different in the bulk of the spacetime.

5.2.7 Vielbein, semimetric, four-legs, and tetrad

5.2.7.1 The work by Møller

In 1961, Møller [30] realized that four-legs or tetrad form of Einstein gravity might help us to get the covariant conserved law for the gravitational field. Different from the action of Einstein by Christoffel symbols in eq.(5.105), the action by Møller is written in the form by Ricci rotation coefficients, see the Lagrangian (5.109) studied by Duan. The action is diffeomorphsim invariant or covariant, and one can get a covariant conservation law for the gravitational field.

However, there are still problems in Møller's vielbein or tetrad formalism. For example, although the final conservation law is diffeomorphsim invariant or covariant, it depends on some internal gauge condition of $SO(1,3)$. So, in some sense, it is still background dependent. The second problem is that the tetrad formalism is not totally equivalent to the metric formalism of the gravity because the global existence of the tetrad needs a topological condition of the spacetime manifold (it is actually the requirement for the existence of spinor structure). Finally, as pointed out by Møller in his paper, in some cases, the canonical energy-momentum tensor (proposed by himself) does not have well defined asymptotic behavior to ensure a finite conserved quantity [30, 33]. Based on the work by Møller, Duan and his collaborator provided a solid analysis on this problem and obtained correct conserved quantities.

5.2.7.2 The work by Duan

Prof. Duan was the first person who introduced the notion of *vielbein* to China. In fact, in 1957, he has obtained general covariant equations for fields of arbitrary spin by using the notion of vielbein [32]. In his paper, he used an old terminology *semimetric* to describe the vielbein, i.e., (to respect the original work by Prof.Duan, here we use i, j instead of μ, ν or a, b.)

$$g_{ij} = \lambda_{i(\alpha)}\lambda_{j(\alpha)}, \tag{5.108}$$

where $\lambda_{i(\alpha)}$ is the so-called semimetric, and (α) denotes the index of $SO(1,3)$. In China, even nowadays, a lot of senior researchers still remember this vivid word. Nearly at the same time, Møller and some western researchers used *four-legs* or *orthogonal tetrad* to describe the vielbein.

At the end of 1950s, several researchers in the world have realized that semimetric can be used to describe the gravitational field. Prof. Duan was one of them. Actually, in 1963, his work on the conserved quantities in general relativity was published on a top journal in China—Acta Physica Sinica [33][4]. The main aim of that work is to establish some covariant conservation law for gravitational field by using semimetric. Here, we give a brief summary of the work.

As mentioned in last subsection, pseudotensor or superpotential are sensitive to the boundary terms. Besides this, there still exist some remanning freedoms to choose the superpotential. Merely replacing the metric with semimetric is not helpful. The difference comes from the boundary terms. By using semimetric representative, the Lagrangian has a form

$$L_\lambda = \frac{1}{16\pi}\left[\eta_{(\alpha)}\eta_{(\alpha)} - \eta_{(\alpha\beta\gamma)}\eta_{(\gamma\beta\alpha)}\right], \tag{5.109}$$

where $\eta_{(\alpha\beta\gamma)}$ is Ricci rotation coefficients, and $\eta_{(\alpha)} = \eta_{(\beta\alpha\beta)}$. The difference between this Lagrangian and the first order Lagrangian by Einstein L_g in eq.(5.105), is a boundary term, i.e.,

$$\sqrt{g}L_\lambda = \sqrt{g}L_g + \Delta, \tag{5.110}$$

where Δ is the boundary term and has a form

$$\Delta = \frac{1}{16\pi}\frac{\partial}{\partial x^i}\left[\sqrt{g}\left(\lambda^i_{(\alpha)}\frac{\partial\lambda^j_{(\alpha)}}{\partial x^j} - \lambda^j_{(\alpha)}\frac{\partial\lambda^i_{(\alpha)}}{\partial x^j}\right)\right]. \tag{5.111}$$

With these Lagrangians in hand, by using variation, one can get correct equations of motion, i.e., the Einstein equations. Considering infinitesimal coordinate translation with a form

$$x^i \to x'^i = x^i + \lambda^i_{(\alpha)}a_{(\alpha)}, \tag{5.112}$$

where $a_{(\alpha)}$ does not depend on coordinates, and can be understood as the translation along the α direction in some inner space. By these, one can obtain

$$\frac{\partial}{\partial x^i}\left[\sqrt{g}(t^i_{(\alpha)} + T^i_{(\alpha)})\right] = 0, \tag{5.113}$$

where $T^i_{(\alpha)} = T^i{}_j\lambda^j_{(\alpha)}$ and the canonical energy-momentum tensor $t^i_{(\alpha)}$ has a form

$$\begin{aligned}
t^i_{(\alpha)} = \frac{1}{16\pi}\Big\{&\lambda^i_{(\alpha)}\left[\eta_{(\beta)}\eta_{(\beta)} - \eta_{(\delta\beta\gamma)}\eta_{(\gamma\beta\delta)}\right]\\
&- 2\lambda^i_{(\beta)}\left[\eta_{(\beta)}\eta_{(\alpha)} - \eta_{(\delta\beta\gamma)}\eta_{(\gamma\alpha\delta)}\right]\\
&- 2\lambda^i_{(\gamma)}\left[\eta_{(\beta)}\eta_{(\alpha\beta\gamma)} + \eta_{(\beta)}\eta_{(\beta\alpha\gamma)}\right]\\
&+ 2\lambda^i_{(\delta)}\eta_{(\alpha\beta\gamma)}\eta_{\gamma\beta\delta}\Big\}.
\end{aligned} \tag{5.114}$$

---

[4]Actually, this work had been completed before October of 1962.

Furthermore, one finds

$$\sqrt{g}\left(t^i_{(\alpha)} + T^i_{(\alpha)}\right) = \frac{\partial}{\partial x^j} v^{ij}_{(\alpha)} \, , \tag{5.115}$$

where $v^{ij}_{(\alpha)} = \sqrt{g} V^{ij}_{(\alpha)}$ with

$$V^{ij}_{(\alpha)} = \frac{1}{8\pi}\left[\lambda^i_{(\beta)}\lambda^j_{(\gamma)}\eta_{(\alpha\beta\gamma)} + (\lambda^i_{(\alpha)}\lambda^j_{(\beta)} - \lambda^j_{(\alpha)}\lambda^i_{(\beta)})\eta_{(\beta)}\right] . \tag{5.116}$$

Obviously, $V^{ij}_{(\alpha)}$ is an antisymmetric tensor on the spacetime, and the above equation is covariant. The 4-momentum $P_{(\alpha)}$ for some region enclosed by some closed two surface is defined by the surface integral of $V^{ij}_{(\alpha)}$.

This covariant conservation law has been further developed by Prof.Duan and his collaborators to discuss the energy and angular momentum of various gravitational systems. See [34, 35, 36, 37, 47, 39, 49, 41, 42] for its modern form and details.

5.2.8 *Gravitational energy-momentum can not be locally defined*

Any energy density is observer dependent. For the matter fields, we have a well defined covariant energy-momentum tensor, and we can get the energy density for an arbitrary observer on the background spacetime where the matter fields live on. Furthermore, for the same observer, one gets the same energy density in any coordinate system. This energy density of the matter fields has direct relation to the geometry of the spacetime by gravitational equations.

However, gravitational field is different. As mentioned in the previous sections, gravitational fields are built into the geometry of the spacetime, and there is no preferred background structure for themselves. Actually, if we assume a covariant energy-momentum tensor for gravitational field, then, in principal, it could not be the source of gravitational equations. Otherwise we have to modify the gravitational equations, such as Einstein equations, or introduce a reference metric into the system and this breaks the covariance or diffeomorphsim invariant of the theory (and the resulting energy-momentum tensor is a pseudotensor). So, in this sense, it could not curve the spacetime like the matter fields [43]. In general relativity, from Raychaudhuri equation, this energy-momentum tensor also can not effect the geodesic deviation of world line of particles, so there is no local observation effect for this energy-momentum tensor [43]. To get the pseudotensor t^{ab}, one has to break the covariance of the theory. Actually, if the background structure is a preferred coordinate system, on the world line of a free falling observer, this tensor might be vanishing in this coordinate system. However, generally, it will not be vanishing in another coordinate system.

Of course, gravity has energy. For example, in linear gravity theory, with the Minkowski spacetime as a background spacetime, one has a well defined energy-momentum tensor for gravitational wave. The problem is that this energy can not be localized without a preferred background structure. Some researchers think that the question to seek for the energy or momentum density of gravitational field is a wrong question [43]

Anybody who looks for a magic formula for "local gravitational energy - momentum" is looking for the right answer to the wrong question. Unhappily, enormous time and effort were devoted in the past to trying to "answer this question" before investigators realized the futility of the enterprise. ......

Nowadays, most of people think that the energy or momentum density of gravitational field can only be defined globally or at most quasilocally (defined inside a two dimensional closed surface in four dimensions). Some people would think that the pseudotensor method has become an old topic in general relativity, and probably it is only interesting to scientific historians. The situation is not so bad, actually, Chern and Nester have shown that a pesudotensor corresponds to a Hamilton boundary term [44]. So they are quasilocal and acceptable; each is the energy-momentum density for a definite physical situation with certain boundary condition [44].

We set aside the discussion on the pseudotensors for a moment, and turn to an important definition of the energy and angular momentum for some special spacetime with isometries (for example, stationary spacetimes). This is the so-called Komar integral in general relativity.

5.2.9   *Komar mass and angular momentum*

Based on Møller's discussion on the pseudotensor and superpotential [29], In 1959, Komar provided a covariant method to define conserved quantities for a spacetime with some timelike isometry in general relativity [45]. This definition has been used to study the mechanics of black hole in the early stage of black hole physics [46], and plays an important role in the study of the conserved quantities in stationary spacetime in general relativity.

Let ξ be a Killing vector field of a spacetime (M, g_{ab}). From Killing equation we have

$$\nabla_a \nabla_b \xi_c = R_{dabc}\xi^d \,, \tag{5.117}$$

we can have

$$\nabla_a \nabla^{[a}\xi^{b]} = -R_a{}^b \xi^a \,, \tag{5.118}$$

and

$$(\mathrm{d}\mathbf{K}_\xi)_{a_1 \cdots a_{n-1}} = 2R_a{}^b \xi^a \epsilon_{ba_1 \cdots a_{n-1}} \,, \tag{5.119}$$

where $\mathbf{K}_\xi$ is an $(n-2)-$form

$$(\mathbf{K}_\xi)_{a_1 \cdots a_{n-2}} = \nabla^{[a}\xi^{c]}\epsilon_{aca_1 \cdots a_{n-2}} \,. \tag{5.120}$$

The Komar conserved quantity for the Killing vector field ξ is defined as an integral on an $(n-2)$−surface of the spacetime, i.e.,

$$K_S[\xi] = -\frac{1}{8\pi} \int_S \mathbf{K}_\xi \,. \tag{5.121}$$

By using Einstein equations, and considering the integral of eq.(5.119) on a spacelike hypersurface Σ with two boundaries S_o and S_i, we have

$$\int_\Sigma \mathrm{d}\mathbf{K}_\xi = 16\pi \int_\Sigma \left(T_{ab} - \frac{1}{n-2}Tg_{ab}\right)\xi^a n^b \epsilon_\Sigma \,, \tag{5.122}$$

where n^a satisfying $n_a n^a = -1$ is the normal vector field of Σ, and ϵ_Σ is the volume element of Σ, and $T = g^{ab}T_{ab}$. By using Stokes theorem, we have

$$K_{S_o}[\xi] - K_{S_i}[\xi] = 2\int_\Sigma \left(T_{ab} - \frac{1}{n-2}Tg_{ab}\right)\xi^a n^b \epsilon_\Sigma \,. \tag{5.123}$$

Hence, for a solution of vacuum Einstein equations with the Killing vector field ξ^a, the Komar integral does not depend on the selection of the codimension-2 surface, and it is conserved in this sense. For a stationary axisymmetric solution of the vacuum Einstein equations, assume t^a is the timelike like Killing vector field which has an unit norm at infinity, and φ^a is the Killing vector field corresponding to the rotating symmetry, one can define the Komar mass and angular momentum for the $(n-2)$ dimensional surface S

$$M_S = K_S[t]\,, \qquad J_S = K_S[\varphi]\,. \tag{5.124}$$

When S approaches the spacelike infinity of an asymptotically flat spacetime, the Komar mass coincides with the ADM mass (see next section for ADM mass).

It should be pointed out here, the Komar integral is based on the Einstein equations. To our knowledge, Komar type conserved quantities in general diffeomorphsim invariant theories have not been found in literatures. However, in general Lovelock gravity theory, one has already found similar integrals [47] (see also [48] for early work). The Lagrangian of the Lovelock gravity theory has a form

$$\mathbf{L} = L\epsilon\,, \tag{5.125}$$

with

$$L = \sum_{k=0}^{p} c_k L_{(k)}\,, \tag{5.126}$$

where $p \leq [(n-1)/2]([N]$ denotes the integral part of the number N), c_k are arbitrary constants with dimension of $[\text{Length}]^{2k-2}$, and L_k can be expressed as

$$L_{(k)} = \frac{1}{2^k}\delta^{a_1 \cdots a_k b_1 \cdots b_k}_{c_1 \cdots c_k d_1 \cdots d_k} R^{c_1 d_1}{}_{a_1 b_1} \cdots R^{c_k d_k}{}_{a_k b_k}\,. \tag{5.127}$$

Here, $\delta^{\cdots}_{\cdots}$ is generalized Kroneck delta tensor. $L_{(0)} = 1$, so the constant c_0 is just the cosmological constant. $L_{(1)}$ gives us the usual curvature scalar term, and for simplicity, we set $c_1 = 1$, while $L_{(2)}$ is just the Gauss–Bonnet term. The equations of motion following from the Lagrangian (5.126) have the form $\mathcal{G}_{ab} = 8\pi T_{ab}$, where

$$\mathcal{G}^a_b = \sum_{k=0}^{p} c_k G_{(k)}{}^a_b \tag{5.128}$$

with

$$G_{(k)}{}^a_b = -\frac{1}{2^{k+1}} \delta^{a a_1 \cdots a_k b_1 \cdots b_k}_{b c_1 \cdots c_k d_1 \cdots d_k} R^{c_1 d_1}{}_{a_1 b_1} \cdots R^{c_k d_k}{}_{a_k b_k} . \tag{5.129}$$

By these, one finds

$$G_{(k)}{}^a_b = R_{(k)}{}^a_b - \frac{1}{2} \delta^a_b L_{(k)} , \tag{5.130}$$

where

$$R_{(k)}{}^a_b = \frac{k}{2^k} \delta^{a_1 a_2 \cdots a_k b_1 \cdots b_k}_{b \, c_2 \cdots c_k d_1 \cdots d_k} R^{a d_1}{}_{a_1 b_1} \cdots R^{c_k d_k}{}_{a_k b_k} , \tag{5.131}$$

and

$$R_{(k)}{}^a_a = k L_{(k)} . \tag{5.132}$$

From the above relations, following the logic of the Komar integral in the Einstein gravity theory, one has

$$\nabla_a B^{ab}_{(k)} = -R_{(k)}{}^b_c \xi^c , \tag{5.133}$$

where ξ^a is a Killing vector field and

$$B^{ab}_{(k)} = \frac{k}{2^k} \delta^{a b a_1 \cdots a_{k-1} b_1 \cdots b_{k-1}}_{c d c_1 \cdots c_{k-1} d_1 \cdots d_{k-1}} (\nabla^c \xi^d) R^{c_1 d_1}{}_{a_1 b_1} \cdots R^{c_{k-1} d_{k-1}}{}_{a_{k-1} b_{k-1}} . \tag{5.134}$$

The Hodge dual of $B^{ab}_{(k)}$ is required Komar like integral for pure Lovelock gravity. Further, it is easy to find

$$J^a_{(k)} = G^{ab}_{(k)} \xi_b \tag{5.135}$$

satisfies $\nabla_a J^a_{(k)} = 0$, so locally one has

$$J^a_{(k)} = \nabla_b \omega^{ba}_{(k)} , \tag{5.136}$$

where $\omega^{ab}_{(k)}$ is an 2-form. By these considerations, Kastor finds that the general integral kernel of the Komar like integral has a form

$$B^{ab} = c_p B^{ab}_{(p)} + \sum c_k \left[\left(\frac{n-2k}{n-2p} \right) B^{ab}_{(k)} + 2 \left(\frac{p-k}{n-2p} \right) \omega^{ab}_{(k)} \right] , \tag{5.137}$$

and one also has

$$\nabla_b B^{ba} = -8\pi \left(T^{ab} - \frac{1}{n-2p} T g^{ab} \right) \xi_b . \tag{5.138}$$

It should be pointed out here that the potential $\omega^{ab}_{(k)}$ is absent in the case of Einstein [47]. To make the above integral include the special case of Einstein gravity, one can rearrange the terms in the above procedure and get a little bit different expression from the above B^{ab}. Another point is: $\omega^{ab}_{(0)}$ is actually the Killing potential of the Killing vector field, and the integral for this term will give a volume like quantity. This volume plays an important role in the study of the thermodynamics of black holes in gravity theories with the cosmological constant, see [47] for details and references therein. Finally, for general diffeomorphsim invariant theories, the Komar like integrals are still absent up to date. It is interesting to generalize the above procedure to obtain the Komar like integrals in these theories.

5.3   Global definitions

Due to the absence of the local energy density of gravitational field, it has been developed to have several global definitions of the energy (and linear momentum) of gravity field. These quantities are defined at the infinities, such as the spacelike infinity and null infinity of the spacetime. For example, the ADM energy and momentum are defined at the spacelike infinity, and Bondi–Sachs energy and momentum are defined at the null infinity.

The definitions of these global quantities can be realized by the Hamiltonian of the gravitational system. The Hamiltonian description comes from the symplectic structure (or presymplectic when constraints are present) of the theory. The traditional way to construct the Hamiltonian structure of general relativity is based on the $1 + 3$ decomposition of the spacetime, and the phase space is spanned by the induced metric h_{ab} of the spacelike hypersurface and the conjugate momentum $\pi^{ab} = \sqrt{h}(K^{ab} - Kh^{ab})$, where K^{ab} is the extrinsic curvature of the hypersurface. This means we have a set of preferred canonical variables for the system. Another way to construct the Hamiltonian is the so-called covariant phase space method. In the discussion of covariant phase space, it is not necessary to separate the generalized coordinates and generalized momentums in advance. In other words, the covariant phase space method does not depend on the coordinates of the phase space, see [49, 50] for the earlier discussion on this method in classical field systems.

5.3.1   *General definition on conserved quantities*

Here, we give a brief review on the covariant phase analysis method developed by Wald *et. al.* [4, 5, 51]. This method is valid for general diffeomorphsim invariant theories and does not depend on some special asymptotical conditions of the spactimes. The method is based on the Lagrangian formalism of the theory. For the general diffeomorphsim invariant Lagrangian as eq.(5.17) in Sec.5.2.2, one has the $(n - 1)-$form $\boldsymbol{\Theta}(\phi, \delta\phi)$. This $(n - 1)-$ form is the so-called symplectic potential form, and the symplectic form is defined as

$$\boldsymbol{\omega}(\phi, \delta_1\phi, \delta_2\phi) = \delta_2\boldsymbol{\Theta}(\phi, \delta_1\phi) - \delta_1\boldsymbol{\Theta}(\phi, \delta_2\phi) \,. \tag{5.139}$$

In this definition, one has to consider two parameter family of the fields in the configuration space, i.e., $\phi(\lambda_1, \lambda_2)$. The variation δ_1 is with respect to λ_1 parameter while λ_2 is fixed. The definition of δ_2 is similar. By this symplectic form, one can define a presymplectic form Ω as

$$\Omega_\Sigma(\phi, \delta_1\phi, \delta_2\phi) = \int_\Sigma \boldsymbol{\omega}(\phi, \delta_1\phi, \delta_2\phi) \,, \tag{5.140}$$

where Σ is an $(n - 1)-$submanifold of the spacetime (sometimes it is chosen to be a Cauchy surface). The fields discussed are assumed to satisfy some smooth condition and asymptotic condition. With the assumption of the asymptotic condition on the fields, Ω_Σ is independent of Σ. All the fields satisfy such conditions form a space, i.e., the configuration space $\mathcal{F}$. For the details of the properties of $\mathcal{F}$, see [5, 51].

Let ξ^a be a vector field on the spacetime, the Hamiltonian conjugate to ξ^a, i.e., $H_\xi : \mathcal{F} \to \mathcal{R}$, is defined as a variation equation

$$\delta H_\xi = \Omega_\Sigma(\phi, \delta\phi, \mathcal{L}_\xi\phi) = \int_\Sigma \omega(\phi, \delta\phi, \mathcal{L}_\xi\phi) \,. \tag{5.141}$$

It is not hard to find the variation of the Noether current (5.46) is given by

$$\delta \mathbf{J}[\xi] = \omega(\phi, \delta\phi, \mathcal{L}_\xi\phi) + \mathrm{d}(\xi \cdot \mathbf{\Theta}(\phi, \delta\phi)) \,. \tag{5.142}$$

For general diffeomorphsim invariant theory, the Neother current locally can be expressed as

$$\mathbf{J}[\xi] = \mathrm{d}\mathbf{Q}[\xi] + \xi^a \mathbf{C}_a \,, \tag{5.143}$$

where $\mathbf{C}_a = 0$ when the fields are on-shell, i.e., ϕ belongs to solutions space $\bar{\mathcal{F}} \subset \mathcal{F}$. Usually, these $\mathbf{C}_a$ are thought as the constraints of the theory [7]. So one gets

$$\omega(\phi, \delta\phi, \mathcal{L}_\xi\phi) = \xi^a \delta\mathbf{C}_a + \mathrm{d}\delta\mathbf{Q} - \mathrm{d}(\xi \cdot \mathbf{\Theta}) \,. \tag{5.144}$$

As a result, when restricted on $\bar{\mathcal{F}}$, the Hamiltonian can be expressed as

$$\delta H_\xi = \int_\Sigma \xi^a \delta\mathbf{C}_a + \int_{\partial\Sigma} (\delta\mathbf{Q} - \xi \cdot \mathbf{\Theta}) \,. \tag{5.145}$$

So when the variation of the fields satisfies linearized field equations, the Hamiltonian is a surface term

$$\delta H_\xi = \int_{\partial\Sigma} (\delta\mathbf{Q} - \xi \cdot \mathbf{\Theta}) \,. \tag{5.146}$$

This is the necessary condition of the existence of the Hamiltonian. The sufficient condition of the existence of the Hamiltonian is given by

$$(\delta_1\delta_2 - \delta_2\delta_1)H_\xi = -\int_{\partial\Sigma} \xi \cdot \omega(\phi, \delta_1\phi, \delta_2\phi) = 0 \,, \tag{5.147}$$

for $\phi \in \bar{\mathcal{F}}$, and all pairs of $\delta_1\phi$ and $\delta_2\phi$ which satisfy linearized field equations, i.e., $\delta_1\phi$ and $\delta_2\phi$ are tangent to $\bar{\mathcal{F}}$.

When the Hamiltonian exists and $\bar{F}$ is simple connected, then for one parameter family of solution $\phi(\lambda)$, one has

$$H_\xi[\phi] = \int_0^1 \mathrm{d}\lambda \int_{\partial\Sigma} [\delta\mathbf{Q}(\lambda) - \xi \cdot \mathbf{\Theta}(\lambda)] \,. \tag{5.148}$$

$\phi(0)$ can be viewed as the reference solution. The simple connected condition ensures the integral is independent of the path. We always assume this condition holds in the following discussions.

To define the conserved quantities at infinity of the spacetime, usually we apply the conformal completion techniques by Penrose [52, 53]. Assume (M, g_{ab}) is a physical spacetime, and $(\tilde{M}, \tilde{g}_{ab})$ is an unphysical spacetime which is the conformal completion of (M, g_{ab}). We assume $\tilde{M} = M \cup B$, where B is an $(n-1)$−dimensional submanifold of $\tilde{M}$. The intersection of Σ with B is the cross section of B. With this structure, one can define the so-called asymptotic symmetry generator as follows. Assume ξ^a is a complete vector field on $M \cup B$, so that ξ^a is tangent to B on B, then ξ^a is said to be a representative of an infinitesimal asymptotic symmetry if the one-parameter group associated with ξ^a maps $\bar{\mathcal{F}}$ to $\bar{\mathcal{F}}$. See [51] for details.

Assume ω can be continuously extended to B, then according the existence condition (5.147) of Hamiltonian, one has two types of definitions of conserved quantities.

Case I. If $\bar{\omega} = 0$, i.e., the pull back of ω to B is vanishing, then by the existence condition (5.147), H_ξ exists for all infinitesimal asymptotic symmetries, and independent of the choice of representative ξ^a. Furthermore, H_ξ is independent of the choice of the section of B, and H_ξ is really conserved.

CaseII. If $\bar{\omega}$ is not vanishing. Generally, H_ξ does not exist. In some special case where ξ is everywhere tangent to $\partial\Sigma$, then the pull back of $\xi \cdot \omega$ to $\partial\Sigma$ is vanishing on $\partial\Sigma$ so that H_ξ exists. However, in general, H_ξ (even in the case it exists) depends on the cross section of B and is not conserved.

For example, at null infinity, the generator corresponding to supertranslation does not tangent to $\partial\Sigma$ (consider the limit to the null infinity), and then H_ξ does not exist.

This is a general discussion on the conserved quantities for general diffeomorphsim invariant gravity theories. The ADM mass and momentum defined at spacelike infinity in general relativity belongs to the case I. While the case II corresponds the "conserved quantities" at the null infinity of the spacetime. Since this method does not rely on some special asymptotical conditions of the spacetime, it can also be used to define the conserved quantities for asymptotically AdS spacetimes, for example, see [54].

5.3.2 *Global quantities at spacelike infinity*

5.3.2.1 *Coordinate dependent definitions*

In general relativity, the spacelike infinity i^0 of a spacetime has a technique definition by Ashtekar and Hansen [55]. With this definition, one can put the ADM mass or momentums [56] into a coordinate free form.

However, a lot of discussions on ADM mass are based on the coordinate dependent description, for example, the proof of the mass positivity by Yau [57]. Here, we give some discussion on this prescription. One considers a spacelike hypersurface Σ

in a asymptotically flat spacetime (M, g_{ab}). Assume U is an open set (of M) and has a well defined time function t such that $\Sigma \cap U$ is the slice with a constant t. Let $\{t, x^i\}$ be a coordinate system covering U, and D_o be a region inside U with

$$r := \sqrt{\delta_{ij} x^i x^j} > r_o \,. \tag{5.149}$$

In the region D_o, one has a reference metric which has a form of Euclidean, i.e., δ_{ij}, and

$$g_{ij} = \delta_{ij} + \mathcal{O}(r^{-\alpha}) \,, \qquad \partial_k g_{ij} = \mathcal{O}(r^{-\alpha-1}) \,, \tag{5.150}$$

where the value of α has an appropriate range such that this asymptotic condition can include most of physically interest spacetimes. For example, in four dimensions, $\alpha = 1$. The function r can be arbitrarily large, and when r approaches infinity, one approaches the spacelike infinity of the spacetime.

It seems that the asymptotic symmetry group of the spacetime near the spacelike infinity should be the usual Poincaré group. However, it turns out not so simple. Actually, one has a larger symmetry group preserving the asymptotical condition, i.e., the so-called SPI group. This group includes the translation depends on the coordinates of the $(n-2)$−sphere, and the its Lie algebra is also larger than the Poincaré Lie algebra [55].

With the ADM decomposition as

$$g = -N^2 dt^2 + h_{ij}(dx^i + N^i dt)(dx^j + N^j dt) \,, \tag{5.151}$$

where N, N^i are lapse and shift vector, respectively, and the usual Hamilton analysis, one finds the on-shell Hamiltonian of the Einstein gravity is a surface term (see for example [58, 59] and [60])

$$H[N, \boldsymbol{N}] = -\frac{1}{8\pi} \int_S \left[N(k - k_0) - N^i(K_{ij} - K h_{ij})n^j \right] \epsilon_S \,, \tag{5.152}$$

where S is an $(n-2)$−dimensional surface embedded in (Σ, h_{ij}), and k is the trace of the extrinsic curvature of S (embedded in Σ), and n^i is the normal vector field of S inside Σ, while k_0 is the trace of the extrinsic curvature of S embedded in the reference space (for example, Euclidean space here). K_{ij} is the extrinsic curvature of Σ and has a form

$$2N K_{ij} = 2h_{k(i} D_{j)} N^k - \mathcal{L}_t h_{ij} \,, \tag{5.153}$$

where D_i is the covariant derivative on (Σ, h_{ij}), and t^a is the time flow vector of the ADM decomposition which can be expressed as

$$t^a = \left(\frac{\partial}{\partial t}\right)^a = N u^a + \boldsymbol{N} \,, \tag{5.154}$$

where

$$\boldsymbol{N} = N^i \left(\frac{\partial}{\partial x^i}\right)^a \,, \tag{5.155}$$

and u^a with $u^a u_a = -1$ is the norm vector field of Σ. So the Hamiltonian can be viewed as the dual of the time flow vector.

Near the spacelike infinity of the spacetime, with the Cartesian coordinates $\{t, x^i\}$, the asymptotic conditions on the components of the metric imply

$$h_{ij} = \delta_{ij} + \mathcal{O}(r^{-\alpha}),$$
$$K_{ij} = 0_{ij} + \mathcal{O}(r^{-\alpha-1}). \tag{5.156}$$

The $(n-2)-$ surface S with constant r can be viewed as the embedded submanifold in Euclidean space $(\mathcal{E}^{n-1}, \delta_{ij})$, and at the same time it is also the embedded submanifold in (Σ, h_{ij}). k_0 is the mean curvature of S embedded in the Euclidean space, while k is the one embedded in (Σ, h_{ij}). By a little complicated calculation, see for example reference [61], one finds

$$k - k_0 = -\frac{1}{2}n^i h^{kj}(\partial_k h_{ij} - \partial_i h_{kj}) + n_i \partial_j \left(n^{[i}\Delta n^{j]}\right) + \cdots, \tag{5.157}$$

where "$\cdots$" are higher order terms compared to $r^{-\alpha}$, and n^i is the unit norm vector field of S embedded in $(\mathcal{E}^{n-1}, \delta_{ij})$. Assume n^i_Σ is the unit norm vector field of S embedded inside Σ, then Δn^i in the above equation is given by $n^i_\Sigma - n^i$. The last term in the above equation corresponds to an exact form, so its integral will be vanishing due to the closeness of S. Considering the time flow vector near spacelike infinity with $N = 1$ and $N^i = 0$, one gets

$$E_{\mathrm{ADM}} = H[1, \mathbf{0}] = \frac{1}{16\pi} \lim_{r \to \infty} \int_S (\partial_j h_{ij} - \partial_i h_{jj}) n^i \epsilon_S. \tag{5.158}$$

Actually, the leading term of Δn^i is the order of $r^{-\alpha}$, the normal vector field n^i can also be understood as the norm vector field of S embedded in (Σ, h_{ij}), i.e., the vector field n^i_Σ.

Here, we have used the Cartesian coordinates for the Euclidean space. For general coordinates of the Euclidean space, the expression for the ADM energy has a form

$$E_{\mathrm{ADM}} = H[1, \mathbf{0}] = \frac{1}{16\pi} \lim_{r \to \infty} \int_S h^{jk}(\bar{D}_j h_{ik} - \bar{D}_i h_{jk}) n^i \epsilon_S, \tag{5.159}$$

where $\bar{D}_i$ is the covariant derivative of the Euclidean space. For more details, see [61] and [10].

Since near the spacelike infinity of the spacetime t^a approaches the time translation vector in Poincaré algebra, E_{ADM} can be understood as the dual of the time translation. So it can be understood as the energy of the system, i.e., ADM energy of the spacetime.

Similar to the ADM energy, for the spacial translation near spacelike infinity, one can define ADM linear momentum as

$$P_{(i)} = H[0, \boldsymbol{\xi}_{(i)}] = \frac{1}{8\pi} \lim_{r \to \infty} \int_S \left[(K_{ij} - K h_{ij}) n^j\right] \epsilon_S, \tag{5.160}$$

where $\xi_{(i)} = \delta^j{}_i(\partial/\partial x^j)$ corresponds to the spacial translations near the spacelike infinity. ADM angular momentum is defined as

$$J_{(i)} = H[0, \varphi_{(i)}] = \frac{1}{8\pi} \lim_{r\to\infty} \int_S \left[\varphi^k_{(i)}(K_{kj} - Kh_{kj})n^j \right] \epsilon_S \,, \tag{5.161}$$

where

$$\varphi_{(i)} = \varphi^j_{(i)}\left(\frac{\partial}{\partial x^j}\right) \tag{5.162}$$

is the rotation generator near the spacelike infinity. From the ADM energy and ADM momentum, it is natural to define an 4-momentum ($n-$momentum here) $P_{(\mu)} = (-E_{\text{ADM}}, P_{(i)})$. Under the Lorentz transformation at the spacelike infinity, this momentum transforms as a Lorentz vector. The invariant

$$M_{\text{ADM}} = \sqrt{(E_{\text{ADM}})^2 - \delta^{ij} P_{(i)} P_{(j)}} \tag{5.163}$$

is the ADM mass of the spacetime. It should be pointed out here, unlike the ADM energy and linear momentum, the ADM angular momentum $J_{(i)}$ is not well defined in the Cartesian coordinates. For example, in four dimensions, with the asymptotic condition of K_{ij}, the integral for $J_{(i)}$ might be divergent [10].

By using the covariant phase space method in the previous section, one can also get the formula for ADM energy. Similar to the Hamiltonian in eq.(5.152), the on-shell Hamiltonian in covariant phase space method is also a surface term, i.e., eq.(5.146). When $\partial\Sigma$ in eq.(5.146) approaches spacelike infinity, by assuming Σ be a constant t slice, and $\partial\Sigma$ be a constant r surface S and

$$\int_S \xi \cdot \boldsymbol{\Theta} = \delta \int_S \xi \cdot \mathbf{B} \,, \tag{5.164}$$

one has

$$H_\xi = \lim_{r\to\infty} \int_S (\mathbf{Q}[\xi] - \xi \cdot \mathbf{B}) \,. \tag{5.165}$$

For the Einstein gravity without cosmological constant, it is not hard to find

$$\big(\boldsymbol{\Theta}(g, \delta g)\big)_{a_1\cdots a_{n-1}} = \frac{1}{16\pi} \epsilon_{aa_1\cdots a_{n-1}} g^{ab} g^{cd} (\nabla_c \delta g_{bd} - \nabla_b \delta g_{cd}) \,, \tag{5.166}$$

and

$$(\mathbf{Q}[\xi])_{a_1\cdots a_{n-1}} = -\frac{1}{16\pi} \epsilon_{a_1\cdots a_{n-2}cd} \nabla^c \xi^d \,. \tag{5.167}$$

So the integral of $\mathbf{Q}[\xi]$ is exactly one half of Komar integral in eq.(5.121) if ξ^a is a Killing vector field. Of course, this cannot give the ADM energy (or momentum). The term with $\mathbf{B}$ in eq.(5.165) will provide another one half. This requires the details of $\mathbf{B}$. By the explicit form of $\boldsymbol{\Theta}$, one can find $\mathbf{B}$ and get the Hamiltonian [5].

Assume the spacetime is asymptotically flat, and near the spacelike infinity the metric components approach the ones of Minkowski metric in the Lorentz coordinates, one has

$$E = H_t = \frac{1}{8\pi} \lim_{r\to\infty} \int_S (\partial_j h_{ij} - \partial_i h_{jj}) n^i \epsilon_S \,, \tag{5.168}$$

where t is the time flow vector. Similarly, linear momentum and angular momentum of the asymptotically flat spacetime can be defined as

$$P_{(i)} = H_{\xi_{(i)}}\,, \qquad J_{(i)} = H_{\varphi_{(i)}}\,. \tag{5.169}$$

The ADM energy in the above eqs.(5.158) or (5.168) can also be written in a more simple form. By using Gauss-Codazzi equation and the asymptotic behavior of the metric at spacelike infinity, one can put the ADM energy in the form [62]

$$E_{\mathrm{ADM}} = \frac{1}{16\pi} \lim_{r\to\infty} \int_S r q^{ik} q^{jl} R_{ijkl}\epsilon_S\,, \tag{5.170}$$

where R_{ijkl} is the components of the Riemann tensor of the spacetime, and q_{ij} is the induced metric on the constant r surface. This expression has a close relation to the definition by Ashtekar and Hansen with the components of Weyl tensors at spacelike infinity [55].

5.3.2.2 *Asymptotical flat at spacelike infinity and SPI group*

All of the above discussions are based on the coordinates (for example, the Cartesian coordinates in Euclidean space) near spacelike infinity. Actually, based on the conformal completion, one can get the ADM energy or momentum in the forms which do not depend on the coordinates. These are the definitions made by Ashtekar and Hansen [55, 63]. Here, we focus on the general relativity case.

A spacetime (M, g_{ab}) will be said to be asymptotically empty and flat at spatial and null infinity if there exists a spacetime $(\tilde{M}, \tilde{g}_{ab})$ which is smooth everywhere except at a point i^0 where $\tilde{g}_{ab}$ is $C^{>0}$, together with an imbedding of M into $\tilde{M}$ (identifying M with its image in $\tilde{M}$) satisfying the following conditions:

(i). $\overline{J(i^0)} = \tilde{M} - M$;

(ii). There exists a function Ω on $\tilde{M}$ such that, on M, $\tilde{g}_{ab} = \Omega^2 g_{ab}$; on $\dot{J}(i^0) - i^0$, $\Omega = 0$, $\tilde{\nabla}_a\Omega \neq 0$; and, at i^0, $\Omega = 0$, $\tilde{\nabla}_a\Omega = 0$, and

$$\lim_{\to i^0} \tilde{\nabla}_a\tilde{\nabla}_b\Omega = 2\tilde{g}_{ab}|_{i^0}\,,$$

(iii). There exists a neighborhood N of $J(i^0)$ in $\tilde{M}$ such that $(N, \tilde{g}_{ab})$ is strongly causal and time orientable, and in $M \cap N$, g_{ab} satisfies Einstein's vacuum equations.

The point i^0 represents the spatial infinity of (M, g_{ab}) while $\mathcal{I} := \dot{J}(i^0) - i^0$ represents the null infinity. Usually, one has

$$\mathcal{I} = \mathcal{I}^+ \cup \mathcal{I}^-\,,$$

where $\mathcal{I}^\pm$ denote the future infinity and past infinity respectively. Once the conformal completion of Penrose is applied, the spacelike infinity is a point, i.e., i^0, and one has to consider the differentiability conditions on this point. The $C^{>0}$ differentiability requirement on g_{ab} ensures only that g_{ab} is C^0, smooth in its "angular dependence" at i^0, and that its derivatives suffer from only finite radial discontinuities at the point i^0.

We have mentioned the asymptotic group below eq.(5.148). SPI group is a kind of asymptotic group which is just the subgroup of diffeomorphism group which preserves the universe structure of the spaclike infinity in the above definition. In [55], to define conserved quantities associated with the single point i^0, the author has used a technique "blowing up" borrowing from algebraic geometry. The point i^0 is blown up to a four dimensional manifold $\mathcal{S}$ which is formed by the equivalence class of some "regular curves" passing through i^0. Two regular curves are equivalent if they have same velocity and acceleration at i^0. The manifold $\mathcal{S}$ has a bundle structure—the base space is the unit timelike hyperboloid $\mathcal{K}$ in the tangent space of i^0, i.e., $T_{i^0}(\tilde{M})$, and the structure group is additive group $\mathcal{R}$. In a word, we have a bundle

$$P(\mathcal{S}, \mathcal{K}, \pi; \mathcal{R}) \,.$$

This $\mathcal{S}$ will be called SPI (spatial infinity). Furthermore, $\mathcal{S}$ has two tensor fields: a covariant, second rank, symmetric, degenerate tensor field $h_{ab} = \pi^* \boldsymbol{h}_{ab}$ which is the pullback of the natural metric $\boldsymbol{h}_{ab}$ on the hyperboloid $\mathcal{K}$; and a vertical vector field v^a, the generator of the natural one parameter transformations on $\mathcal{S}$ induced by its structure group. The diffeomorphsims which preserve the bundle structure of $\mathcal{S}$, h_{ab}, and v^a form a subgroup, denoted by $\mathcal{G}$, of diffeomorphsim group. This $\mathcal{G}$ is the so-called SPI group. With this definition, let ξ be the generator of the diffeomorphsim in $\mathcal{G}$, then one has

$$\mathcal{L}_\xi h_{ab} = 0 \,, \qquad \mathcal{L}_\xi v^a = 0 \,. \tag{5.171}$$

Assume ξ satisfies the above condition, then the corresponding one parameter family of diffeomorphsims automatically preserves the fibre bundle structure of $\mathcal{S}$. All of such ξ form the Lie algebra of SPI group, denoted by $\mathbf{L}_\mathcal{G}$.

The generators which have vanishing projection to $\mathcal{K}$ (i.e., $\pi_* \xi = 0$) form an Abelian sub Lie algebra of $\mathbf{L}_\mathcal{G}$. It has been proved in [55] that there is a one-to-one correspondence between (arbitrary) scalar fields on $\mathcal{K}$ and elements of $\mathbf{L}_\mathcal{G}$ whose projection on $\mathcal{K}$ vanishes. The elements in this sub Lie algebra, denoted by $\mathbf{L}_\mathcal{S}$, are called infinitesimal SPI supertranlations. Furthermore, it is proved $\mathbf{L}_\mathcal{S}$ is an ideal of $\mathbf{L}_\mathcal{G}$, and the quotient algebra $\mathbf{L}_\mathcal{G}/\mathbf{L}_\mathcal{S}$ is Lorentz algebra.

Consider the functions f of the type $f(k) = k_a \eta^a$ where k_a is any covector at i^0 and η^a is the position vector of points on the hyperboloid $\mathcal{K}$ in $T_{i^0}(\tilde{M})$. The infinitesimal supertranslations of the type $f(k)v^a$ on $\mathcal{S}$ form a four-dimensional Abelian Lie algebra, denoted by $\mathbf{L}_\mathcal{T}$. It can be proved that $\mathbf{L}_\mathcal{T}$ is an ideal of $\mathbf{L}_\mathcal{G}$. The elements in $\mathbf{L}_\mathcal{T}$ are called infinitesimal SPI-translalions

In Minkowski spacetime, the spacetime translations are precisely the elements of $\mathbf{L}_{\mathcal{T}}$. It is also shown that every Killing field in the physical spacetime (M, g_{ab}) gives rise to an unique element of $\mathbf{L}_{\mathcal{G}}$ [55].

5.3.2.3 *Coordinate free definitions*

To define conserved quantities, one has to consider the limits of physical fields and their derivatives at i^0. These limits are direction dependent when the fields have no usual limits at i^0. Let γ be a differentiable spacelike curve passing through i^0 in $\tilde{M}$ and η^a be the unit tangent vector at i^0, then, the limit of a tensor field $\tau^{a_1 \cdots a_r}{}_{b_1 \cdots b_s}$ along γ at i^0

$$\lim_{\to i^0} \tau^{a_1 \cdots a_r}{}_{b_1 \cdots b_s} = \boldsymbol{\tau}^{a_1 \cdots a_r}{}_{b_1 \cdots b_s}(\eta)$$

defines a mapping from $\mathcal{K}$ to tensors at i^0 with the same tensor type of $\tau^{a_1 \cdots a_r}{}_{b_1 \cdots b_s}$. This mapping is also smooth, i.e., one can define the derivatives $\boldsymbol{\partial}_e \boldsymbol{\tau}^{a_1 \cdots a_r}{}_{b_1 \cdots b_s}(\eta)$ respect to η^e (to arbitrary order). The limits of the derivatives of fields with respect to $\tilde{\nabla}_a$ have relation to $\boldsymbol{\partial}_e$ as follows

$$\lim_{\to i^0} \Omega^{\frac{1}{2}} \tilde{\nabla}_e \tau^{a_1 \cdots a_r}{}_{b_1 \cdots b_s} = \boldsymbol{\partial}_e \boldsymbol{\tau}^{a_1 \cdots a_r}{}_{b_1 \cdots b_s}(\eta) \,,$$

where $\tilde{\nabla}_a$ is the derivative operator associated with $\tilde{g}_{ab}$.

Since $\tilde{g}_{ab}$ has an usual limit at i^0, i.e., $\lim_{\to i^0} \tilde{g}_{ab} = \boldsymbol{g}_{ab}$, so $\boldsymbol{g}_{ab}$ does not depend on the directions. So we have $\boldsymbol{\partial}_c \boldsymbol{g}_{ab}(\eta) = 0$. The induced metric $\boldsymbol{h}_{ab}(\eta) = \boldsymbol{g}_{ab}(\eta) - \eta_a \eta_b$ on $\mathcal{K}$ is obviously direction dependent.

Assume $\boldsymbol{\tau}^{a_1 \cdots a_r}{}_{b_1 \cdots b_s}(\eta)$ is orthogonal to η^a for all indices, the covariant derivative on $\mathcal{K}$ is defined as

$$\boldsymbol{D}_e \boldsymbol{\tau}^{a_1 \cdots a_r}{}_{b_1 \cdots b_s}(\eta) = \boldsymbol{h}_e{}^f(\eta) \boldsymbol{h}_{c_1}{}^{a_1}(\eta) \cdots \boldsymbol{h}_{c_r}{}^{a_r}(\eta) \boldsymbol{h}_{d_1}{}^{b_1}(\eta)$$
$$\cdots \boldsymbol{h}_{d_s}{}^{b_s}(\eta) \boldsymbol{\partial}_f \boldsymbol{\tau}^{c_1 \cdots c_r}{}_{d_1 \cdots d_s}(\eta) \,. \tag{5.172}$$

With this preliminary, for physical fields, the asymptotic data is encoded in the fields on $\mathcal{K}$. For example, by the assumption on g_{ab}, one finds that $\Omega^{\frac{1}{2}} C_{abc}{}^d$ has direction dependent limit at i^0, where $C_{abc}{}^d$ is the Weyl tensor of the spacetime, and the asymptotic behavior of Weyl tensor is determined by their "electric" and "magnetic" parts, i.e.,

$$\boldsymbol{E}_{ab}(\eta) = \lim_{\to i^0} C_{acbd}(\eta) \eta^c \eta^d \,, \tag{5.173}$$

$$\boldsymbol{B}_{ab}(\eta) = \lim_{\to i^0} C^*_{acbd}(\eta) \eta^c \eta^d \,. \tag{5.174}$$

The asymptotic gravitational equations are given by the equations on $\mathcal{K}$ as follows

$$\boldsymbol{D}_{[a} \boldsymbol{E}_{b]c} = 0 \,, \qquad \boldsymbol{D}_{[a} \boldsymbol{B}_{b]c} = 0 \,. \tag{5.175}$$

From these one also has $D^a E_{ab} = 0 = D^a B_{ab}$. Since E_{ab} is trace free and divergence free, one finds two form

$$Q[\boldsymbol{\xi}]_{cd} = E_{ab}\xi^a \epsilon^b{}_{cd} \tag{5.176}$$

is a closed form if ξ^a is a conformal Killing vector field of $(\mathcal{K}, h_{ab})$. Where ϵ_{abc} is the Levi-Civita tensor on $\mathcal{K}$. Hence, the integral of $Q[\boldsymbol{\xi}]$ on the two dimensional spacelike cross section C of $\mathcal{K}$ is a conserved quantity, i.e.,

$$P[\boldsymbol{\xi}] = \int_C Q = \int_C E_{ab}\xi^a u^b \epsilon_C \tag{5.177}$$

is independent of C. Here ϵ_C is the volume element of C, and u^b is the unit norm of C in $\mathcal{K}$. The cross section C can be viewed as the limit (at i^0) of a sequence of two dimensional closed surfaces in the physical spacetime. $(\mathcal{K}, h_{ab})$ admits ten conformal Killing vector fields. Six of them are Killing vectors. However, these Killing vectors do not have nontrivial conserved quantities because Q is exact in this case.

The four conformal Killing (but not Killing) vectors ξ^a of $(\mathcal{K}, h_{ab})$ actually correspond to the infinitesimal translations in $\mathbf{L}_\mathcal{T}$. In fact, the functions of $\mathcal{K}$ correspond to infinitesimal supertranslation, and the function with the form $f(k) = k_a \eta^a$ corresponds to infinitesimal translation as mentioned before. From the definition, it is not hard to prove that the derivative $D_a f(k)$ is a conformal Killing vector field of $(\mathcal{K}, h_{ab})$. So every element of $\mathbf{L}_\mathcal{T}$ has a conserved quantity, and $P[\boldsymbol{\xi}]$ is just the momentum associated with the translation. Further, $P[\boldsymbol{\xi}]$ can be expressed as

$$P[k] = \int_C E_{ab} D^a f(k) u^b \epsilon_C \tag{5.178}$$

because we have $D^a f(k) = h^a{}_b k^b$. Hence, the conserved quantities can be viewed as a mapping from $T_{i^0}\tilde{M}$ to $\mathcal{R}$, and belong to the algebraic dual space of $T_{i^0}\tilde{M}$, and can be viewed as covectors. Denoting P as P_a, then one gets

$$P_a k^a = \int_C E_{ab} k^a u^b \epsilon_C . \tag{5.179}$$

Based on the calculation for Kerr spacetime, Ashtekar and Hansen has shown this momentum P_a is exactly the ADM linear momentum.

To define the angular momentum for the spacetime, one has to consider the Lorentz algebra at the spacelike infinity. However, SPI algebra $\mathbf{L}_\mathcal{G}$ includes infinity versions of Lorentz algebra (as many as supertranslations). Additional conditions have to be imposed to define angular momentum in the usual sense. One of these conditions is the vanishing of B_{ab}. This means the magnetic part of Weyl tensor decays faster than the electric part. The next order of Weyl tensor is defined as

$$\beta_{ab} = \lim_{\to i^0} C^*_{acbd}(\tilde{\nabla}^c \Omega^{\frac{1}{2}})(\tilde{\nabla}^d \Omega^{\frac{1}{2}}) . \tag{5.180}$$

For an arbitrary skew tensor F_{ab} at i^0, one defines the angular momentum as an antisymmetric tensor M^{ab} at i^0

$$M^{ab} F_{ab} = \frac{1}{2} \int_C \boldsymbol{\beta}_{ab} \boldsymbol{\xi}^a \boldsymbol{u}^b \boldsymbol{\epsilon}_C , \tag{5.181}$$

where $\boldsymbol{\xi}^a$ is a Killing vector field on $(\mathcal{K}, \boldsymbol{h}_{ab})$ induced by $\epsilon^{abcd} F_{bc} \eta_d$ at i^0. See [55] for details.

5.3.2.4 *Further development*

To avoid the awkward technique to study the differentiability of the fields at the single point i^0, Ashtekar and Romano proposed another completion of the spacetimes, see [63] and reference therein.

Definition 1. A spacetime (M, g_{ab}) will be said to possess an asymptote at spacial infinity if there exists a manifold $\tilde{M}$ with boundary $\mathcal{H}$, a smooth function Ω defined on $\tilde{M}$, and a diffeomorphism from M to $\tilde{M} - \mathcal{H}$ (M is identified with its image in $\tilde{M}$) such that ($l = g^{ab} \tilde{\nabla}_a \Omega \tilde{\nabla}_b \Omega$)

   (i). $\Omega \hat{=} 0$, and $\tilde{\nabla}_a \Omega \hat{\neq} 0$, where "$\hat{=}$" stands for equals when evaluated at points of $\mathcal{H}$;
   (ii). $q_{ab} := \Omega^2 (g_{ab} - l^{-1} \tilde{\nabla}_a \Omega \tilde{\nabla}_b \Omega)$ and $n^a := \Omega^{-4} g^{ab} \tilde{\nabla}_b \Omega$ admit smooth limits on $\mathcal{H}$, such that q_{ab} has a signature $(-, +, +)$ on $\mathcal{H}$;
   (iii). $\lim_{\Omega \to 0} G_{ab} = 0$, where $G_{ab} := R_{ab} - \frac{1}{2} R g_{ab}$ is the Einstein tensor of g_{ab}.

Definition 2. If (M, g_{ab}) admits an asymptote at spatial infinity in which $\mathcal{H}$ has topology $S^2 \times \mathcal{R}$ then (M, g_{ab}) will be said to be asymptotically flat at spacial infinity (AFSI).

Definition 3. An AFSI spacetime (M, g_{ab}) will be said to be asymptotically Minkowskian (AM) if the boundary $\mathcal{H}$ is geodesically complete with respect to the metric q_{ab} thereon.

The above definition is not the standard conformal completion by Penrose [52, 53], so the spacial infinity is not a single point. With the definition of AM spacetime, one can also study the asymptotic group as before and get SPI group. The analysis of the asymptotic behavior of the fields looks simpler than those by using the i^0 treatment.

Two completions (M, Ω) and (M, Ω') of (M, g_{ab}) are said to be smooth related if $\Omega' = \alpha \Omega$ for some smooth function α and $\alpha \hat{=}$ constant. Defining F as

$$F = \mathcal{L}_n \Omega = \Omega^{-4} l ,$$

then for two smooth related completions, we have

$$F' = \mathcal{L}_{n'} \Omega' \hat{=} \alpha^{-2} F .$$

The field equations implies $F \mathrel{\hat{=}} \text{constant}$ and $\mathcal{R}_{ab} \mathrel{\hat{=}} 2F q_{ab}$. Where $\mathcal{R}_{ab}$ is the Ricci tensor of $\Omega = \text{constant}$ surface. The completion with $F \mathrel{\hat{=}} 1$ and $\mathcal{R}_{ab} \mathrel{\hat{=}} 2 q_{ab}$ is called "unit hyperboloid completion". The pull back of q_{ab} and $\mathcal{R}_{ab}$ to $\mathcal{H}$ are denoted by $\boldsymbol{q}_{ab}$ and $\boldsymbol{\mathcal{R}}_{ab}$. Since $\boldsymbol{\mathcal{R}}_{ab} = 2\boldsymbol{q}_{ab}$, so under the unit hyperboloid completion, $\mathcal{H}$ is isometric to the $\mathcal{K}$ in the previous discussion. Hence, now $\mathcal{H}$ will play the role of $\mathcal{K}$, and all asymptotic data of fields are fields on $\mathcal{H}$. The relevant electric and magnetic parts of the Weyl tensor now can be expressed as

$$E_{ab} = l^{-1}\Omega^{5} C_{acbd} n^{c} n^{d} , \qquad B_{ab} = l^{-1}\Omega^{5} C_{acbd}^{*} n^{c} n^{d} , \tag{5.182}$$

and they satisfy

$$D_{[a}E_{b]c} \mathrel{\hat{=}} 0 , \qquad D_{[a}B_{b]c} \mathrel{\hat{=}} 0 . \tag{5.183}$$

With these, as the previous discussion, one can define conserved quantities as

$$P[\boldsymbol{\xi}] = \frac{1}{8\pi} \int_{C} \boldsymbol{E}_{ab} \boldsymbol{\xi}^{a} \boldsymbol{u}^{b} \epsilon_{C} . \tag{5.184}$$

where $\boldsymbol{E}_{ab}$ is the pull back of E_{ab} to $\mathcal{H}$, and $\boldsymbol{\xi}$ is the conformal Killing vector (but not Killing) for $\mathcal{H}$, and $\boldsymbol{u}$ is the unit norm of C in $\mathcal{H}$. As before, $\boldsymbol{\xi}$ corresponds to the SPI translation. To define the angular momentum, one has to set $B_{ab} \mathrel{\hat{=}} 0$ and consider the next order, i.e., β_{ab} as in the i^{0} treatment, see [63] for details.

In a conclusion, the ADM energy and momentum can be put into a coordinate free form. The definitions of the conserved quantities are a little bit complicated. However, the final results are very clear—the ADM momentum can be viewed as the dual of infinitesimal SPI translations. Of course, to really understand these surface integrals at infinity, one has to construct Hamiltonian of the theory (or others) to get these integrals. Some derivation of these conserved quantities from Hamiltonian formalism can be found in [64, 70].

5.3.3 *Global quantities at null infinity*

The spacelike infinity has no causal contact to the bulk events of the spacetime. So the ADM energy or momentum defined at the spacelike infinity characterizes the initial data on constant t slice in some sense. ADM mass is a constant which does not change with the time t. The null infinity is different from the spacelike infinity because it has causal connection with the bulk events. For example, due to the existence of the flux of outgoing gravitational radiation, the Bondi–Sachs energy is not a constant and depends on the retarded time u or the cross sections of the null infinity. So the Bondi–Sachs energy (if it can be constructed from the Hamiltonian formalism) belongs to the case II in the covariant phase space method mentioned at the beginning of this section.

To study the gravitational wave near infinity, Bondi *et. al.* proposed a definition of asymptotic flat (near the null infinity) by using a special coordinate [65]. The original definition of Bondi is based on axisymmetric spacetime, and was generalized by Sachs to the spacetime without this symmetry [66]. The definition by Bondi and Sachs was later generalized and put into the form of "conformal infinity" by applying the conformal completion technique of Penrose [52, 53]. Here, we follow the notations by Ashtekar [67].

5.3.3.1 *Asymptotical flat spacetime at null infinity*

A spacetime (M, g_{ab}) is asymptotically flat at null infinity if there exists a spacetime $(\tilde{M}, \tilde{g}_{ab})$ with boundary $\mathcal{I}$ and a diffeomorphism from M onto $\tilde{M} - \mathcal{I}$ (M and $\tilde{M} - \mathcal{I}$ are identified) such that:

(i). There exists a smooth function Ω on $\tilde{M}$ with $\tilde{g}_{ab} = \Omega^2 g_{ab}$ on M; $\Omega = 0$ on $\mathcal{I}$; and $n_a := \tilde{\nabla}_a \Omega$ is nowhere vanishing on $\mathcal{I}$;

(ii). $\mathcal{I}$ is topologically $S^2 \times \mathcal{R}$;

(iii). g_{ab} satisfies Einstein equations $R_{ab} - \frac{1}{2} R g_{ab} = 8\pi T_{ab}$, where $\Omega^{-2} T_{ab}$ has a smooth limit to $\mathcal{I}$.

From the Einstein equations and the condition on the energy-momentum tensor, it is not hard to find

$$\tilde{g}^{ab} \tilde{\nabla}_a \Omega \tilde{\nabla}_b \Omega = \tilde{g}^{ab} n_a n_b \,\hat{=}\, 0 \,, \tag{5.185}$$

where $\hat{=}$ means that the limit to $\mathcal{I}$ is taken. So $\mathcal{I}$ is null hypersurface in $(\tilde{M}, \tilde{g}_{ab})$. Again, considering Einstein equations and the relations between the Ricci tensors of (M, g_{ab}) and $(\tilde{M}, \tilde{g}_{ab})$, we have

$$\tilde{\nabla}_a n_b \,\hat{=}\, \frac{1}{2} \left[\lim_{\to \mathcal{I}} \Omega^{-1} n_c n^c \right] \tilde{g}_{ab} \,\hat{=}\, \frac{1}{4} (\tilde{\nabla}_c n^c) \tilde{g}_{ab} \,. \tag{5.186}$$

Hence, on $\mathcal{I}$, $\tilde{\nabla}_a n_b$ is determined by its divergence.

Of course, the conformal completion is not unique. Assume $\Omega' = \omega \Omega$ with positive smooth function on $\tilde{M}$ is also a well defined conformal factor in (i) in the above definition. It is easy to find

$$\tilde{\nabla}_a n_b' \,\hat{=}\, \frac{1}{4} \omega (\tilde{\nabla}_c n^c) \tilde{g}_{ab} + (n^c \tilde{\nabla}_c \omega) \tilde{g}_{ab} \,. \tag{5.187}$$

where $n_a' = \tilde{\nabla}_a \Omega'$. So, with an appropriate ω, we have

$$\tilde{\nabla}_a n_b \,\hat{=}\, 0 \,. \tag{5.188}$$

From (5.186) we know: if $\tilde{\nabla}_c n^c \,\hat{=}\, 0$, we have the above relation automatically. For this reason, some people call the conformal factor with $\tilde{\nabla}_c n^c \,\hat{=}\, 0$ as divergence free conformal frame [67]. The conformal factor which satisfies (5.188) is also said to satisfy Bondi gauge. Of course, there still remains gauge freedoms to select the conformal factor after the Bondi gauge. It is easy to find $\Omega' = \omega \Omega$ also satisfies the Bondi gauge if $\mathcal{L}_n \omega \,\hat{=}\, 0$.

5.3.3.2 *BMS group and algebra*

Definition: An asymptotically flat space-time is said to be asymptotically Minkowski if $\mathcal{I}$ is complete in any divergence free conformal frame.

This definition ensures that $\mathcal{I}$ has the same global structure as it has in the completion of Minkowski space-time.

Assume the induced metric of $\mathcal{I}$ is $\tilde{h}_{ab}$, then it is degenerate with signature $(0,+,+)$. It naturally satisfies $n^a \tilde{h}_{ab} \,\hat{=}\, 0$, and $\mathcal{L}_n \tilde{h}_{ab} \,\hat{=}\, 0$ if the Bondi gauge is satisfied (at present, we choose the Bondi gauge). The pair $(n^a, \tilde{h}_{ab})$ on $\mathcal{I}$ represents the asymptotic behavior of the metric. The pair

$$(n'^a = \omega^{-1} n^a, \quad \tilde{h}'_{ab} = \omega^2 \tilde{h}_{ab})$$

with $\mathcal{L}_n \omega = 0$ plays the same role. As suggested by Geroch [68], the pair can also be recast into a tensor

$$\Gamma^{ab}{}_{cd} = n^a n^b \tilde{h}_{cd}.$$

Obviously, this tensor does not depend on the selection of ω.

The asymptotic symmetry should preserve this universal structure, i.e., the pair or the tensor $\Gamma^{ab}{}_{cd}$. A vector field ξ^a preserves the universal structure if $\mathcal{L}_\xi \Gamma^{ab}{}_{cd} = 0$ or equivalently

$$\mathcal{L}_\xi \tilde{h}_{ab} = 2\alpha \tilde{h}_{ab}, \qquad \mathcal{L}_\xi n^a = -\alpha n^a \tag{5.189}$$

for some function α on $\mathcal{I}$ satisfying $\mathcal{L}_n \alpha = 0$. This ξ^a is called the representative of the asymptotic symmetry. All such vectors form a Lie algebra $\mathbf{L}_B$, which is just the so-called BMS algebra.

It is not hard to find that $\xi^a = f n^a$ with $\mathcal{L}_n f = 0$ for some function f on $\mathcal{I}$ is a member of $\mathbf{L}_B$. This kind of representative of asymptotic symmetry is called supertranslation, and they form a subalgebra, denoted by $\mathbf{L}_S$, of $\mathbf{L}_B$. Furthermore, it is easy to find $\mathbf{L}_S$ is also an ideal of $\mathbf{L}_B$. Hence, we can define the quotient algebra. This quotient algebra is nothing but the Lorentz algebra, i.e., dimension of

$$\mathbf{L}_B / \mathbf{L}_S \tag{5.190}$$

is finite although $\mathbf{L}_B$ and $\mathbf{L}_S$ both have infinity dimensions. In the case of finite transformations, the BMS group $\mathcal{B}$ is the semi-product of Lorentz group and super-tranlation group $\mathcal{S}$.

The BMS generator ξ^a can be projected to the two dimensional spacelike sub-mifold (S, q_{ab}) because ξ^a preserves the direction of n^a. Assume the projection is given by $\bar{\xi}^a$, then one can prove that $\bar{\xi}^a$ is a conformal Killing vector field on (S, q_{ab}). One can further select the conformal factor such that q_{ab} is the metric of a sphere by using remanning freedom after the Bondi gauge (usually called Bondi frame). With this, a BMS supertranslation $\xi^a = f n^a$ is a translation (corresponds to the translation in Poincaré algebra) when f is the linear combination of spherical harmonic functions with $l = 0, 1$. In a word, we can find a translation subalgebra, denoted by $\mathbf{L}_T$, from the supertranslation algebra $\mathbf{L}_S$.

5.3.3.3 *Conserved quantities at null infinity*

Roughly speaking, the ADM energy or momentum is associated with the spacelike hypersurface which asymptotically anchored at spacelike infinity i^0. Bondi–Sachs energy, however, is defined as the one for the hypersurface Σ which is asymptotically approaches a null hypersurface near the null infinity. In other words, in asymptotic region with Cartesian coordinates $\{t, x^i\}$, the hypersurface Σ for Bondi–Sachs energy approaches

$$u = t - r = \text{constant}\,, \qquad r = \sqrt{\delta_{ij} x^i x^j}\,.$$

Here u is the so-called retarded time. The cross section of this hypersurface with $\mathcal{I}$, denoted by C_u, is a codimension-2 surface of the unphysical spacetime. The Bondi–Sachs energy and momentum are defined as integrals on C_u.

A simple way to get the Bondi–Sachs momentum is the utilization of the Komar integral (5.121) in the previous section. The logic is like this: although the Komar integral depends on the timelike Killing vector field and we have no Killing vector field for a general spacetime, the vector field ξ^a which approaches asymptotic symmetry has some property similar to a Killing vector field especially when we take a limit to the null infinity. Assume $\xi^a_{(\mu)}$, $\mu = 0, \cdots 3$ are vector fields on the spacetime, and they correspond to the four translation generators (the elements in $\mathbf{L}_{\mathcal{T}}$) of $\mathcal{I}$ when they are extended to $\mathcal{I}$. It should be pointed out here, the selection of $\xi^a_{(\mu)}$ is not unique, there exist a lot of $\xi^a_{(\mu)}$ which correspond to the same translations on $\mathcal{I}$, and they form an equivalent class (for their common translations on $\mathcal{I}$). The Komar like integral depends on this selection, so one has to impose some condition to cure this problem. Geroch and Winicour proposed a condition to select $\xi^a_{(\mu)}$, i.e., the Geroch–Winicour condition [69]

$$\nabla_a \xi^a_{(\mu)} \,\hat{=}\, 0\,,$$

and defined the Bondi–Sachs momentum as

$$P_{(\mu)} = -\frac{1}{8\pi} \lim_{S_u \to C_u} \int_{S_u} \epsilon_{abcd} \nabla^c \xi^d_{(\mu)}\,, \tag{5.191}$$

where S_u is a sequence of two dimensional surface (in Σ) which approaches C_u. It should be point out here, usually, the Bondi–Sachs momentum depends on the retarded time u or the cross section C_u. For example, the Bondi–Sachs energy $E = P_{(0)}$ generally depends on u, i.e., we have $E = E(u)$. So a different Σ has a different Bondi–Sachs energy. The physical reason is simply the existence of the gravitational radiation flux to null infinity. Bandi and his collaborators showed

$$\frac{\mathrm{d}E(u)}{\mathrm{d}u} = -\int_{C_u} n^2\, \epsilon_{C_u}\,, \tag{5.192}$$

where $n^2 \geq 0$ called news functions, and ϵ_{C_u} is the volume element of C_u. This suggests that outgoing gravitational radiation carries positive energy, and the Bondi–Sachs energy is always non-increasing.

By using covariant phase space analysis, Wald and Zoupas have studied the conserved quantities associated to the null infinity [51]. Here, we give a brief summary on their analysis and results.

Assume the pull back of symplectic current form $\boldsymbol{\omega}$ to B (now B is the future null infinity $\mathcal{I}^+$), denoted by $\bar{\boldsymbol{\omega}}$, does not vanish on B. On B, assume $\bar{\boldsymbol{\Theta}}$ be a symplectic potential for $\bar{\boldsymbol{\omega}}$ such that

$$\bar{\boldsymbol{\omega}}(\delta\phi, \delta_1\phi, \delta_2\phi) = \delta_1\bar{\boldsymbol{\Theta}}(\phi, \delta_2\phi) - \delta_2\bar{\boldsymbol{\Theta}}(\phi, \delta_1\phi). \tag{5.193}$$

The symplectic potential form $\bar{\boldsymbol{\Theta}}$ is assumed to be locally constructed from dynamical fields ϕ and their derivatives. Other requirement on $\bar{\boldsymbol{\Theta}}$ can be found in [51]. Let $\mathcal{H}_\xi$ satisfy

$$\delta\mathcal{H}_\xi = \int_{\partial\Sigma} [\delta\mathbf{Q}[\xi] - \xi \cdot \boldsymbol{\Theta}] + \int_{\partial\Sigma} \xi \cdot \bar{\boldsymbol{\Theta}}, \tag{5.194}$$

Then, this $\mathcal{H}_\xi$ satisfies the existence condition (5.147). So one gets conserved quantity by $\mathcal{H}_\xi$ after selecting an appropriate reference solution ϕ_0. However, it should be noted that $\mathcal{H}_\xi$ is very different from the real Hamiltonian H_ξ. So by considering $\mathcal{H}_\xi$, one solves the existence problem for H_ξ. The next problem, i.e., the conservation problem of $\mathcal{H}_\xi$, is explained as follows. Generally $\mathcal{H}_\xi$ is not conserved due to the existence of the gravitational radiation on B. So there must be a nonzero flux $\mathbf{F}_\xi$ on B such that

$$\delta\mathcal{H}_\xi|_{\partial\Sigma_1} - \delta\mathcal{H}_\xi|_{\partial\Sigma_2} = -\int_{B_{12}} \delta\mathbf{F}_\xi, \tag{5.195}$$

where B_{12} is the portion between $\partial\Sigma_1$ and $\partial\Sigma_2$. Of course, in stationary case, this flux must be vanishing. From eq.(5.194), it is easy to find

$$\delta\mathbf{F}_\xi = \delta\bar{\boldsymbol{\Theta}}(\phi, \mathcal{L}_\xi\phi), \tag{5.196}$$

and, by requiring $\bar{\boldsymbol{\Theta}} = 0$ for stationary spacetime, one finds

$$\mathbf{F}_\xi = \bar{\boldsymbol{\Theta}}(\phi, \mathcal{L}_\xi\phi). \tag{5.197}$$

To get a well defined Hamiltonian, the selection of reference solution ϕ_0 (assumed to be stationary) also has some requirement: For all representatives ξ^a and η^a of infinitesimal asymptotic symmetries and for all cross section $\partial\Sigma$, ϕ_0 has to satisfy

$$0 = \int_{\partial\Sigma} \left(\mathcal{L}_\eta \mathbf{Q}[\xi] - \mathbf{Q}[\mathcal{L}_\eta\xi] - \xi \cdot \boldsymbol{\Theta}(\phi_0, \mathcal{L}_\eta\phi_0) \right). \tag{5.198}$$

In general relativity, with physically reasonable asymptotical conditions of the fields, Wald and Zoupas has proved that $\bar{\boldsymbol{\omega}}$ really does not vanish on the null infinity, and $\bar{\boldsymbol{\Theta}}$ has an unique form

$$\bar{\boldsymbol{\Theta}} = -\frac{1}{32\pi} N_{ab}\tau^{ab}\epsilon_{\mathcal{I}}, \tag{5.199}$$

where $\epsilon_{\mathcal{I}}$ is the volume element of the null infinity, and τ_{ab} (has smooth limit to $\mathcal{I}$) is defined as $\delta\tilde{g}_{ab} = \Omega\tau_{ab}$, and N_{ab} is the so-called news tensor which is defined as

$$N_{ab} = \bar{S}_{ab} - \rho_{ab}, \tag{5.200}$$

where $\bar{S}_{ab}$ is the pull back of the Schouten tensor for g_{ab} to the null infinity, and ρ_{ab} is chosen to be $(1/2)k\tilde{h}_{ab}$ [68]. Actually, one has applied the gauge freedom after the Bondi gauge to set the unphysical metric $\tilde{h}_{ab}$ (the restriction of $\tilde{g}_{ab}$ to $\mathcal{I}$) on $\mathcal{I}$ such that the induced metric on all the cross sections of $\mathcal{I}$ is that of a round sphere with a constant curvature k.

By these results, for any $\xi^a \in \mathbf{L}_{\mathcal{B}}$, for each cross section of $\mathcal{I}$, the conserved quantity $\mathcal{H}_\xi$ satisfies

$$\delta\mathcal{H}_\xi = \int_{\partial\Sigma} [\delta\mathbf{Q}[\xi] - \xi \cdot \mathbf{\Theta}] - \frac{1}{32\pi} \int_{\partial\Sigma} N_{ab} \cdot \mathcal{T}^{ab} \xi \cdot \epsilon_{\mathcal{I}}. \qquad (5.201)$$

Assume for Minkowski spacetime, one has $\mathcal{H}_\xi = 0$, principally, one can get the explicit form of $\mathcal{H}_\xi$. $\mathbf{\Theta}$ and $\mathbf{Q}$ can be found in eqs.(5.166) and (5.167). For ξ^a which is tangent to $\partial\Sigma$ everywhere, we have a real Hamiltonian and the term including $\mathbf{Q}$ in (5.201) gives correct Komar like integrals for angular momentums. Together with the Geroch–Winicour condition on ξ, one gets the expressions for angular momentums at the null infinity. For supertranlation, i.e., $\xi^a \in \mathbf{L}_{\mathcal{S}}$ (which is not tangent to $\partial\Sigma$), one finds $\mathcal{H}_\xi$ agrees with that proposed by Dray and Streubel [70, 71, 72, 73].

The study of the ADM momentum can be straightforwardly generalized to the Einstein gravity theory in higher dimensions. However, in a higher dimensions, the null infinity and Bondi–Sachs momentum are not so simple, see [74] for details. The study of the Bondi–Sachs energy and momentum in general diffeomorphsim invariant gravity theories is still absent up to date.

5.4 Quasi-local definitions

As we stressed in the above, there does not exist well defined locally energy-momentum tensor for gravitational field. On the other hand, we have seen from the previous section that the energy, momentum, angular momentum, and even super-translation momentum (at the null infinity) can be well defined globally. Whether the energy (and momentum) can be defined on some patch of the spacetime? This is the so-called quasi-local energy. There are a lot of reasons to define the energy and angular momentum quasi-locally. Some are from the viewpoint of physical practice, and some are from deep understanding of the nature of gravity. For example, one may argue that no physical system is truly isolated, and the asymptotic conditions to define the global conserved quantities are just theoretical idealizations.

The quasi-local quantities are defined on the subset of the spacetime. This subset can be a globally hyperbolic domain with compact closure in the spacetime; the compact spacelike hypersurface with boundary; or a closed, orientated spacelike codimension-2 surface embedded in the spacetime [2]. Here, we only focus on the last case, i.e., the quasi-local quantities defined on the closed, orientated spacelike codimension-2 surface.

There are several ways to construct the quasi-local quantities. For example, they can be defined from the Noether theorem based on the Lagrangian of the theory in the first section; They can also be defined from the Hamiltonian structure of the theory [2]; Some quasi-local quantities are directly defined from the geometry and symmetry of the codimension-2 surface with some physical consideration; and some quasi-local quantities are defined even without any systematic procedure. In any approach, there are ambiguities of uniqueness in the definition of these quantities. Due to the work by Nester and Chen [44], the situation is not better than the pseudotensor method—there are infinite types of definitions for these quasi-local quantities.

In this section, we only list several quasi-local defined quantities, and the detailed procedure to get these quantities will be omitted. Before introducing the quasi-local quantities, we firstly give a brief review on the geometry of the codimension-2 surface.

5.4.1　*Geometry of codimension-2 surface*

The codimension-2 spacelike surface plays an important role in the definition of quasi-local quantities in gravity theories. To describe the geometry of the codimension-2 surface, in this section, we list some important formulas in the submanifold theory. More details can be found in the papers [75, 76, 77, 78], and some text books for differential geometry, for example, the well-known book by Kobayashi and Nomizu [79]. The readers who are familiar with the submanifold theory can skip this subsection.

On any point p of a spacelike submanifold S embedded in a spacetime (M, g_{ab}), the tangent space $T_p M$ can be expressed as

$$T_p M = T_p S \oplus T_p^{\perp} S\,.$$

Any vector $v^a \in T_p M$ can be decomposed as $v^a = v_{\parallel}^a + v_{\perp}^a$, where $v_{\parallel}^a \in T_p S$, $v_{\perp}^a \in T_p^{\perp} S$, and $g_{ab}(p)v_{\parallel}^a v_{\perp}^a = 0$. Hence, the tangent bundle TM has a decomposition $TM = TS \oplus T^{\perp} S$. Since $T_p M$ has the decomposition, we can select a basis in $T_p M$ and its algebraic dual basis as

$$\{(e_i)^a, (\theta_I)^a\}, \quad \{(e^{*i})_a, (\theta^{*I})_a\}, \tag{5.202}$$

where $i = 1, \cdots, m\,; I = m + 1, \cdots, n$, satisfying

$$(e_i)^a = (e_i)_{\parallel}^a\,, \quad (\theta_I)^a = (\theta_I)_{\perp}^a\,,$$

and (according to the definition of the dual basis)

$$(e^{*j})_a(e_i)^a = \delta_i{}^j\,, \quad (\theta^{*J})_a(\theta_I)^a = \delta_I{}^J\,.$$

Since we have $g_{ab}(e_i)^a(\theta_I)^b = 0$, if we define

$$g_{ab}(e_i)^a(e_j)^b = q_{ij}\,, \quad g_{ab}(\theta_I)^a(\theta_J)^b = h_{IJ}\,, \tag{5.203}$$

then g_{ab} can be expressed as

$$g_{ab} = q_{ij}(e^{*i})_a(e^{*j})_b + h_{IJ}(\theta^{*I})_a(\theta^{*J})_b \,,$$

or

$$g_{ab} = h_{ab} + q_{ab} \,, \tag{5.204}$$

where

$$q_{ab} = q_{ij}(e^{*i})_a(e^{*j})_b \,, \qquad h_{ab} = h_{IJ}(\theta^{*I})_a(\theta^{*J})_b \,.$$

It is easy to find that the inverse metric of g^{ab} can also be decomposed as

$$g^{ab} = h^{ab} + q^{ab} \,, \tag{5.205}$$

where

$$q^{ab} = q^{ij}(e_i)^a(e_j)^b \,, \qquad h^{ab} = h^{IJ}(\theta_I)^a(\theta_J)^b \,.$$

Here, q^{ij} and h^{IJ} are the inverses of q_{ij} and h_{IJ} respectively, or

$$q^{ij}q_{jk} = \delta^i{}_k \,, \qquad h^{IJ}h_{JK} = \delta^I{}_K \,.$$

From the above relations, it is easy to show

$$h_{ac}q^{cb} = q_{ac}h^{cb} = 0 \,. \tag{5.206}$$

From the expression of h_{ab} and q_{ab}, we have

$$h_a{}^b = h_{ac}g^{cb} = h_{ac}h^{cb} = \delta^I{}_J(\theta^{*J})_a(\theta_I)^b \,,$$

$$q_a{}^b = q_{ac}g^{cb} = q_{ac}q^{cb} = \delta^i{}_j(e^{*j})_a(e_i)^b \,,$$

and

$$\delta_a{}^b = h_a{}^b + q_a{}^b \,. \tag{5.207}$$

It is easy to find

$$h_a{}^c h_c{}^b = h_a{}^b \,, \qquad q_a{}^c q_c{}^b = q_a{}^b \,, \qquad h_a{}^c q_c{}^b = q_a{}^c h_c{}^b = 0 \,. \tag{5.208}$$

So $h_a{}^c$ and $q_a{}^c$ have the properties of projection operators. They are $(1,1)-$ tensors defined on $S \subset M$. At any point $p \in S$, $\delta_a{}^b$ is the identity mapping from T_pM to itself. $q_a{}^b$ is the projection from T_pM to T_pS, and $h_a{}^b$ is the projection from T_pM to $T_p^\perp S$. For examples, we have

$$v^a = v^b\delta_b{}^a = v^b h_b{}^a + v^b q_b{}^a = v_\parallel^a + v_\perp^a \,,$$

where

$$v_\parallel^a = v^b q_b{}^a = v^i(e_i)^a \,, \qquad v^i = (e^{*i})_a v^a \,,$$

$$v_\perp^a = v^b h_b{}^a = v^I(\theta_I)^a \,, \qquad v^I = (\theta^{*I})_a v^a \,.$$

The symmetric tensor q_{ab} is the induced metric on the surface S. This induced metric q_{ab} is Riemannian because S is spacelike (now we set $m = n - 2$), while the transverse part, i.e., h_{ab}, has a Lorentzian signature. So it's natural to introduce two future directed null vector fields ℓ^a and n^a, and h_{ab} can be expressed as

$$h_{ab} = -\ell_a n_b - n_a \ell_b = \varepsilon_{IJ}(\theta^{*I})_a(\theta^{*J})_b \,. \tag{5.209}$$

where I and J take values $\{1,2\}$, and $(\theta^{*1})_a = \ell_a$, $(\theta^{*2}) = n_a$. The symbol ε_{IJ} represents a constant matrix given by $\varepsilon_{11} = \varepsilon_{22} = 0$, $\varepsilon_{12} = \varepsilon_{21} = -1$. According to pointing to center or not, the vectors n^a and ℓ^a are called inward (ingoing) and outward (outgoing) respectively. Obviously, there are some freedoms to choose ℓ^a and n^a. However, if we require $\ell_a n^a = -1$, the only remanning freedom is the rescaling of the null vectors, i.e., $\ell \to \lambda \ell$, $n \to n/\lambda$ with some positive regular function λ. Certainly, one can also introduce an orthogonal frame such that h_{ab} can be expressed as

$$h_{ab} = -u_a u_b + v_a v_b \,, \tag{5.210}$$

where u_a and v_a satisfy: $u^a u_a = -1$, $v^a v_a = 1$ and $u^a v_a = 0$. Now, in eq.(5.209), we can take $(\theta^{*1})_a = u_a$, $(\theta^{*2})_a = v_a$, $\varepsilon_{11} = -\varepsilon_{22} = -1$ and $\varepsilon_{12} = \varepsilon_{21} = 0$. Similar to the null frame, there are also some freedoms (for example, the $SO(1,1)$ rotation of the frame) to take different u_a and v_a.

Assuming the covariant derivative of the spacetime (M, g_{ab}) is given by ∇_a, then, the second fundamental form is defined as a mapping

$$K : \Gamma(TS) \times \Gamma(TS) \to \Gamma(T^\perp S)\,, \quad (u,v) \mapsto K(u,v)\,,$$

with (the abstract index is added)

$$K^c(u,v) = h_b{}^c u^a \nabla_a v^b \,.$$

It is easy to find that the mapping K is bilinear, i.e., we have

$$K(f_1 u, f_2 v) = f_1 f_2 K(u,v)\,,$$

where f_1, f_2 are smooth functions defined on S. The mapping can be trivially extended to $\Gamma(TM) \times \Gamma(TM)$(by requiring $K : (u,v) \mapsto 0$ if $u^a = u_\perp^a$ or $v^a = v_\perp^a$), so that K is a bilinear mapping from $\Gamma(TM) \times \Gamma(TM)$ to $\Gamma(TM)$. Thus K can be understood as a $(1,2)-$type tensor of the spacetime defined on the submanifold S. Hence, it can be denoted by $K_{ab}{}^c$. From the definition, we have

$$K_{ab}{}^c u^a v^b = h_b{}^c u^a \nabla_a v^b \,, \qquad \forall u^a, v^a \in T\Sigma \,.$$

It is obvious

$$\begin{aligned}
K_{ab}{}^c u^a v^b &= h_b{}^c u^a \nabla_a v^b = h_b{}^c u^a \nabla_a (q_d{}^b v^d) = h_b{}^c u^a v^d \nabla_a q_d{}^b \\
&+ h_b{}^c u^a q_d{}^b \nabla_a v^d = q_d{}^b u^a v^d \nabla_a q_b{}^c = (q_d{}^b q_e{}^a \nabla_a q_b{}^c) u^e v^d \\
&= (q_a{}^d q_b{}^e \nabla_d q_e{}^c) u^a v^b \,.
\end{aligned} \tag{5.211}$$

So one can define the second fundamental tensor $K_{ab}{}^c$ as

$$K_{ab}{}^c = q_a{}^d q_b{}^e \nabla_d q_e{}^c \,. \tag{5.212}$$

This is an important extrinsic quantity of the surface S, and it can be defined without introducing any local frame of the spacetime actually. From Frobenius condition for the existence of the submanifold, it is easy to find $K_{[ab]}{}^c = 0$. From the definition, it is also easy to show that

$$h_a{}^d K_{db}{}^c = K_{ab}{}^d q_d{}^c = 0 \, . \tag{5.213}$$

The second fundamental tensor can be decomposed into a traceless part $(C_{ab}{}^c)$ and a trace part (K^c), i.e.,

$$K_{ab}{}^c = \frac{1}{n-2} q_{ab} K^c + C_{ab}{}^c \, , \tag{5.214}$$

where $K^c = g^{ab} K_{ab}{}^c$ is called extrinsic curvature vector or mean curvature vector of the codimension-2 surface. By using the null frame, one gets these extrinsic quantities along the directions of ℓ and n:

$$K_{ab}^{(\ell)} = -K_{ab}{}^c \ell_c = q_a{}^c q_b{}^d \nabla_c \ell_d \, , \qquad K_{ab}^{(n)} = -K_{ab}{}^c n_c = q_a{}^c q_b{}^d \nabla_c n_d \, . \tag{5.215}$$

Similarly, the extrinsic vector is also decomposed as

$$\theta^{(\ell)} = -K^c \ell_c = q^{ab} \nabla_a \ell_b \, , \qquad \theta^{(n)} = -K^c n_c = q^{ab} \nabla_a n_b \, . \tag{5.216}$$

These two quantities are called the expansions along ℓ and n respectively. The traceless part is decomposed as

$$\sigma_{ab}^{(\ell)} = -C_{ab}{}^c \ell_c = \left(q_a{}^c q_b{}^d - \frac{1}{n-2} q_{ab} q^{cd} \right) \nabla_c \ell_d \, ,$$

$$\sigma_{ab}^{(n)} = -C_{ab}{}^c n_c = \left(q_a{}^c q_b{}^d - \frac{1}{n-2} q_{ab} q^{cd} \right) \nabla_c n_d \, . \tag{5.217}$$

These are just the usual shear tensors along the directions of ℓ and n. For the orthogonal frame $\{u, v\}$, we can also get $K_{ab}^{(u)} = -K_{ab}{}^c u_c$, $K_{ab}^{(v)} = -K_{ab}{}^c v_c$ and the corresponding expansions and shear tensors. Actually, for an arbitrary normal vector X, we can define

$$K_{ab}^{(X)} = -K_{ab}{}^c X_c = q_a{}^c q_b{}^d \nabla_c X_d \, , \tag{5.218}$$

and the expansion and the shear tensor are respectively given by

$$\theta^{(X)} = -K^c X_c \, , \qquad \sigma_{ab}^{(X)} = -C_{ab}{}^c X_c \, .$$

To study the intrinsic geometry of S, it's necessary to introduce the corresponding connection or covariant derivative D_a on S. For (r, s)–type tensor $\tau^{a_1 \cdots a_r}{}_{b_1 \cdots b_s}$ on S satisfying

$$\tau^{a_1 \cdots a_r}{}_{b_1 \cdots b_s} = q_{c_1}{}^{a_1} \cdots q_{c_r}{}^{a_r} q_{b_1}{}^{d_1} \cdots q_{b_s}{}^{d_s} \tau^{c_1 \cdots c_r}{}_{d_1 \cdots d_s} \, , \tag{5.219}$$

the covariant derivative is defined as

$$D_a \tau^{a_1 \cdots a_r}{}_{b_1 \cdots b_s} = q_{c_1}{}^{a_1} \cdots q_{c_r}{}^{a_r} q_{b_1}{}^{d_1} \cdots q_{b_s}{}^{d_s} q_a{}^b \nabla_b \tau^{c_1 \cdots c_r}{}_{d_1 \cdots d_s} \, . \tag{5.220}$$

It is easy to check this definition satisfies all of the requirements for a derivative operator [3]. With this definition of the covariant derivative, for vectors ξ and η

which are tangent to S (invariant under the projection operator $q_a{}^b$), it's easy to find

$$\eta^c \nabla_c \xi^b = \eta^c D_c \xi^b + K_{ac}{}^b \eta^a \xi^c .$$
(5.221)

This is just Gauss's formula. By using this covariant derivative D_a, from the usual definition

$$\mathcal{R}_{abcd}\xi^d = (D_a D_b - D_b D_a)\xi_c ,$$
(5.222)

one gets the intrinsic Riemann curvature tensor $\mathcal{R}_{abcd}$. The relation between this intrinsic curvature of S and the curvature of the spacetime is encoded in Gauss equation:

$$\mathcal{R}_{abcd} = K_{ca}{}^e K_{bde} - K_{cb}{}^e K_{ade} + q_a{}^e q_b{}^f q_c{}^g q_d{}^h \, R_{efgh} ,$$
(5.223)

where R_{abcd} is the Riemann curvature of the spacetime. This equation can be easily found from the definitions of D_a and R_{abcd}. In the case of codimension-1, it can be shown

$$\mathcal{R}_{abcd} = (K_{ca}K_{bd} - K_{cb}K_{ad}) + q_a{}^e q_b{}^f q_c{}^g q_d{}^h \, R_{efgh} .$$
(5.224)

In the case of codimension-2, from the Gauss equation, one can have

$$\mathcal{R} + K_{abc}K^{abc} - K^e K_e = R - 2R_{ab}h^{ab} + h^{ac}h^{bd}R_{abcd} ,$$
(5.225)

or

$$\mathcal{R} + K_{abc}K^{abc} - K^e K_e = R - R_{ab}q^{ab} + q^{ac}q^{bd}R_{abcd} .$$
(5.226)

Similar to the covariant D_a for the intrinsic geometry of S, for $(r+p, s+q)-$type tensors $T^{a_1 \cdots a_r a_{r+1} \cdots a_{r+p}}{}_{b_1 \cdots b_s b_{s+1} \cdots b_{s+q}}$ defined on S satisfying

$$T^{a_1 \cdots a_r a_{r+1} \cdots a_{r+p}}{}_{b_1 \cdots b_s b_{s+1} \cdots b_{s+q}} = q_{c_1}{}^{a_1} \cdots q_{c_r}{}^{a_r} h_{c_{r+1}}{}^{a_{r+1}} \cdots h_{c_{r+p}}{}^{a_{r+p}}$$
$$q_{b_1}{}^{d_1} \cdots q_{b_s}{}^{d_s} h_{b_{s+1}}{}^{d_{s+1}} \cdots h_{b_{s+q}}{}^{d_{s+q}} T^{c_1 \cdots c_r c_{r+1} \cdots c_{r+p}}{}_{d_1 \cdots d_s d_{s+1} \cdots d_{s+q}} ,$$
(5.227)

we can define $\tilde{D}_a$ as

$$\tilde{D}_a T^{a_1 \cdots a_r a_{r+1} \cdots a_{r+p}}{}_{b_1 \cdots b_s b_{s+1} \cdots b_{s+q}} = q_{c_1}{}^{a_1} \cdots q_{c_r}{}^{a_r} h_{c_{r+1}}{}^{a_{r+1}} \cdots h_{c_{r+p}}{}^{a_{r+p}}$$
$$q_{b_1}{}^{d_1} \cdots q_{b_s}{}^{d_s} h_{b_{s+1}}{}^{d_{s+1}} \cdots h_{b_{s+q}}{}^{d_{s+q}} q_a{}^b \nabla_b T^{c_1 \cdots c_r c_{r+1} \cdots c_{r+p}}{}_{d_1 \cdots d_s d_{s+1} \cdots d_{s+q}} .$$
(5.228)

It is easy to find $\tilde{D}_a f = q_a{}^b \nabla_b f$, and $\tilde{D}_a$ is linear and satisfies Leibnitz rule. Obviously D_a is just the special case of $\tilde{D}_a$. For tensors with all indices are "normal" to S, sometimes $\tilde{D}_a$ is denoted by $D_a^\perp$ (i.e., the normal connection).

Obviously, for tangent tensor (which is invariant under $q_a{}^b$), this covariant derivative reduces to the derivative D_a. For an arbitrary normal vector X and a tangent vector ξ, by using above definition, it's easy to find

$$\xi^c \nabla_c X^b = -K_a{}^b{}_c \xi^a X^c + \xi^c \tilde{D}_c X^b .$$
(5.229)

This is just Weingarten's formula. So, for normal vectors, $\tilde{D}_a$ is just the usual normal covariant derivative $D_a^\perp$. Based on this covariant derivative, by calculating

$$\Omega_{abcd}X^d = (\tilde{D}_a\tilde{D}_b - \tilde{D}_b\tilde{D}_a)X_c \tag{5.230}$$

for an arbitrary normal vector X, we get the corresponding curvature tensor Ω_{abcd}, which has form

$$\Omega_{abcd} = q_a{}^e q_b{}^f h_c{}^g h_d{}^h R_{efgh} + K_{aed}K_b{}^e{}_c - K_{bed}K_a{}^e{}_c. \tag{5.231}$$

This is Ricci equation. Obviously, this curvature has property of the Weyl tensor. In fact, after some rearrangement, it can be expressed as

$$\Omega_{abcd} = q_a{}^e q_b{}^f h_c{}^g h_d{}^h C_{efgh} + C_{aed}C_b{}^e{}_c - C_{bed}C_a{}^e{}_c, \tag{5.232}$$

where C_{efgh} is the Weyl tensor of the spacetime, while C_{abc} is the traceless part of the second fundamental tensor.

Further, in our codimension-2 case, from eq.(5.228), for an arbitrary normal vector $X_a = \alpha\ell_a + \beta n_a$, it's easy to find

$$\tilde{D}_a X_b = (D_a\alpha + \omega_a\alpha)\ell_b + (D_a\beta - \omega_a\beta)n_b, \tag{5.233}$$

where ω_a is defined as

$$\omega_a = -q_a{}^e n_d \nabla_e \ell^d. \tag{5.234}$$

This is just the normal covariant derivative given in some references (for example see [81]). Sometimes, ω_a is called the $SO(1,1)$ connection of the $SO(1,1)$ normal bundle (see, for example, [80]). Of course, this definition of the connection on the normal bundle depends on the null frame (so it's gauge dependant). Similarly, if we consider the orthogonal frame $\{u, v\}$, for $X_a = \alpha u_a + \beta v_a$, it's easy to find

$$\tilde{D}_a X_b = (D_a\alpha + \omega_a\beta)u_b + (D_a\beta + \omega_a\alpha)v_b, \tag{5.235}$$

and now ω_a is defined as

$$\omega_a = -q_a{}^c u^b \nabla_c v_b. \tag{5.236}$$

It should be noted here: we have used the same notation ω_a as in the case of the null frame, but their values are usually different from each other.

Generally, the connection is defined to be

$$\omega_{abc} = \varepsilon_{IJ}(\theta^{*I})_c \tilde{D}_a(\theta^{*J})_b = \omega_a \epsilon_{bc}, \tag{5.237}$$

where $\epsilon_{ab} = n_a\ell_b - \ell_a n_b$ for the null frame $\{\ell, n\}$, and ω_a is given in eq.(5.234). While for the orthogonal frame $\{u, v\}$, $\epsilon_{ab} = u_a v_b - v_a u_b$ and ω_a can be found in eq. (5.236). It's easy to see this ω_{abc} satisfies standard relations

$$\omega_{abc}\ell^c = \tilde{D}_a\ell_b, \qquad \omega_{abc}n^c = \tilde{D}_a n_b, \tag{5.238}$$

for the null frame, and

$$\omega_{abc}u^c = \tilde{D}_a u_b, \qquad \omega_{abc}v^c = \tilde{D}_a v_b, \tag{5.239}$$

for the orthogonal frame. For the null frame $\{\ell, n\}$ or the orthogonal frame $\{u, v\}$, from eq.(5.230) and above two relations, it's easy to find that the curvature tensor Ω_{abcd} now can be put into

$$\Omega_{abcd} = \tilde{D}_a \omega_{bcd} - \tilde{D}_b \omega_{acd} + \omega_{ac}{}^e \omega_{bed} - \omega_{bc}{}^e \omega_{aed} \,. \tag{5.240}$$

Considering the definition of $\tilde{D}_a$, this curvature is just the one proposed by Carter [75, 76, 77]:

$$\Omega_{abcd} = \left(h_c{}^e h_d{}^f q_a{}^g q_b{}^h \nabla_g \omega_{hef} + \omega_{ac}{}^e \omega_{bed} \right) - (a \leftrightarrow b) \,. \tag{5.241}$$

Actually, the expressions (5.240) and (5.241) are valid in the case with an arbitrary codimension. In the special case of codimension-2 in this paper, after a short calculation, we find

$$\Omega_{abcd} = (D_a \omega_b - D_b \omega_a)\epsilon_{cd} \,. \tag{5.242}$$

Here, $\Omega_{ab} = D_a \omega_b - D_b \omega_a$ is the curvature associated with the $SO(1,1)$ connection (5.234).

Another important formula is Codazzi equation. This equation can be obtained by applying the covariant derivative $\tilde{D}_a$ on the second fundamental tensor. From equation

$$\tilde{D}_d(K_{abc} X^c) = \tilde{D}_d(K_{abc}) X^c + K_{abc} \tilde{D}_d X^c \,, \tag{5.243}$$

and the relation $K_{ab}{}^c X_c = -q_a{}^e q_b{}^f \nabla_e X_f$, one can show:

$$\tilde{D}_a K_{bcd} - \tilde{D}_b K_{acd} = -q_a{}^e q_b{}^f q_c{}^h h_d{}^g R_{efhg} \,. \tag{5.244}$$

This is just the Codazzi equation. For an arbitrary normal vector Y, it gives

$$\left(\frac{n-3}{n-2} \right) D_a \theta^{(Y)} - D_b \sigma^{(Y)b}_a + K_d \tilde{D}_a Y^d - K_a{}^b{}_d \tilde{D}_b Y^d = q_a{}^e q^{bc} Y^d R_{ebcd} \,. \tag{5.245}$$

In the case of codimension-1, this equation is just the so called momentum constraint equation in Hamiltonian formalism in general relativity if we select $Y^a = u^a$, where u^a is the unit normal vector of some spacelike hypersurface. In the case of codimension-2, by using eq.(5.238), immediately, we get

$$\left(\frac{n-3}{n-2} \right) (D_a - \omega_a) \theta^{(\ell)} - (D_b - \omega_b) \sigma^{(\ell)b}_a = q_a{}^e q^{bc} \ell^d R_{ebcd} \tag{5.246}$$

and

$$\left(\frac{n-3}{n-2} \right) (D_a + \omega_a) \theta^{(n)} - (D_b + \omega_b) \sigma^{(n)b}_a = q_a{}^e q^{bc} n^d R_{ebcd} \,. \tag{5.247}$$

The right hand sides of above equations can also be transformed into the form composed by the Weyl tensor C_{abcd} and Einstein tensor G_{ab}:

$$q_a{}^e q^{bc} Y^d R_{ebcd} = q_a{}^e q^{bc} Y^d C_{ebcd} - \left(\frac{n-3}{n-2} \right) q_a{}^e Y^b G_{eb} \,. \tag{5.248}$$

We can get similar equations

$$\left(\frac{n-3}{n-2}\right)\left(D_a\theta^{(u)} - \omega_a\theta^{(v)}\right) - \left(D_b\sigma^{(u)b}_{\ \ a} - \omega_b\sigma^{(v)b}_{\ \ a}\right) = q_a^{\ e}q^{bc}u^d R_{ebcd} , \qquad (5.249)$$

$$\left(\frac{n-3}{n-2}\right)\left(D_a\theta^{(v)} - \omega_a\theta^{(u)}\right) - \left(D_b\sigma^{(v)b}_{\ \ a} - \omega_b\sigma^{(u)b}_{\ \ a}\right) = q_a^{\ e}q^{bc}v^d R_{ebcd} \qquad (5.250)$$

by using eq.(5.239) if the orthogonal frame is considered. Eqs.(5.223), (5.231) and (5.244) are important relations in the submanifold theory. They are valid in the case with an arbitrary codimension. In the four dimensional case, some of the formulas for the codimension-2 surfaces we have listed here can also be found in [81].

5.4.2 Misner–Sharp energy

Generally, the metric of an n-dimensional spherically symmetric spacetime can be written as

$$g = h_{\mu\nu}(y)dy^\mu dy^\nu + r(y)^2\gamma_{ij}(z)dz^i dz^j , \qquad (5.251)$$

where $\gamma_{ij}dz^i dz^j$ is the standard metric of an $(n-2)$-dimension sphere of radius one. The function r is a geometric quantity which can be defined from the area of the $(n-2)$-dimension orbit sphere of $SO(n-1)$.

For the Einstein gravity in n dimensions, the so-called Misner–Sharp energy is defined as

$$E_{\text{MS}} = \frac{(n-2)\Omega_{n-2}}{16\pi}r^{n-3}\left(1 - \nabla_a r\nabla^a r\right) , \qquad (5.252)$$

where Ω_{n-2} is the area of the sphere with unit radius. In four dimensions, this energy is exactly the form given by [82]. Assuming the codimension-2 surface is just the $(n-2)$-sphere with radius r, then we have

$$h_{ab} = h_{\mu\nu}(dy^\mu)_a(dy^\nu)_b , \qquad q_{ab} = r^2\gamma_{ij}(dz^i)_a(dz^j)_b . \qquad (5.253)$$

With these identifications, we have

$$K_{abc} = -\frac{1}{r}q_{ab}\nabla_c r = \frac{1}{n-2}q_{ab}K_c , \qquad K_c = -\frac{(n-2)}{r}\nabla_c r . \qquad (5.254)$$

Thus, the Misner–Sharp energy (5.252) can be expressed as

$$E_{\text{MS}} = \frac{\Omega_{n-2}}{16\pi(n-3)}r^{n-1}\left[\mathcal{R} - \left(\frac{n-3}{n-2}\right)K^c K_c\right] , \qquad (5.255)$$

where

$$\mathcal{R} = \frac{(n-2)(n-3)}{r^2}$$

is the scalar curvature of the $(n-2)$-sphere. By using the area, $\int \epsilon_q$, of the closed codimension-2 surface, we can transform it into a more general form:

$$E_{\text{MS}} = \frac{\left(\int \epsilon_q\right)^{\frac{n-3}{n-2}}}{16\pi\left(\Omega_{n-2}\right)^{\frac{1}{n-2}}(n-3)}\left\{\frac{\int \epsilon_q\mathcal{R}}{\left(\int \epsilon_q\right)^{\frac{n-4}{n-2}}} - \left(\frac{n-3}{n-2}\right)\frac{\int \epsilon_q K_c K^c}{\left(\int \epsilon_q\right)^{\frac{n-4}{n-2}}}\right\} . \qquad (5.256)$$

The above discussion is independent of the null frame. If we introduce the null frame $\{\ell, n\}$, then the mean curvature vector can be expressed as

$$K_a = \theta^{(\ell)} n_a + \theta^{(n)} \ell_a = -\frac{(n-2)}{r} \nabla_a r \,, \tag{5.257}$$

and the Misner–Sharp energy has a form

$$E_{\text{MS}} = \frac{(n-2)\Omega_{n-2}}{16\pi} r^{n-3} \left[1 + \frac{2\, r^2}{(n-2)^2} \theta^{(\ell)} \theta^{(n)} \right] . \tag{5.258}$$

In four dimensions, this expression is the special case of Hawking energy (see the following subsection). The relation between the Misner–Sharp energy and ADM energy has been studied by Hayward [83]. It was shown: in general relativity, for asymptotic flat spacetime (both at spacelike infinity and null infinity) one has

$$E_{\text{BS}} = \lim_{\to \mathcal{I}} E_{\text{MS}} \,, \tag{5.259}$$

and

$$E_{\text{ADM}} = \lim_{\to i^0} E_{\text{MS}} \,, \tag{5.260}$$

where E_{BS} is Bondi–Sachs energy and E_{ADM} is ADM energy.

For general Lovelock gravity theory (5.126), one can also obtain a generalized Misner–Sharp energy [84], which has a form

$$E_{\text{MS}} = \frac{\Omega_{k,(n-2)}}{16\pi} \sum_{i=0}^{p} c_i \frac{(n-2)!}{(n-1-2i)!} r^{n-1-2i} \left(k - \nabla_a r \nabla^a r \right)^i \,, \tag{5.261}$$

where k is the sectional curvature of $(n-2)$−surface S with maximal symmetry, and $\Omega_{k,(n-2)}$ is the area of S with an unit radius.

The Misner–Sharp energy can be constructed from the Noether currents [84] associated with the Kodama vector [85]

$$k^a = -\epsilon^{ab} \nabla_b r \,,$$

where

$$\epsilon_{ab} = \epsilon_{\mu\nu} (dy^\mu)_a (dy^\nu)_b$$

is the Levi-Civita tensor for the Lorentian part with metric h_{ab}, and $\epsilon_{\mu\nu} = \sqrt{-h}\varepsilon_{\mu\nu}$. Here, $\varepsilon_{\mu\nu}$ is the alternating symbol in the two dimensional space.

One can also construct the Misner–Sharp energy in other gravity theories, for example, $f(R)$ gravity [86] and two dimensional dilaton gravity theory [87]. This energy plays an important role in studying the mechanics of some quasi-local horizons, see, for example, [88, 89, 90, 91, 92].

5.4.3 *Hawking mass*

To study the gravitational radiation in an expanding universe, in 1968, Hawking introduced an mass which is interpreted as the total mass of the source and disturbance in the dust-filled Friedmann universe [93]. This mass monotonically decreases as gravitational radiation is emitted.

In section (5.3.3), the study of the relation between the outgoing gravitational radiation and the conserved quantities at the null infinity of an asymptotically flat spacetime is an ideal and reasonable approximation when the regions involved are small compared to the Hubble radius of the universe. However, when the gravitational wave propagates a large distance, this approximation is not appropriate, and one has to study the gravitational radiation on the background of an expanding universe. This is an important motivation in [93]. Of course, in this study, the quasi-local definition of energy or mass is necessary. Here, we give a brief review on this mass.

5.4.3.1 *The Hawking mass*

In general relativity, for a closed two surface S, the Hawking energy is defined as

$$E_H(S) = \sqrt{\frac{A(S)}{16\pi}} \left[\frac{1}{2}\chi(S) - \frac{1}{16\pi} \int_S (K_a K^a) \epsilon_S \right],$$ (5.262)

where $A(S)$ is the area of the surface S, $\chi(S)$ is the Euler number of S, and K^a is the mean curvature of S. By choosing the null frame $\{\ell, n\}$, one gets

$$E_H(S) = \sqrt{\frac{A(S)}{16\pi}} \left[\frac{1}{2}\chi(S) + \frac{1}{8\pi} \int_S \theta^{(\ell)} \theta^{(n)} \epsilon_S \right].$$ (5.263)

Note that in the original work by Hawking [93], Newman-Penrose formalism has been used, and the expression given by Hawking is a little bit different from the one in (5.263).

In the case with spherical symmetry, the Hawking energy reduces to the Misner–Sharp energy in four dimensions. It is easy to see that the Hawking energy is the just the special case of (5.256) with $n = 4$. However, the energy (5.256) is not the most general generalization of the Hawking energy to higher dimensions. Probably, it might be valid only when the surface S is an Einstein manifold.

When S is a sphere, the Hawking mass gives the Bondi–Sachs energy if S approaches to the null infinity, and tends to ADM energy when S approaches to the spacelike infinity [93]. Namely we have

$$E_{\mathrm{BS}} = \lim_{\to \mathcal{I}} E_{\mathrm{H}},$$ (5.264)

and

$$E_{\mathrm{ADM}} = \lim_{\to i^0} E_{\mathrm{H}} \,.$$ (5.265)

The Hawking mass is widely used in studying the mechanics of general dynamical black hole. Relevant discussions on the mechanics of dynamical black hole can be found in $[81, 95, 96, 97, 98, 99, 100, 101, 102, 103]$, or in the review $[104]$.

Here let us mention that there are several other energy definitions which are similar to the Hawking energy, for example, the Geroch–Hawking energy and Hayward energy. The Geroch–Hawking energy depends on some hypersurface in the spacetime. For more details for these energy definitions and associated properties, see $[2]$ and references therein.

5.4.3.2 *Hayward mass*

The Hawking mass (and Geroch–Hawking mass) is not vanishing for Minkowski spacetime. This inspires Hayward to introduce a new mass. The definition has a close relation to the geometry of the codimension-2 surface.

Note that the contracted Gauss equations (5.225) can be rewritten as

$$\mathcal{R} + C_{abc}C^{abc} - \frac{n-3}{n-2}K_c K^c = R - q^{ab}R_{ab} + q^{ac}q^{bd}R_{abcd}\,,$$ (5.266)

or

$$\mathcal{R} + C_{abc}C^{abc} - \frac{n-3}{n-2}K_c K^c = 2(n-3)q^{ab}S_{ab} + q^{ac}q^{bd}C_{abcd}\,,$$ (5.267)

where S_{ab} is Schouten tensor defined as

$$S_{ab} = \frac{1}{n-2}\left[R_{ab} - \frac{1}{2(n-1)}Rg_{ab}\right]\,.$$ (5.268)

In four dimensions, the integral of the left hand side of eq.(5.266) or (5.267) on the codimension-2 spacelike surface S is nothing but just the Hayward energy (up to a coefficient) $[94]$, i.e.,

$$E_{\mathrm{Hayward}}[S] = \sqrt{\frac{A(S)}{16\pi}}\frac{1}{8\pi}\int_S \left[\mathcal{R} + C_{abc}C^{abc} - \frac{1}{2}K_c K^c\right]\epsilon_S\,.$$ (5.269)

This expression does not depend on any frame of the normal bundle of S. If we choose the null frame $\{\ell, n\}$, and consider the definition of Euler number for the two dimensional closed surface, we have

$$E_{\mathrm{Hayward}}[S] = \sqrt{\frac{A(S)}{16\pi}}\left\{\frac{1}{2}\chi(S) + \frac{1}{8\pi}\int_S \left[\theta^{(\ell)}\theta^{(n)} - 2\sigma^{(\ell)}_{ab}\sigma^{(n)ab}\right]\epsilon_S\right\}\,.$$ (5.270)

From the contracted Gauss equation (5.266) or (5.267), it is easy to find that Hayward energy is vanishing for Minkowski spacetime. However, the Hawking energy

(5.263) does not share this property because the shear tensor C_{abc} usually is non-vanishing for an arbitrary S.

At spacelike infinity, assume the matter fields satisfy appropriate fall off condition, the Hayward mass has a limit

$$\lim_{S \to i^0} E_{\text{Hayward}}[S] = \lim_{S \to i^0} \sqrt{\frac{A(S)}{16\pi}} \frac{1}{8\pi} \int_S (q^{ac} q^{bd} R_{abcd}) \epsilon_S \,. \tag{5.271}$$

With eq.(5.266), it can be seen that this is very similar to the one in eq.(5.170). Actually, the Hayward energy really gives ADM energy at spacelike infinity [2, 94]. Note that any contraction of indices of Weyl tensor gives a vanishing value and considering eq.(5.267), the above limit can be put into a form as

$$\lim_{S \to i^0} E_{\text{Hayward}}[S] = \lim_{S \to i^0} \sqrt{\frac{A(S)}{16\pi}} \frac{1}{8\pi} \int_S (h^{ac} h^{bd} C_{abcd}) \epsilon_S \,. \tag{5.272}$$

This expression has a form of the ADM energy defined by Ashtekar *et. al.* [see eq.(5.184)].

However, if we take the limit to the null infinity, the Hayward energy cannot reduce to the Bondi–Sachs energy. The reason is quite simple: since the term with shear tensor, i.e., $C_{abc}C^{abc}$, represents the energy density of gravitational radiation in some sense, and this is already included in the Hayward energy. The Hawking energy obviously does not have the contribution from this part. The Hawking energy can reduce to the Bondi–Sachs energy at null infinity. Thus, in general, it is impossible for the Hayward energy. In fact, the Hayward energy will reduce to the so-called Newman-Unti energy at null infinity (see [2] and references therein).

5.4.3.3 *The case with a cosmological constant*

The causal structure of a spacetime drastically changes when the cosmological constant is presented. The conformal infinity of the spacetime is very different from the one without the cosmological constant. So the presence of the cosmological constant forces us to find new global defined conserved quantities[5]. The quasi-locally defined quantity has to match the global one when the codimension-2 surface approaches the infinity of the spacetime, so we have to modify the Hawking mass or Hayward mass in the previous to adapt the change of the global quantity.

When cosmological constant Λ is present, one has a generalized Hawking mass as [108]

$$E_H(S) = \sqrt{\frac{A(S)}{16\pi}} \left\{ \frac{1}{2}\chi(S) - \frac{1}{16\pi} \int_S \left[K_a K^a + \frac{4}{3}\Lambda \right] \epsilon_S \right\} \,. \tag{5.273}$$

---

[5]One can find the definition of an asymptotically (A)dS spacetime and associated conserved quantities in the appendix of this paper.

Similarly, from eq.(5.267), we have

$$\mathcal{R} - \frac{n-3}{n-2} K_c K^c - 2\frac{n-3}{n-1}\Lambda =$$

$$2(n-3)q^{ab}\tilde{S}_{ab} - C_{abc}C^{abc} + q^{ac}q^{bd}C_{abcd}, \tag{5.274}$$

where we have defined

$$\tilde{S}_{ab} = \frac{8\pi}{n-2}\left[T_{ab} - \frac{1}{n-1}Tg_{ab}\right], \tag{5.275}$$

where $T = g^{ab}T_{ab}$. Actually, from the Einstein equations, we have

$$S_{ab} = \tilde{S}_{ab} + \frac{\Lambda}{(n-1)(n-2)}g_{ab}. \tag{5.276}$$

In four dimensions, the integration of the left hand side of eq.(5.274) just gives the Hawking mass (5.273) up to a constant factor

$$\sqrt{\frac{A(S)}{16\pi}} \cdot \frac{1}{8\pi}.$$

So the energy (5.273) is not vanishing even for pure AdS(dS) spacetime. To cure this problem, following the Hayward energy, we propose a definition of mass for the spacetime with a cosmological constant as

$$E_{\text{New}}(S) = \sqrt{\frac{A(S)}{16\pi}}\left\{\frac{1}{2}\chi(S) - \frac{1}{16\pi}\int_S \left[K_a K^a - 2C_{abc}C^{abc} + \frac{4}{3}\Lambda\right]\epsilon_S\right\}. \tag{5.277}$$

Here some remarks are in order.

(i). Obviously, from the contracted Gauss equation (5.274), one can see that this mass is identically vanishing for the pure AdS (dS) spacetime. Actually, from (5.274), we have

$$E_{\text{New}}(S) = \sqrt{\frac{A(S)}{16\pi}}\int_S \left[q^{ab}T_{ab} - \frac{2}{3}T + \frac{1}{8\pi}q^{ac}q^{bd}C_{abcd}\right]\epsilon_S. \tag{5.278}$$

(ii). From eq.(5.278), and assume that energy-momentum tensor T_{ab} decays quickly enough near the infinities, we can arrive at

$$\lim_{S\to\mathcal{I}} E_{\text{New}}(S) = \lim_{S\to\mathcal{I}} \sqrt{\frac{A(S)}{16\pi}}\frac{1}{8\pi}\int_S (h^{ac}h^{bd}C_{abcd})\epsilon_S. \tag{5.279}$$

Hence, the limit (to infinity) of the mass formula is similar to the global energy defined by Ashtekar, Magnon, and Das at the infinity $\mathcal{I}$ of an asymptotically AdS spacetime [105] (see also [106] for higher dimensional case). See appendix A for detailed discussion.

(iii). Of course, our definition is quasi-local and is independent of the detailed asymptotic behavior of the spacetime in some sense. So it is also valid for a positive cosmological constant. For a weaker fall-off condition of the matter field, this new mass formula approaches to the conserved quantities defined by Ashtekar, Bonga, and Kesavan in [107]. See appendix C for details.

(iv). It should be stressed here that this definition has included some possible gravitational radiation represented by the shear tensor C_{abc} as in the case without the cosmological constant. For AdS spacetimes, the existence of the timelike boundary is equivalent to a confining box in some sense, so it is reasonable to include the contribution from the radiation. This is different from the asymptotically flat case.

(v). In the case with spherical symmetry (or the symmetry of a two dimensional maximally symmetric space), if the codimension-2 surface is just the surface with constant $r = \sqrt{A(S)/4\pi}$, then the shear tensor C_{abc} automatically vanishes, and we get the Misner–Sharp energy with a cosmological constant. See eq.(5.261), and consider $n = 4$.

(vi). It is interesting to study the properties of this definition. For example, the positivity, monotonicity (as in [108] by Bray *et.al.*). And, it is also interesting to find its usage in the study of the mechanics of a dynamical black hole with a cosmological constant.

(vii). To study the gravitational radiation in an expanding universe, Hawking proposed the Hawking mass. By the same logic, based on the new mass formula, it is interesting to study the gravitational radiation on some Friedmann–Robertson–Walker universe with the cosmological constant. According to the observation in astronomy, the cosmological constant plays an important role in modern cosmology. For the source which is far from us, to study the gravitational radiation, principally, it is better to use the Friedmann–Robertson–Walker universe as the background spacetime. Probably, the new mass formula can provide some clue to study this problem.

At the end of the subsection, we should point out that these definitions of energy, however, are not so easy to understand from the Hamiltonian formulation of general relativity, and some of them even have no direct relation to the Noether currents. The natural question one may ask is whether these quasi-local energies can be derived from the Hamiltonian formulation of the Einstein gravity theory. The quasi-local energy definition by Brown and York provides some clues to understand this question.

5.4.4 *Brown–York quasi-local energy-momentum tensor*

Roughly speaking, in the Brown–York formalism, the energy-momentum tensor of the gravitational field in the Einstein gravity is realized by a symmetric rank-2 tensor on the timelike boundary of a patch of the spacetime [58]. The symmetric rank-2 tensor is just the so-called Brown–York energy-momentum tensor. Brown–York formalism has a close relation to the usual Hamiltonian formulation of general relativity. So the foliation of the spacetime is required. Actually, to get the Brown–York energy-momentum tensor, the Hamilton–Jacobi analysis is enough.

5.4.4.1 *Quasi-local region*

For a quasi-local region of the spacetime, to get well defined variation problem, one has to consider the boundary conditions of the fields involved and appropriate boundary terms. Consider a quasi-local region $\mathcal{M}$ with a timelike boundary $\mathcal{B}$ and two spacelike boundaries $\Sigma_{\pm}$ in a spacetime (M, g_{ab}). The "$\pm$" in $\Sigma_{\pm}$ means future and past respectively. So the boundary of $\mathcal{M}$ is given by

$$\partial \mathcal{M} = \mathcal{B} \cup \Sigma_+ \cup \Sigma_- .$$

The intersections of $\mathcal{B}$ and $\Sigma_{\pm}$, i.e., the corners, are denoted by $B_{\pm}$, see Fig.(5.2).

Assume the quasi-local region is foliated by $\{\Sigma_t, t \in [t_-, t_+]\}$. For each slice Σ_t, its (unit) normal vector, induced metric, and extrinsic curvature are denoted by u^a, h_{ab}, and K_{ab} respectively. Similarly, the unit normal vector, induced metric, and extrinsic curvature on the timelike boundary $\mathcal{B}$ are denoted by $\bar{n}^a$, γ_{ab}, and Θ_{ab} respectively. The intersection of Σ_t and $\mathcal{B}$ is denoted by B. The unit normal vectors of B in Σ_t and $\mathcal{B}$ are denoted by n^a and $\bar{u}^a$ respectively. Obviously, on B, we have $u_a n^a = 0 = \bar{u}_a \bar{n}^a$. However, in general $u_a \bar{u}^a \neq 0$. Furthermore, since u^a and $\bar{u}^a$ are both normal to B and can be viewed as the four velocities of two observers meeting at the point on B, the 3-velocity of u^a measured by $\bar{u}^a$ is simply given by (u^a and $\bar{u}^a$ are both future pointing)

$$v = \frac{u_a \bar{n}^a}{u_a \bar{u}^a} . \tag{5.280}$$

So we have $u_a \bar{n}^a = -\gamma v$,

$$u^a = \gamma \bar{u}^a - \gamma v \bar{n}^a ,$$
$$n^a = -\gamma v \bar{u}^a + \gamma \bar{n}^a , \tag{5.281}$$

and inverse transformation

$$\bar{u}^a = \gamma u^a + \gamma v n^a ,$$
$$\bar{n}^a = \gamma v u^a + \gamma n^a . \tag{5.282}$$

where $\gamma = 1/\sqrt{1 - v^2}$.

5.4.4.2 *Variation problem, boundary terms, and Brown–York energy-momentum tensor*

As mentioned above, to have a well defined variation problem, one has to add some boundary terms to the Einstein–Hilbert action, and consider the following action

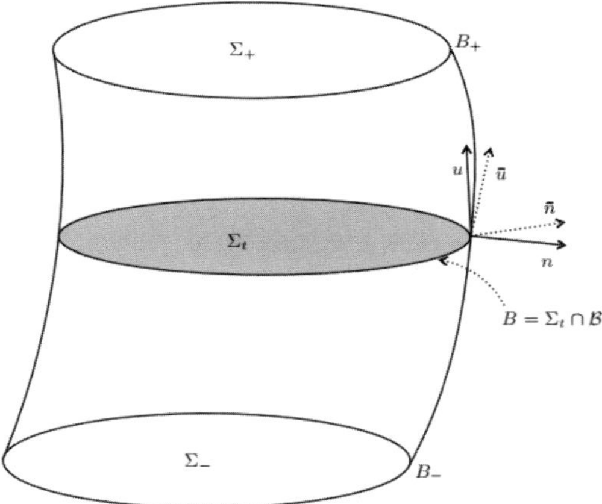

Fig. 5.2   The quasi-local region in the spacetime.

$$I = -\frac{1}{8\pi}\int_{B_+}\theta\epsilon_{B_+} + \frac{1}{8\pi}\int_{B_-}\theta\epsilon_{B_-}$$

$$+ \frac{1}{8\pi}\int_{\Sigma_+}K\epsilon_{\Sigma_+} - \frac{1}{8\pi}\int_{\Sigma_-}K\epsilon_{\Sigma_-}$$

$$- \frac{1}{8\pi}\int_{B}\Theta\epsilon_{B}$$

$$+ \frac{1}{16\pi}\int_{\mathcal{M}}R\epsilon\,, \tag{5.283}$$

where $\theta = -u_a\bar{n}^a$, and ϵ, ϵ_B, $\epsilon_{\Sigma_\pm}$, and $\epsilon_{B_\pm}$ are the natural volume elements of spacetime, $\mathcal{B}$, $\Sigma_\pm$, and $B_\pm$ respectively. When the metric g_{ab} is on-shell, one has

$$\delta I = -\frac{1}{16\pi}\int_{B_+}(\theta q^{ab}\delta q_{ab})\epsilon_{B_+} + \frac{1}{16\pi}\int_{B_-}(\theta q^{ab}\delta q_{ab})\epsilon_{B_-}$$

$$+ \frac{1}{2}\int_{\Sigma_+}(\pi^{ab}\delta h_{ab})\epsilon_{\Sigma_+} - \frac{1}{2}\int_{\Sigma_-}(\pi^{ab}\delta h_{ab})\epsilon_{\Sigma_-} - \frac{1}{2}\int_{\mathcal{B}}(\tau^{ab}\delta\gamma_{ab})\epsilon_{\mathcal{B}}\,,$$

$$\tag{5.284}$$

where

$$8\pi\tau_{ab} = \Theta_{ab} - \Theta\gamma_{ab}\,, \tag{5.285}$$

$$8\pi\pi_{ab} = K_{ab} - Kh_{ab}\,. \tag{5.286}$$

The symmetric rank-2 tensor τ_{ab} is divergence free according to the covariant derivative D_a induced on $\mathcal{B}$, i.e., we have $D_a\tau^{ab} = 0$. This tensor is the so-called Brown–York energy-momentum tensor or surface stress energy tensor for the timelike boundary $\mathcal{B}$. Formally it can be written as

$$\epsilon_{\mathcal{B}}\tau^{ab} = -2\frac{\delta I}{\delta\gamma_{ab}}\,. \tag{5.287}$$

Since $\bar{u}^a$ can be viewed as the velocity field for a reference frame on $(\mathcal{B}, \gamma_{ab})$, then γ_{ab} can be decomposed as $\gamma_{ab} = -\bar{u}_a\bar{u}_b + q_{ab}$. The energy density $\bar{\varepsilon}$, momentum density $\bar{j}_a$, and stress tensor $\bar{s}_{ab}$ observed by the observer with the velocity $\bar{u}^a$ have forms

$$8\pi\bar{\varepsilon} = -8\pi\tau_{ab}\bar{u}^a\bar{u}^b = \gamma k + \gamma v l \,,$$
$$8\pi\bar{j}_a = 8\pi q_a{}^c\tau_{bc}\bar{u}^b = q_a{}^c n^b K_{bc} - d_a\theta \,,$$
$$8\pi\bar{s}_{ab} = 8\pi q_a{}^c q_b{}^d\tau_{cd} = \gamma(k_{ab} - kq_{ab})$$
$$+ (\bar{n}_c\bar{a}^c)q_{ab} + \gamma v(l_{ab} - lq_{ab}) \,. \tag{5.288}$$

In the above the following relation has been used

$$\Theta_{ab} = \gamma k_{ab} + \gamma v l_{ab} + (\bar{n}_c\bar{a}^c)\bar{u}_a\bar{u}_b + 2q_{(a}{}^c\bar{u}_{b)}(K_{cd} - \nabla_c\theta) \,, \tag{5.289}$$

where k_{ab} and l_{ab} are the extrinsic curvature of B (embedded in the spacetime) along n^a and u^a respectively, i.e.,

$$k_{ab} = -q_a{}^c q_b{}^d\nabla_c n_b \,, \qquad l_{ab} = -q_a{}^c q_b{}^d\nabla_c u_b \,. \tag{5.290}$$

In addition, in the above equations,

$$k = q^{ab}k_{ab} \,, \qquad l = q^{ab}l_{ab} \,,$$

and $\bar{a}^c$ is the acceleration of $\bar{u}^a$, and d_a is the covariant derivative on (B, q_{ab}).

On $\Sigma_\pm$, we have $h_{ab} = n_a n_b + q_{ab}$, the symmetric tensor π_{ab} can also be decomposed as

$$8\pi\jmath_\vdash = 8\pi\pi_{ab}n^a n^b = -l \,,$$
$$8\pi\jmath_a = 8\pi q_a{}^c\pi_{bc}n^b = q_a{}^c n^b K_{bc} \,,$$
$$8\pi t_{ab} = -8\pi q_a{}^c q_b{}^d\pi_{cd} = -\gamma(l_{ab} - lq_{ab}) + (u_c b^c)q_{ab} \,, \tag{5.291}$$

where $b^c = n^b\nabla_b n^c$ and

$$K_{ab} = l_{ab} + (u_c b^c)n_a n_b + 2q_{(a}{}^c n_{b)}K_{cd}n^d \tag{5.292}$$

has been used.

In the above discussion, the quasi-local region is foliated as $\{\Sigma_t\}$. Certainly, one can foliate this region by $\{\bar{\Sigma}_t\}$ such that $\bar{\Sigma}_t$ is orthogonal to $\mathcal{B}$, i.e., the normal vector of $\bar{\Sigma}_t$ on B is $\bar{u}^a$. With this foliation, one has a new set $(\bar{K}_{ab}, \bar{h}_{ab})$ on $\bar{\Sigma}_t$, and then new $\bar{\pi}_{ab}$. Of course, one also has $\bar{h}_{ab} = \bar{n}_a\bar{n}_b + q_{ab}$. Considering eqs.(5.282), we have

$$8\pi\bar{\jmath}_\vdash = 8\pi\bar{\pi}_{ab}\bar{n}^a\bar{n}^b = -\bar{l} = -\gamma l - \gamma v k \,,$$
$$8\pi\bar{\jmath}_a = 8\pi q_a{}^c\bar{\pi}_{bc}\bar{n}^b = q_a{}^c\bar{n}^b\bar{K}_{bc} = q_a{}^c n^b K_{bc} - d_a\theta \,,$$
$$8\pi\bar{t}_{ab} = -8\pi q_a{}^c q_b{}^d\bar{\pi}_{cd} = -\bar{l}_{ab} + \bar{l}q_{ab} + (\bar{u}_c\bar{b}^c)q_{ab}$$
$$= -\gamma(l_{ab} - lq_{ab}) + \bar{u}_c\bar{b}^c q_{ab} - \gamma v(k_{ab} - kq_{ab}) \,, \tag{5.293}$$

where

$$\bar{l}_{ab} = -q_a{}^c q_b{}^d\nabla_c\bar{u}_d \,, \qquad \bar{l} = q^{ab}\bar{l}_{ab} \,,$$

and $\bar{b}^a = \bar{n}^a \nabla_a \bar{n}^b$.

Σ_+ and $\mathcal{B}$ joint at the corner B_+. On B_+, from the above equations, it is easy to find

$$\bar{\varepsilon} = \gamma\varepsilon - \gamma v \jmath_+ \,, \qquad \bar{\jmath}_+ = \gamma\jmath_+ - \gamma v \varepsilon \,. \tag{5.294}$$

where $8\pi\varepsilon = k$. So one gets

$$\varepsilon u^a + \jmath_+ n^a = \bar{\varepsilon}\bar{u}^a + \bar{\jmath}_+ \bar{n}^a \,. \tag{5.295}$$

This means $\varepsilon u^a + \jmath_+ n^a$ does not depend on the $SO(1,1)$ rotations of the normal bundle of B. Obviously we also have

$$8\pi\bar{\jmath}_a = 8\pi\jmath_a - d_a\theta \,,$$

This is similar to the gauge transformation of the gauge potential on the normal bundle of B. More detailed discussion on the transformations of the quantities defined by the Brown–York tensor can be found in the paper by Brown, Lau, and York [62].

The above calculation shows that the Brown–York energy-momentum tensor depends on the foliation of the quasi-local region. However, one can construct various $S(1,1)$ boost invariant quantities from the quantities calculated in the above.

5.4.4.3 *Background contribution and subtraction*

In the simple case with the orthogonal foliation $\{\Sigma_t\}$ (From now on we only consider this kind of foliation, and omit the bar since the bar frame coincides with the unbar frame by this selection. Actually this reduce to the simple case considered by Brown and York in 1992 [58]), from the Brown–York tensor, on $\mathcal{B}$, one has the energy density, momentum density, and stress tensor. They have simple forms

$$8\pi\varepsilon = k \,,$$
$$8\pi\jmath_a = q_a{}^c n^b K_{bc} \,,$$
$$8\pi s_{ab} = k_{ab} - k q_{ab} + (n_c a^c) q_{ab} \,. \tag{5.296}$$

In principal, with this energy density, we should get the total energy associated with the surface B. However, the situation is not so simple because this energy density is nonvanishing even for the Minkowski spacetime. In general relativity, the strategy to cure this problem is the so-called "background subtraction" which removes the contribution from some reference solution, for example, the Minkowski spacetime. Usually, a term $-I_0$ will be added to the action (5.283) such that the variation of

$$I - I_0 \tag{5.297}$$

provides a renormalized Brown–York energy-momentum tensor which is vanishing for the reference solution. The term I_0 depends on the boundary data only, so it does not change the equations of motion.

By this consideration, one has an energy

$$E_{\text{BY}} = \int_B \varepsilon \epsilon_B = \frac{1}{8\pi} \int_B (k - k_0) \epsilon_B \,, \qquad (5.298)$$

where k_0 represents the contribution from the reference spactime. Since τ_{ab} is divergence free on $(\mathcal{B}, \gamma_{ab})$, so for a Killing vector field ξ on $(\mathcal{B}, \gamma_{ab})$, one can define a conserved quantity as

$$Q_{\text{BY}}[\xi] = \int_B \tau_{ab} \xi^a u^b \epsilon_B \,. \qquad (5.299)$$

By this, one can define quasi-local conserved quantities, such as the angular momentum for a quasi-local region of the spacetime. It should be noted here that in this expression, τ_{ab} has been renormalized, i.e., one has applied background subtraction method to remove the contribution from the reference background.

5.4.4.4 *Further development*

For a timelike boundary $\mathcal{B}$ of a quasi-local region, we can establish the Brown–York energy-momentum tensor τ_{ab} (assume renormalization has been done if necessary). However, it is obvious that the definition of τ_{ab} depends on the boundary $\mathcal{B}$ although the final result is an integral on a codimension-2 surface. This dependence implies the Brown–York energy depends on the observer. The situation is similar to the ADM energy which changes if we do a Lorentz transformation at the spacelike infinity. To get an ADM mass type quantity, some proposals have been made, for example, Liu–Yau mass [109], Kijowski mass [110], Epp mass [111], Wang-Yau mass [112], and for more definitions see [2]. For example, the Liu-Yau mass is defined as

$$E_{\text{LY}} = \frac{1}{8\pi} \int_B \left(\sqrt{k^2 - l^2} - k_0 \right) \epsilon_B \,. \qquad (5.300)$$

Since the mean curvature vector K^a of B (embedded in the spacetime) is simply

$$K^a = kn^a + lu^a \,,$$

so the square root in the above integral is actually the Lorentzian norm of the mean curvature. Obviously, this term does not depend on the frame of the normal bundle, so it is gauge invariant.

In the Einstein gravity theory with a negative cosmological constant, the definition of Brown–York energy-momentum tensor plays an important role. With the development of AdS/CFT correspondence [113, 114, 115, 116], this surface energy-momentum tensor is extensively used in a lot of investigations relevant to the AdS/CFT correspondence and its applications in various fields. According to the AdS/CFT correspondence, or more general holographic properties of gravity [117, 118], AdS gravity in the bulk is equivalent to a field theory on its timelike boundary.

Some techniques in the boundary field theory bring some ideas to solve the problems in gravity theory. This includes a method to renormalize the Brown–York energy-momentum tensor. The idea is to introduce some counter-terms on the boundary [119, 120]. These surface counter-terms are constructed from the intrinsic geometric quantities, for example, Riemann tensor of the boundary manifold, such that the action I is finite when the quasi-local region approaches to the full space-time. Consequently, the Brown–York energy-momentum tensor is also finite when the boundary approaches to the AdS boundary. Usually, the surface counter-terms consist of the higher order terms of the Riemann tensor of the boundary. This corresponds to the ultraviolet divergence of the boundary fields. Hence, the renormalization of the Brown–York energy-momentum tensor in some sense corresponds to the short scale renormalization of the boundary field theory.

5.4.5 *Other quasi-local definitions*

Besides the above quasi-local definitions, there are a lot of other quasi-local definitions of energy and angular momentums in general relativity. For example, the definition by Penrose based on twistor theory, and the way to get the conserved quantities from the integrals of Nester–Witten two form. Almost all of the quasi-local definitions and relevant references can be found in [2].

5.5 Conclusion and discussion

In this paper, we have reviewed some aspects of the conserved quantities in gravity theories. As claimed at the beginning of this paper, this is not a thorough review on this topic due to the space limitation.

The Noether theorem is the basis to discuss conserved quantities. Based on the Lagrangian formalism of the Noether theorem, the conservation laws in field theories have been established in Minkowski spacetime. The Minkowski spacetime have ten independent infinitesimal isometries. These symmetries actually are the relativistic principle of the special relativity. This sense, the conservation laws of the energy, linear momentum, and angular momentum are the natural results of relativistic principle of special relativity. In general relativity, however, the spacetime has no isometries in general, and we have no conservation law for the matter fields without the isometries. This can be found in eq.(5.61). Once we believe the general relativity is a geometry theory of the spacetime, the usual conservation law in the Minkowski spacetime does not exist any more. Some readers might think the diffeomorphsim group of the spacetime is large enough, why this group could not provide the symmetry for the Noether theorem? Of course, we can use the infinitesimal diffeomorphsim to construct a Noether current. However, this Noether current can not be simply used to define the energy and momentum for a general spacetimes (as in the Minkowski spacetime), see [121] for detailed discussion. In fact, for pure Einstein–Hilbert action, one gets the contracted Bianchi identity from the Noether theorem. This actually leads us to the a deep question of general relativity—whether general covariance is the fundamental principle of general relativity, and whether general relativity is really a theory of relativity [122]. The topic is beyond this paper, we will not discuss it here, and only view that gravity is the geometry of the spacetime without any nondynamical background. To introduce the canonical energy-momentum tensor for gravitational field, one has to introduce some background structure to the spacetime, and arrives at the pseudotensor description.

The covariant phase space analysis by Wald and his collaborators provides a framework to study the conservation law or conserved quantities not only in Einstein gravity but also in general diffeomorphsim invariant gravity theories. Based on this analysis, we can construct the pseudotensor and superpotential by introducing background metric, and we have constructed the canonical energy-momentum tensor and canonical spin tensor in a simple way, see sec.(5.2.5). By considering detailed analysis near infinities, one can also get the ADM energy at a spacelike ininity and the conserved quantities dual to the BMS supertranslations at a null infinity. It is interesting to study the conserved quantities in other diffeomorphsim invariant gravity theories by this method. The definition of Brown–York energy-momentum tensor is based on the usual Hamiltonian analysis. To get a covariant phase space version Hamiltonian for a quasi-local region, one has to study the boundary data and to find the existence condition for the Hamiltonian, i.e., eq.(5.147).

The study of the globally defined conserved quantities are important to black hole physics. The hairs of a black hole in general relativity is determined by the conserved quantities. So the classical black hole is characterized by these conserved quantities. These conserved quantities are also the variables in the thermodynamic phase space of the black hole thermodynamics. So it is also important in the study of the semi-classical behavior of quantum gravity. For stationary black hole spacetimes which have well defined asymptotical behavior, these globally conserved quantities are also well defined, and one can establish the laws of thermodynamics [123]. For black holes in our universe, of course, under certain approximation, the black hole system can still be viewed as an isolated system. However, the stationary condition cannot be fulfilled in general, and we do not know their asymptotical behavior in principle, and we do not know how to study their thermodynamics. Probably, to overcome these difficulties, the quasi-local definition of black hole is necessary [104]. In these quasi-local definitions of the black holes, the quasi-local energy and angular momentums are important to study the mechanics or thermodynamics of the system.

There are two byproducts of this review: (i). By the analysis of the detailed structure of the Noether current, we provide a simple way to get the canonical energy-momentum tensor and canonical spin tensor on general curved spacetimes in the subsubsection (5.2.5.2). The canonical energy-momentum tensor and canonical spin tensor satisfy the equations which have been given in [2]. (ii). Following the Hayward energy, we provide a new mass definition for the case with a cosmological constant. This mass definition has some interesting properties. It includes the contribution of the gravitational radiation and approaches to the Ashtekar–Magnon–Das conformal mass at the infinity of an asymptotical AdS spacetime. For the asymptotically Schwarzschild-de Sitter spacetimes, this mass approaches to the definition by Ashtekar, Bonga, and Kesavan.

Acknowledgement

RGC was supported in part by the National Natural Science Foundation of China with grants No. 11690022, No.11375247, No.11435006, No.11447601 and No.11647601, and by the Strategic Priority Research Program of CAS Grant No.XDB23030100 and by the Key Research Program of Frontier Sciences of CAS. LMC was supported in part by the National Natural Science Foundation of China with grants No.11622543 and No.11235010.

Appendix

In this appendix, we give a brief review on the definition of an asymptotical (A)dS spacetime by Ashtekar *et.al.* and the global defined conserved quantities for this spacetime [105, 106, 107]. We show that new mass formula (5.277) tends to the Ashtekar–Magnon–Das conformal energy of the asymptotically AdS spacetime when the codimension-2 closed surface approaches to the infinity of the spacetime.

5.6 Asymptotically AdS spacetimes

Definition 1: An 4-dimension spacetime (M, g_{ab}) is said to be *weakly asymptotically anti de Sitter* if there exists a manifold $\tilde{M}$ with boundary $\mathcal{I}$, equipped with a metric $\tilde{g}_{ab}$ and a diffeomorphism from M onto $\tilde{M} - \mathcal{I}$ (M and $\tilde{M} - \mathcal{I}$ are identified by this mapping), such that:

(i). there exists a function Ω on $\tilde{M}$ such that $\tilde{g}_{ab} = \Omega^2 g_{ab}$ on $\tilde{M}$; and Ω is vanishing on $\mathcal{I}$ and $\tilde{n}_a = \tilde{\nabla}_a \Omega$ is nowhere vanishing on $\mathcal{I}$;

(ii). $\mathcal{I}$ is topologically $\mathcal{R} \times S^2$;

(iii). g_{ab} satisfies $G_{ab} + \Lambda g_{ab} = 8\pi T_{ab}$ with $\Lambda < 0$, where $\Omega^{-1} T_{ab}$ admits a smooth limit to $\mathcal{I}$.

Based on this definition, one can define the conserved quantity at the infinity of the spacetime (This will be summarized soon.). However, this definition can not grantee that the asymptotic symmetry group is the expected anti-de Sitter group. Actually, it is the diffeomorphsim group of $\mathcal{I}$, i.e., $\mathrm{Diff}(\mathcal{I})$, which is infinity dimension and larger than the anti-de Sitter group [105]. To reduce $\mathrm{Diff}(\mathcal{I})$ to (a finite dimension) asymptotic symmetry group, one has to strength the asymptotic condition as follows

Definition 2: A spacetime (M, g_{ab}) is said to be *asymptotically anti de Sitter* if in additional to the definition 1, one has

(iv). The conformal group of $(\mathcal{I}, \gamma_{ab})$ is the anti-de Sitter group (or covering group thereof), where γ_{ab} is the induced metric on $\mathcal{I}$.

It has been shown in [105] that the vanishing of the Bach tensor of $(\mathcal{I}, \gamma_{ab})$ is equivalent to the vanishing of the magnetic part of the leading order of the Weyl tensor

$$\boldsymbol{B}_{ab} \;\hat{=}\; \ell^2 \boldsymbol{K}^*_{acbd} \tilde{n}^c \tilde{n}^d \,.$$

So, when $\boldsymbol{B}_{ab} \;\hat{=}\; 0$, $(\mathcal{I}, \gamma_{ab})$ is conformally flat and admits ten conformal Killing vectors. This additional condition also implies that on $\mathcal{I}$ one has a global coordinate system $\{t, \theta, \phi\}$ such that γ_{ab} is conformally related to

$$\overset{o}{\gamma} \;=\; -\,\mathrm{d}t^2 + \ell^2(\mathrm{d}\theta^2 + \sin^2\theta \mathrm{d}\phi^2)\,, \tag{5.301}$$

where $\ell^{-2} = -\Lambda/3$. Hence, by the condition (iv) in the definition 2, the conformal class of γ_{ab} has been restricted to the one in which all metrics are conformally related to $\overset{o}{\gamma}$.

The above definition has been extended to higher dimensions [106].

Definition 3: A spacetime (M, g_{ab}) is said to be asymptotically anti-de Sitter if there exists a manifold $\tilde{M}$ with boundary $\mathcal{I}$, equipped with a metric $\tilde{g}_{ab}$ and a diffeomorphism from M onto $\tilde{M} - \mathcal{I}$ (M and $\tilde{M} - \mathcal{I}$ are identified by this diffeomorphsim), such that:

(i). there exists a function Ω on $\tilde{M}$ such that $\tilde{g}_{ab} = \Omega^2 g_{ab}$ on $\tilde{M}$;

(ii). $\mathcal{I} = \partial\tilde{M}$ is topologically $S^{n-2} \times \mathcal{R}$, and $\Omega = 0$ on $\mathcal{I}$ and $\tilde{n}_a = \tilde{\nabla}_a\Omega$ is nowhere vanishing on $\mathcal{I}$;

(iii). g_{ab} satisfies $G_{ab} + \Lambda g_{ab} = 8\pi T_{ab}$, with $\Lambda < 0$, where $\Omega^{2-n} T_{ab}$ admits a smooth limit to $\mathcal{I}$;

(iv). The Weyl tensor of $\tilde{g}_{ab}$ is such that $\Omega^{4-n} \tilde{C}_{abcd}$ is smooth on $\tilde{M}$ and vanishes at $\mathcal{I}$.

Obviously, compared to the case of four dimension, the matter field is required to satisfy a faster fall-off condition in (iii). Actually, for usual matter fields, this condition can be fulfilled even in four dimension. The condition (iv) on the Weyl tensor in definition 3 implies the boundary $(\mathcal{I}, \gamma_{ab})$ is conformally flat when $n > 4$. So the conformal group of $(\mathcal{I}, \gamma_{ab})$ is the anti-de Sitter group. In the case of $n = 4$, the condition in (iv) in definition 3 is trivially satisfied. To get the anti-de Sitter group, one has to consider the condition (iv) in the definition 2. This is different from the case in a higher dimension.

Now we give a brief review on the conserved quantities defined at the infinities of the above spactimes. Since we will use $n^a = g^{ab}\nabla_b\Omega$ for a constant Ω surface, to make the discussion transparent, we have introduced $\tilde{n}_a = \tilde{\nabla}_a\Omega = \nabla_a\Omega = n_a$ in the above definitions, and consequently we get

$$\tilde{n}^a = \tilde{g}^{ab}\tilde{\nabla}_b\Omega = \tilde{g}^{ab}\nabla_b\Omega = \Omega^{-2}g^{ab}\nabla_b\Omega = \Omega^{-2}n^a \,.$$

From the Einstein equations and the fall off conditions on the energy-momentum tensor of the matter fields, it is easy to find

$$\tilde{n}_a\tilde{n}^a \mathrel{\hat{=}} \frac{1}{\ell^2} \,, \tag{5.302}$$

where $\hat{=}$ denotes equality restricted to $\mathcal{I}$, and $\ell^{-2} = -2\Lambda/(n-1)(n-2)$. This means the boundary $\mathcal{I}$ is a timelike hypersurface of the unphysical spacetime, and the induced metric on $\mathcal{I}$, denoted by γ_{ab}, has a form

$$\gamma_{ab} \mathrel{\hat{=}} \tilde{g}_{ab} - \frac{1}{\ell^2}\tilde{n}_a\tilde{n}_b \,. \tag{5.303}$$

The Ashtekar–Magnon–Das conserved quantity is defined as

$$Q_{\mathrm{AMD}}[\boldsymbol{\xi}] = -\frac{1}{8\pi}\frac{\ell}{n-3}\int_C E_{ab}\xi^a\tau^b\epsilon_C \,, \tag{5.304}$$

where C is the cross section of $\mathcal{I}$, τ^a is the unit normal vector of C inside $\mathcal{I}$, ξ^a is the representative of some infinitesimal asymptotic symmetry (Actually, it is a conformal Killing vector of $(\mathcal{I}, \gamma_{ab})$), and $\boldsymbol{E}_{ab}$ corresponds to the electric part of the Weyl tensor $\tilde{C}_{abcd}$, i.e,

$$\boldsymbol{E}_{ab} \mathrel{\hat{=}} \ell^2\boldsymbol{K}_{abcd}\tilde{n}^b\tilde{n}^d \,, \tag{5.305}$$

and

$$\boldsymbol{K}_{abcd} = \lim_{\to\mathcal{I}}\Omega^{3-n}\tilde{C}_{abcd} \,. \tag{5.306}$$

The covariant divergence of $\boldsymbol{E}_{ab}$ is given by

$$\boldsymbol{D}^a\boldsymbol{E}_{ab} = -8\pi(n-3)\boldsymbol{T}_{ac}\tilde{n}^a\gamma^c{}_b \,, \tag{5.307}$$

where

$$\boldsymbol{T}_{ab} = \lim_{\to\mathcal{I}}\Omega^{2-n}T_{ab} \,,$$

and the covariant derivative $\boldsymbol{D}_a$ is compatible with γ_{ab}. The definition of the conserved quantity $Q_{\mathrm{AMD}}[\boldsymbol{\xi}]$ depends on the divergence free condition and trace-free condition of $\boldsymbol{E}_{ab}$. So, to get absolutely conserved quantities, the energy-momentum

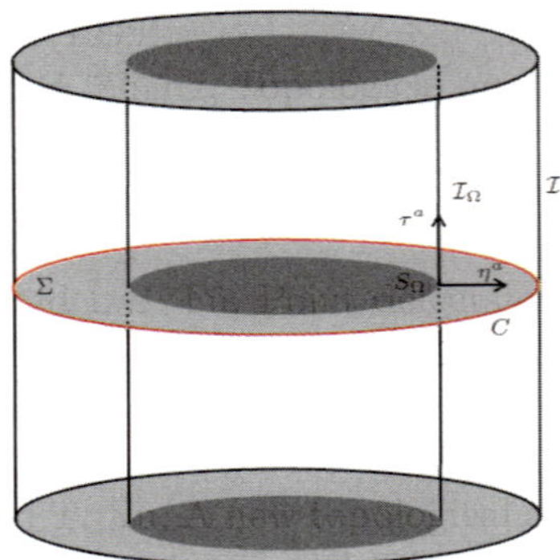

Fig. 5.3   A sequence of S_Ω.

tensor of matter fields has to satisfy some fall off condition. We assume this condition will be fulfilled in the following discussion.

Near the infinity $\mathcal{I}$, the constant Ω surface, denoted by $\mathcal{I}_\Omega$, is a timelike hypersurface. Now, let us consider a spacelike hypersurface $\Sigma \subset \tilde{M}$ which meets $\mathcal{I}$ at the cross section C. The intersection of Σ and $\mathcal{I}_\Omega$ is denoted by S_Ω. Then, near the infinity, with different value of Ω, we have a sequence of codimension-2 surface S_Ω, and $S_\Omega \to C$ when Ω approaches zero (see fig.5.3). We assume Σ is orthogonal to $\mathcal{I}_\Omega$. Let the unit normal vector fields of Σ and $\mathcal{I}_\Omega$ in physical spacetime (M, g_{ab}) be τ^a and η^a respectively, then, we have

$$g^{ab}\tau_a\tau_b = -1\,, \qquad g^{ab}\eta_a\eta_b = 1\,, \qquad g^{ab}\tau_a\eta_b = 0\,, \tag{5.308}$$

With the assumption in the above, obviously, we have

$$\eta^a = (n_c n^c)^{-\frac{1}{2}} n^a = \Omega(\tilde{n}_c\tilde{n}^c)^{-\frac{1}{2}}\tilde{n}^a = \Omega\,\tilde{\eta}^a\,,$$
$$\eta_a = (n_c n^c)^{-\frac{1}{2}} n_a = \Omega^{-1}(\tilde{n}_c\tilde{n}^c)^{-\frac{1}{2}}\tilde{n}_a = \Omega^{-1}\tilde{\eta}_a\,, \tag{5.309}$$

where

$$\tilde{\eta}^a = (\tilde{n}_c\tilde{n}^c)^{-\frac{1}{2}}\tilde{n}^a$$

is the unit normal vector of $\mathcal{I}_\Omega$ embedded in the unphysical spacetime $(\tilde{M}, \tilde{g}_{ab})$. By the same logic, we have

$$\tau^a = \Omega\,\tilde{\tau}^a\,,$$
$$\tau_a = \Omega^{-1}\tilde{\tau}_a\,, \tag{5.310}$$

In the unphsical spacetime, $\tilde{\tau}^a$ can be smoothly extend to the boundary $\mathcal{I}$, so we have $\tilde{\tau}^a \to \tau^a$ when Ω approaches zero.

Now, on the codimension-2 surface S_Ω, we have

$$g_{ab} = -\tau_a\tau_b + \eta_a\eta_b + q_{ab}\,, \tag{5.311}$$

and

$$\tilde{g}_{ab} = -\tilde{\tau}_a\tilde{\tau}_b + \tilde{\eta}_a\tilde{\eta}_b + \tilde{q}_{ab} \,. \tag{5.312}$$

These of course imply $\tilde{q}_{ab} = \Omega^2 q_{ab}$. With these assumptions, the conserved quantity (5.304) can be written as a limit

$$\begin{aligned}
Q_{\mathrm{AMD}}[\xi] &= -\frac{1}{8\pi}\frac{\ell^3}{n-3}\lim_{\Omega\to 0}\int_{S_\Omega}\Omega^{3-n}\tilde{C}_{abcd}\tilde{n}^b\tilde{n}^d\xi^a\tilde{\tau}^c\tilde{\epsilon}_{S_\Omega}\\
&= -\frac{1}{8\pi}\frac{\ell^3}{n-3}\lim_{\Omega\to 0}\int_{S_\Omega}\Omega^{3-n}(\tilde{n}_e\tilde{n}^e)\tilde{C}_{abcd}\tilde{\eta}^b\tilde{\eta}^d\xi^a\tilde{\tau}^c\tilde{\epsilon}_{S_\Omega}\,,
\end{aligned} \tag{5.313}$$

where ξ^a is a vector field in the unphysical spacetime which represents the infinitesimal symmetry and approaches the conformal Killing vector field $\boldsymbol{\xi}^a$ of $(\mathcal{I},\boldsymbol{\gamma}_{ab})$, and $\tilde{\epsilon}_{S_\Omega}$ is the induced volume element of S_Ω (embedded in the unphysical spacetime). Actually, for the infinitesimal symmetry, we have an expansion in the frame $\{\tilde{\tau}^a,\tilde{\eta}^a\}$ as

$$\xi^a = \xi_{(\tau)}\tilde{\tau}^a + \xi_{(\eta)}\tilde{\eta}^a + \tilde{q}^a{}_b\xi^b\,, \tag{5.314}$$

and near the infinity

$$\begin{aligned}
\xi_{(\tau)} &\sim \boldsymbol{\xi}_{(\tau)} + \mathcal{O}(\Omega)\,,\\
\xi_{(\eta)} &\sim 0 + \mathcal{O}(\Omega)\,,\\
\tilde{q}^a{}_b\tilde{\xi}^b &\sim \boldsymbol{q}^a{}_b\boldsymbol{\xi}^a + \mathcal{O}(\Omega)\,,
\end{aligned} \tag{5.315}$$

where $\boldsymbol{\xi}_{(\tau)} = -\boldsymbol{\xi}^a\boldsymbol{\tau}_a$, and $\boldsymbol{q}^a{}_b = \boldsymbol{\gamma}^a{}_b + \boldsymbol{\tau}^a\boldsymbol{\tau}_b$ which is the projection to the cross section of $\mathcal{I}$. By this consideration, we find, in the frame $\{\tau^a,\eta^a\}$ of the physical spacetime, the infinitesimal symmetry can be expressed as

$$\xi^a = \Omega^{-1}\big[\xi_{(\tau)}\tau^a + \xi_{(\eta)}\eta^a\big] + q^a{}_b\xi^b\,. \tag{5.316}$$

Since we have $\Omega^2 C_{abcd} = \tilde{C}_{abcd}$, the conserved quantity (5.304) can also be expressed as

$$Q_{\mathrm{AMD}}[\xi] = -\frac{1}{8\pi}\frac{\ell^3}{n-3}\lim_{\Omega\to 0}\int_{S_\Omega}(\tilde{n}_e\tilde{n}^e)C_{abcd}\eta^b\eta^d\xi^a\tau^c\epsilon_{S_\Omega}\,. \tag{5.317}$$

So we find

$$Q_{\mathrm{AMD}}[\xi] = Q_{\mathrm{AMD}}[\xi_\perp] + Q_{\mathrm{AMD}}[\xi_\parallel]\,, \tag{5.318}$$

where

$$Q_{\mathrm{AMD}}[\xi_\perp] = -\frac{1}{8\pi}\frac{\ell}{n-3}\lim_{\Omega\to 0}\int_{S_\Omega}\Omega^{-1}C_{abcd}\eta^b\eta^d\tau^a\tau^c\epsilon_{S_\Omega}\,, \tag{5.319}$$

and

$$Q_{\mathrm{AMD}}[\xi_\parallel] = -\frac{1}{8\pi}\frac{\ell}{n-3}\lim_{\Omega\to 0}\int_{S_\Omega}C_{abcd}\eta^b\eta^d\xi_\parallel^a\tau^c\epsilon_{S_\Omega}\,. \tag{5.320}$$

The perpendicular part comes from $\xi_\perp^a = \Omega^{-1}\tau^a$, and, similarly, the parallel part corresponds to $\xi_\parallel^a = q^a{}_b\xi^b$. Here, we have chosen $\xi_{(\tau)} = 1$ and $\xi_{(\eta)} = 0$ for simplicity.

The parallel part obviously corresponds to the angular momentum of the spacetime, and the perpendicular part corresponds to the energy (Since the integral is on the surface $\mathcal{I}_\Omega$ with fixed Ω, the limit and integral in eq.(5.317) can be exchanged, and the factor $\tilde{n}_e \tilde{n}^e$ can be replaced by $1/\ell^2$ in the limit procedure.).

Since the Weyl tensor is traceless, we get

$$E[C] = Q_{\mathrm{AMD}}[\xi_\perp] = \frac{1}{16\pi} \frac{\ell}{n-3} \lim_{\Omega \to 0} \int_{S_\Omega} \Omega^{-1} C_{abcd} q^{ac} q^{bd} \epsilon_{S_\Omega} , \qquad (5.321)$$

or

$$E[C] = Q_{\mathrm{AMD}}[\xi_\perp] = \frac{1}{16\pi} \frac{\ell}{n-3} \lim_{\Omega \to 0} \int_{S_\Omega} \Omega^{-1} C_{abcd} h^{ac} h^{bd} \epsilon_{S_\Omega} , \qquad (5.322)$$

where

$$h_{ab} = -\tau_a \tau_b + \eta_a \eta_b .$$

From now on, let us focus on the case of dimension four. In the neighbourhood of the timelike infinity $\mathcal{I}$, the unphysical metric can be written as (as in the Gauss normal coordinates near a hypersurface)

$$\tilde{g}_{ab} = \ell^2 \tilde{n}_a \tilde{n}_b + \tilde{\gamma}_{ab} . \qquad (5.323)$$

Here $\tilde{\gamma}_{ab}$ satisfies $\tilde{\gamma}^{ab} \tilde{n}_b = 0$. Actually, from the discussion before, we have

$$\tilde{\gamma}_{ab} = -\tilde{\tau}_a \tilde{\tau}_b + \tilde{q}_{ab} .$$

Now let us choose the conformal factor Ω such that (this is allowed for the vanishing of the Bach tensor of the boundary)

$$\tilde{\gamma}_{ab} \stackrel{\wedge}{=} \overset{o}{\gamma}_{ab} = -\tau_a \tau_b + \ell^2 \overset{o}{q}_{ab} ,$$

where $\overset{o}{q}_{ab}$ is the standard metric of S^2, and $\tau_a = (dt)_a$ corresponds to the normal vector field for the constant t slice in $\mathcal{I}$. So, on the timelike boundary $\mathcal{I}$, $\tilde{\gamma}_{ab}$ is the standard metric of the so-called Einstein static universe. With this selection of the conformal factor, in the neighbourhood of $\mathcal{I}$, one has an expansion [54]

$$\tilde{g}_{ab} = -\left(1 + \frac{1}{2}\Omega^2\right)(dt)_a (dt)_b + \ell^2\left[(d\Omega)_a (d\Omega)_b + \left(1 - \frac{1}{2}\Omega^2\right) \overset{o}{q}_{ab} \cdots \right]. \qquad (5.324)$$

Based on these, on the codimension-2 surface S_Ω, we obtain

$$q_{ab} = \frac{\ell^2}{\Omega^2}\left(1 - \frac{1}{2}\Omega^2\right) \overset{o}{q}_{ab} \cdots , \qquad (5.325)$$

and the area of S_Ω also has an expansion

$$A(S_\Omega) = 4\pi \frac{\ell^2}{\Omega^2} + \cdots . \qquad (5.326)$$

The "$\cdots$" in the above equations denotes higher order terms of Ω. By the above discussion, we finally arrive at

$$E[C] = \frac{1}{8\pi} \lim_{\Omega \to 0} \sqrt{\frac{A(S_\Omega)}{16\pi}} \int_{S_\Omega} (C_{abcd} q^{ac} q^{bd}) \epsilon_{S_\Omega} . \qquad (5.327)$$

or

$$E[C] = \frac{1}{8\pi} \lim_{\Omega \to 0} \sqrt{\frac{A(S_\Omega)}{16\pi}} \int_{S_\Omega} (C_{abcd} h^{ac} h^{bd}) \epsilon_{S_\Omega} . \qquad (5.328)$$

The later is exactly the expression in eq.(5.279).

5.7 An example: Kerr–AdS spacetime

Here, we calculate the energy of Kerr–AdS spacetime in four dimension. The metric has a form

$$
g = -\frac{\Delta_r}{\rho^2}\left[\mathrm{d}t - \frac{a}{\Xi}\sin^2\theta\,\mathrm{d}\phi\right]^2 + \frac{\rho^2}{\Delta_r}\mathrm{d}r^2 + \frac{\rho^2}{\Delta_\theta}\mathrm{d}\theta^2 + \frac{\Delta_\theta\sin^2\theta}{\rho^2}\left[a\mathrm{d}t - \frac{r^2+a^2}{\Xi}\mathrm{d}\phi\right]^2 ,
\tag{5.329}
$$

where

$$
\rho^2 = r^2 + a^2\cos^2\theta ,
$$

$$
\Delta_r = (r^2+a^2)\left(1+\frac{r^2}{\ell^2}\right) - 2mr ,
$$

$$
\Delta_\theta = 1 - \frac{a^2}{\ell^2}\cos^2\theta ,
$$

$$
\Xi = 1 - \frac{a^2}{\ell^2} .
\tag{5.330}
$$

To get the standard metric of AdS in static coordinates when $m = 0$, one can do coordinates transformations as [124, 125]

$$
T = t ,
$$

$$
\Phi = \phi - \frac{a}{\ell^2}t ,
$$

$$
y\cos\Theta = r\cos\theta ,
$$

$$
y^2\Xi = r^2\Delta_\theta + a^2\sin^2\theta .
\tag{5.331}
$$

The natural bases of these two coordinate systems are related by following equations

$$
\frac{\partial}{\partial T} = \frac{\partial}{\partial t} + \frac{a}{\ell^2}\frac{\partial}{\partial\phi} ,
$$

$$
\frac{\partial}{\partial y} = \frac{r}{y}\frac{r^2+a^2}{r^2+a^2\cos^2\theta}\frac{\partial}{\partial r} + \frac{a^2\sin\theta\cos\theta}{y(r^2+a^2\cos^2\theta)}\frac{\partial}{\partial\theta} ,
$$

$$
\frac{\partial}{\partial\Theta} = -y\sin\Theta\frac{a^2\cos\theta(r^2+\ell^2)}{\ell^2(r^2+a^2\cos^2\theta)}\frac{\partial}{\partial r} + \frac{yr\sin\Theta\Delta_\theta}{\sin\theta(r^2+a^2\cos^2\theta)}\frac{\partial}{\partial\theta} ,
$$

$$
\frac{\partial}{\partial\Phi} = \frac{\partial}{\partial\phi} ,
\tag{5.332}
$$

where

$$
y^2\sin^2\Theta = \frac{(r^2+a^2)\sin^2\theta}{\Xi} .
\tag{5.333}
$$

By using the relations between the natural bases, one can get the components of the metric in the new coordinates $\{T, y, \Theta, \Phi\}$. Unfortunately, the expression of the metric is quite complicated in this new coordinate system. The asymptotical expansion near the infinity can be written as [124] (see [126] for the expansions in higher dimensions)

$$
g = -\left(1+\frac{y^2}{\ell^2}\right)\mathrm{d}T^2 + \left(1+\frac{y^2}{\ell^2} - \frac{2m}{y\Delta_\Theta^{3/2}}\right)^{-1}\mathrm{d}y^2 + y^2(\mathrm{d}\Theta^2 + \sin^2\Theta\mathrm{d}\Phi^2)
$$

$$
+ \frac{2m}{y}\Delta_\Theta^{-5/2}(\mathrm{d}T - a\sin\Theta\mathrm{d}\Phi)^2 + \mathcal{O}(y^{-3}) ,
\tag{5.334}
$$

where

$$\Delta_\Theta = 1 - \frac{a^2}{\ell^2} \sin^2 \Theta \,. \tag{5.335}$$

When y approaches infinity, this metric approaches the pure AdS spacetime. For the two dimensional surface S with constant T and y, we find

$$\epsilon_S = y^2 \sin \Theta \, d\Theta d\Phi + \mathcal{O}(y^{-1}) \,, \tag{5.336}$$

and

$$C_{abcd} q^{ac} q^{bd} = 2m \left(2 + \frac{a^2}{\ell^2} \sin^2 \Theta \right) \cdot \Delta_\Theta^{-5/2} \cdot \frac{1}{y^3} + \mathcal{O}(y^{-4}) \,. \tag{5.337}$$

Substituting the above results into eq.(5.327) or (5.328), and after completing the integration on Θ, we get

$$E[C] = \frac{m}{\Xi^2} \,. \tag{5.338}$$

This is exactly the results obtained in [124] or [126].

5.8   Asymptotically dS spacetimes

Similar to the asymptotically anti de Sitter spacetimes, one can also define a spacetime which has some asymptotic behavior of pure de Sitter spacetime [107].

Definition 1: An 4-dimension spacetime (M, g_{ab}) is said to be *weakly asymptotically de Sitter* if there exists a manifold $\tilde{M}$ with boundary $\mathcal{I}$, equipped with a metric $\tilde{g}_{ab}$ and a diffeomorphism from M onto $\tilde{M} - \mathcal{I}$ (M and $\tilde{M} - \mathcal{I}$ are identified by this mapping), such that:

(i). there exists a function Ω on $\tilde{M}$ such that $\tilde{g}_{ab} = \Omega^2 g_{ab}$ on $\tilde{M}$; and Ω is vanishing on $\mathcal{I}$ and $\tilde{n}_a = \tilde{\nabla}_a \Omega$ is nowhere vanishing on $\mathcal{I}$;

(ii). g_{ab} satisfies $G_{ab} + \Lambda g_{ab} = 8\pi T_{ab}$ with $\Lambda > 0$, where $\Omega^{-1} T_{ab}$ admits a smooth limit to $\mathcal{I}$.

From the Einstein with the positive cosmological constant, it is easy to find

$$\tilde{n}_a \tilde{n}^a \,\hat{=}\, -\frac{1}{\ell^2} \,, \qquad \ell^2 = \frac{\Lambda}{3} \,. \tag{5.339}$$

So $\mathcal{I}$ is spacelike. As in the case of asymptotically flat, the boundary $\mathcal{I}$ usually has two components, i.e., the future infinity $\mathcal{I}^+$ and the past infinity $\mathcal{I}^-$. Hence, usually one has

$$\mathcal{I} = \mathcal{I}^+ \cup \mathcal{I}^- \,.$$

To describe a black hole or something else, the infinity $\mathcal{I}$ is required to be geodesically complete. This inspires a definition

Definition 2: A weakly asymptotically de Sitter space-time (M, g_{ab}) is said to be *asymptotically de Sitter* if $(\mathcal{I}, \gamma_{ab})$ is geodesically complete with respect to the induced metric γ_{ab} (of $\tilde{g}_{ab}$).

Although $(\mathcal{I}, \gamma_{ab})$ is required to be geodesically complete in the definition 2, there are no restrictions on the topology of $\mathcal{I}$. Generally, the topology of $\mathcal{I}$ is given by

$$S^3 - \{p_1, \cdots, p_k\}, \tag{5.340}$$

where $\{p_1, \cdots, p_k\}$ is a set of punctures, and $k = 0, 1, \cdots$. According to the punctures, the spacetimes have following classification:

$k = 0.$ The spacetime (M, g_{ab}) without any puncture is called *globally asymptotically de Sitter* spacetime. In this case, the boundary $\mathcal{I}$ has a topology of S^3;

$k = 1.$ In the case with one puncture, the topology of $\mathcal{I}$ is $\mathcal{R}^3$, and the spacetime is said to be *asymptotically de Sitter in a Poincaré patch*. The puncture corresponds to the spacelike infinity i^0 of the spacetime;

$k = 2.$ In the case with two punctures, $\mathcal{I}$ has a topology of $\mathcal{R} \times S^2$, and the spacetime is called *asymptotically Schwarzschild-de Sitter* spacetime. One puncture represents the spacelike infinity, and another represents the future (or past) timelike infinity i^+ (or i^-) of the spacetime.

The detailed examples for $k = 0, 1, 2$ and the explanation of the case with more punctures can be found in [107].

As in the anti-de Sitter spacetime, to further require the spacetime has the asymptotic symmetry group of de Sitter group (or the subgroup of de Sitter group), one has to impose an additional condition on the conformal class of the boundary spacetime. As in the case of anti-de Sitter, this is equivalent to the condition with a vanished magnetic part of the leading order of the Weyl tensor. This suggests a strong version of the definition.

Definition 3: A space-time (M, g_{ab}) is said to be *strongly asymptotically de Sitter* if in a conformal completion satisfying conditions of Definition 2, the intrinsic metric γ_{ab} on $\mathcal{I}$ is conformally flat.

By this definition, the metric γ_{ab} of the boundary $\mathcal{I}$ is conformally related to the standard metric $\overset{o}{\gamma}$ of S^3, $\mathcal{R}^3$, or $\mathcal{R} \times S^3$, i.e.,

$$\overset{o}{\gamma} = \begin{cases} \mathrm{d}\chi^2 + \sin^2\chi(\mathrm{d}\theta^2 + \sin^2\theta\mathrm{d}\phi^2)\,, & k = 0\,, \\ \mathrm{d}x^2 + \mathrm{d}y^2 + \mathrm{d}z^2\,, & k = 1\,, \\ \mathrm{d}t^2 + \ell^2(\mathrm{d}\theta^2 + \sin^2\theta\mathrm{d}\phi^2)\,, & k = 2\,. \end{cases} \tag{5.341}$$

In the case of $k = 0$, the asymptotic symmetry group is the de Sitter group which has ten generators. However, in the cases of $k = 1$ and $k = 2$, the asymptotic symmetry group is merely the subgroup of the de Sitter group. This is because that the global condition on $(\mathcal{I}, \gamma_{ab})$, i.e., the requirement of the completeness of $(\mathcal{I}, \gamma_{ab})$, precludes some conformal Killing vectors of $(\mathcal{I}, \gamma_{ab})$, see [107] for details.

By the analysis of the asymptotic behavior of the fields and the equations of motion, Ashtekar, Bonga, and Kesavan define the conserved quantities as [107]

$$Q_{\mathrm{ABK}}[\boldsymbol{\xi}] = -\frac{\ell}{8\pi} \int_C \left[\boldsymbol{E}_{ab} + \frac{8\pi}{3} \boldsymbol{T} \gamma_{ab} \right] \xi^a \tau^b \epsilon_C \,, \tag{5.342}$$

where $\boldsymbol{\xi}^a$ is a conformal Killing vector field of $(\mathcal{I}, \gamma_{ab})$, $\boldsymbol{E}_{ab}$ is the electric part of the leading term of the Weyl tensor, ϵ_C is the volume element of the cross section C (of $\mathcal{I}$), τ^a is the unit normal vector of C, and $\boldsymbol{T}$ is the trace of $\boldsymbol{T}^a{}_b$ which has a smooth limit on $\mathcal{I}$ and has a form

$$\boldsymbol{T}^a{}_b = \Omega^{-3} T^a{}_b \,.$$

Similar to the discussion in the case of AdS, this conserved quantity can also be expressed as

$$Q_{\mathrm{ABK}}[\xi] = -\frac{\ell}{8\pi} \lim_{\Omega \to 0} \int_{S_\Omega} \Omega^{-1} \left[C_{abcd} \eta^b \eta^d + \frac{8\pi}{3} T \gamma_{ac} \right] \xi^a \tau^c \epsilon_{S_\Omega} \,, \tag{5.343}$$

where ξ^a is the representative of the asymptotic symmetry and approaches to $\boldsymbol{\xi}^a$ at the infinity, η^a is the unit normal vector field of the constant Ω hypersurface, τ^a is the unit normal vector field of S_Ω (embedded in the constant Ω hypersurface).

For $k = 2$, i.e., for the asymptotically Schwarzschild-de Sitter spacetime, similar to the discussion in the case of AdS, we can choose a conformal factor such that $\tilde{\gamma}_{ab}$ has the standard form $\overset{o}{\gamma}$ on $\mathcal{I}$, and τ^a is a conformal Killing vector field which represents a "time translation" (of course it is not a real time translation because it is spacelike). By these consideration, we have an energy

$$E[C] = Q_{\mathrm{ABK}}[\xi] = \frac{1}{8\pi} \lim_{\Omega \to 0} \sqrt{\frac{A(S_\Omega)}{16\pi}} \int_{S_\Omega} \left[-\frac{16\pi}{3} T + C_{abcd} q^{ac} q^{bd} \right] \epsilon_{S_\Omega} \,. \tag{5.344}$$

Since $\Omega^{-1} T_{ab}$ is assumed to has a smooth limit on $\mathcal{I}$, it is not hard to prove that only $\Omega^{-1} T_{ab} \tilde{n}^a \tilde{n}^b$ has a nontrivial limit on $\mathcal{I}$ by using Einstein equations and Bianchi identity (see [107] for detailed discussion.), so the limit of $q^{ab} T_{ab}$ on $\mathcal{I}$ is trivial. Considering eq.(5.278), we find the energy (5.344) is nothing but the limit

$$E[C] = \lim_{S \to C} E_{\mathrm{New}}(S) \,. \tag{5.345}$$

For the Kerr-de Sitter spacetime, similar to the calculation of Kerr–AdS spacetime, it is not hard to find

$$\lim_{S \to C} E_{\mathrm{New}}(S) = \frac{m}{\Xi^2} \,, \tag{5.346}$$

where $\Xi = 1 + a^2/\ell^2$.

It should be noted here: the fall-off condition, i.e., $\Omega^{-1} T_{ab}$ has a smooth limit on $\mathcal{I}$, is weak enough to include some important spacetimes, for example, the

Friedmann–Robertson–Walker universe filled by dust. Hence, it is reasonable to include the term T in (5.344). Another point is that the Bondi news tensor is always vanishing for the asymptotically de Sitter spacetimes defined in the above. This means that the difference between the energy of two different cross sections C and C', i.e., $E[C] - E[C']$, is merely supported by the matter flux. This is very different from the asymptotically flat case [107].

Bibliography

[1] E. Noether, Gott. Nachr. **1918**, 235 (1918) [Transp. Theory Statist. Phys. **1**, 186 (1971)] [physics/0503066].

[2] L. B. Szabados, Living Rev. Rel. **12**, 4 (2009).

[3] R. M. Wald, *General Relativity*, University of Chicago Press, Chicago, 1984.

[4] J. Lee and R. M. Wald, J. Math. Phys. **31**, 725 (1990).

[5] V. Iyer and R. M. Wald, Phys. Rev. D **50**, 846 (1994) [gr-qc/9403028].

[6] R. M. Wald, J. Math. Phys. **31**, 10, (1990).

[7] V. Iyer and R. M. Wald, Phys. Rev. D **52**, 4430 (1995) [gr-qc/9503052].

[8] V. Iyer, Phys. Rev. D **55**, 3411 (1997) [gr-qc/9610025].

[9] S. Hollands, Rev. Math. Phys. **20**, 1033 (2008) [arXiv:0705.3340 [gr-qc]].

[10] J. L. Jaramillo and E. Gourgoulhon, Fundam. Theor. Phys. **162**, 87 (2011) [arXiv:1001.5429 [gr-qc]].

[11] L. B. Szabados 1991. Canonical pseudotensors, Sparling's form and Noether currents, KFKI Report29/B.

[12] L. B. Szabados, Class. Quant. Grav. **9**, 2521 (1992).

[13] F. J. Belinfante, "On the spin angular momentum of mesons". Physica. **6**, Pages 887-898, (1939),

[14] F. J. Belinfante, "On the current and the density of the electric charge, the energy, the linear momentum and the angular momentum of arbitrary fields". Physica. **7**, Pages 449-474, (1940).

[15] L. Rosenfeld, "Sur le tenseur d'impulsion-energie". Acad. Roy. Belg. Memoirs de classes de Science. (Cl. Sciences) 18, fasc. 6, 2–3 (1940).

[16] T. N. Palmer, Phys. Rev. D **18**, 4399 (1978).

[17] T. N. Palmer, Gen. Relativ. Gravit. **12**, 149(1980).

[18] J. Katz, Class. Quant. Grav. **2**, 423 (1985).

[19] N. Rosen, Found. Phys. **15**, 997 (1986).

[20] J. Katz, and A. Ori, Class. Quant. Grav. **7**, 787 (1990).

[21] J. Katz, J. Bicak and D. Lynden-Bell, Phys. Rev. D **55**, 5957 (1997) [gr-qc/0504041].

[22] J. Katz and D. Lerer, Class. Quant. Grav. **14**, 2249 (1997) [gr-qc/9612025].

[23] A. Einstein, Sitzungsber. Preuss. Akad. Wiss. Berlin (Math. Phys.) **1915**, 778 (1915) Addendum: [Sitzungsber. Preuss. Akad. Wiss. Berlin (Math. Phys.) **1915**, 799 (1915)].

[24] A. Einstein, Annalen Phys. **49**, no. 7, 769 (1916) [Annalen Phys. **14**, 517 (2005)].

[25] A. Einstein, Berlin Ber., 448 (1918).

[26] A. Papapetrou, Proc. Roy. Irish Acad. (Sect. A) **52A**, 11 (1948).

[27] P. G. Bergmann and R. Thomson, Phys. Rev. **89**, 400 (1953).

[28] L. Landau and E. Lifshitz, The Classical Theory of Fields, Addison Wesley, Cambridge, Massachusetts, 1951.

[29] C. Møller, Ann. Phys. **4**, 347 (1958).

[30] C. Møller, Annals Phys. **12**, 118 (1961).

[31] J. N. Goldberg, Phys. Rev. **111**, 315 (1958).

[32] Y. S. Duan, Sov. Phys. JETP **34**, 632 (1958) [arXiv:1707.02217 [gr-qc]].

[33] Y. S. Duan, J. Y. Zhang, Acta Physica Sinica **11**, 19, 689 (1963).

[34] Y. S. Duan and Y. T. Wang, Sci. Sin. **4A**, 343 (1983).

[35] Y. S. Duan, J. C. Liu and X. G. Dong, Acta Physica Sinica **36**, 760 (1987).

[36] Y. S. Duan, J. C. Liu and X. G. Dong, Gen. Rel. Grav. **20**, 485 (1988).

[37] Y. S. Duan and S. S. Feng, Acta Physica Sinica **44**, 1373 (1995) .

[38] S. S. Feng and Y. S. Duan, Gen. Rel. Grav. **27**, 887 (1995).

[39] S. S. Feng and Y. S. Duan, Commun. Theor. Phys. **25**, 485 (1996).

[40] S. S. Feng and Y. S. Duan, Class. Quant. Grav. **16**, 3237(1999) [hep-th/9902096].

[41] Y. S. Duan, Y. X. Liu and L. J. Zhang, Mod. Phys. Lett. A **22**, 2855 (2007) [gr-qc/0508113].

[42] Y. S. Duan, Y. X. Liu, Y. Q. Wang and L. J. Zhang, Mod. Phys. Lett. A **23**, 769 (2008) [gr-qc/0508103].

[43] C. W. Misner, K. S. Thorne and J. A. Wheeler, San Francisco, 1973.

[44] C. C. Chang, J. M. Nester and C. M. Chen, Phys. Rev. Lett. **83**, 1897 (1999) [gr-qc/9809040].

[45] A. Komar, Phys. Rev. **113**, 934 (1959).

[46] J. M. Bardeen, B. Carter and S. W. Hawking, Commun. Math. Phys. **31**, 161 (1973).

[47] D. Kastor, Class. Quant. Grav. **25**, 175007 (2008) [arXiv:0804.1832 [hep-th]].

[48] B. Whitt, Phys. Rev. D **38**, 3000 (1988).

[49] G. J. Zuckerman, Conf. Proc. C **8607214**, 259 (1986).

[50] C. Crnkovic and E. Witten, In Hawking, S.W. (ed.), Israel, W. (ed.): Three hundred years of gravitation, 676–684 and Preprint - Crnkovic, C. (86,rec.Dec.) 13 p

[51] R. M. Wald and A. Zoupas, Phys. Rev. D **61**, 084027 (2000) [gr-qc/9911095].

[52] R. Penrose, Phys. Rev. Lett. **10**, 66 (1963).

[53] R. Penrose, Riv. Nuovo Cim. **1**, 252 (1969) [Gen. Rel. Grav. **34**, 1141 (2002)].

[54] S. Hollands, A. Ishibashi and D. Marolf, Class. Quant. Grav. **22**, 2881 (2005) [hep-th/0503045].

[55] A. Ashtekar and R. O. Hansen, J. Math. Phys. **19**, 1542 (1978).

[56] R. Arnowitt, S. Deser, and C. W. Misner, "The dynamics of General Relativity", edited by L. Witten, Gravitation: an introduction to current research, John Wiley & Sons, Inc., New York · London, 1962.

[57] R. Schon and S. T. Yau, Commun. Math. Phys. **65**, 45 (1979).

[58] J. D. Brown and J. W. York, Phys. Rev. D **47**, 1407 (1993) [arXiv:gr-qc/9209012].

[59] S. W. Hawking and G. T. Horowitz, Class. Quant. Grav. **13**, 1487 (1996) [gr-qc/9501014].

[60] S. W. Hawking and C. J. Hunter, Class. Quant. Grav. **13**, 2735 (1996) [gr-qc/9603050].

[61] D. Baskaran, S. R. Lau and A. N. Petrov, Annals Phys. **307**, 90 (2003) [gr-qc/0301069].

[62] J. D. Brown, S. R. Lau and J. W. York, Jr., gr-qc/0010024.

[63] A. Ashtekar and J. D. Romano, Class. Quant. Grav. **9**, 1069 (1992).

[64] A. Ashtekar, L. Bombelli and O. Reula, PRINT-90-0318 (SYRACUSE).

[65] H. Bondi, Mon. Not. Roy. Astron. Soc. **107**, 410 (1947).

[66] R. K. Sachs, Proc. Roy. Soc. Lond. A **270**, 103 (1962).

[67] A. Ashtekar, arXiv:1409.1800 [gr-qc].

[68] R. P. Geroch, "Asymptotic structure of space-time" 1977.

[69] R. P. Geroch and J. Winicour, J. Math. Phys. **22**, 803 (1981).

[70] A. Ashtekar and M. Streubel, Proc. Roy. Soc. Lond. A **376**, 585 (1981).

[71] W. T. Shaw, Class. Quant. Grav. **1**, no. 4, L33 (1984).

[72] T. Dray, Class. Quant. Grav. **2**, L7 (1985).

[73] T. Dray and M. Streubel, Class. Quant. Grav. **1**, no. 1, 15 (1984).

[74] A. Ishibashi, Class. Quant. Grav. **25**, 165004 (2008) [arXiv:0712.4348 [gr-qc]].

[75] B. Carter, J. Geom. Phys. **8**, 53 (1992).

[76] B. Carter, arXiv:hep-th/9705172.

[77] B. Carter, Int. J. Theor. Phys. **40**, 2099 (2001) [arXiv:gr-qc/0012036].

[78] L. M. Cao, JHEP **1103**, 112 (2011) [arXiv:1009.4540 [gr-qc]].

[79] S. Kobayashi, K. Nomizu. Foundations of Differential Geometry (Wiley Classics Library) Volume 2.

[80] L. B. Szabados, Class. Quant. Grav. **11**, 1833 (1994) [arXiv:gr-qc/9402001].

[81] I. Booth and S. Fairhurst, Phys. Rev. D **75**, 084019 (2007) [arXiv:gr-qc/0610032].

[82] C. W. Misner and D. H. Sharp, Phys. Rev. **136**, B571 (1964).

[83] S. A. Hayward, Phys. Rev. D **53**, 1938 (1996) [gr-qc/9408002].

[84] H. Maeda and M. Nozawa, Phys. Rev. D **77**, 064031 (2008) [arXiv:0709.1199 [hep-th]].

[85] H. Kodama, Prog. Theor. Phys. **63**, 1217 (1980).

[86] R. G. Cai, L. M. Cao, Y. P. Hu and N. Ohta, Phys. Rev. D **80**, 104016 (2009) [arXiv:0910.2387 [hep-th]].

[87] R. G. Cai and L. M. Cao, Fundam. Theor. Phys. **187**, 31 (2017) [arXiv:1609.08306 [gr-qc]].

[88] S. A. Hayward, Class. Quant. Grav. **15**, 3147 (1998) [gr-qc/9710089].

[89] R. G. Cai and S. P. Kim, JHEP **0502**, 050 (2005) [arXiv:hep-th/0501055].

[90] M. Akbar and R. G. Cai, Phys. Rev. D **75**, 084003 (2007) [arXiv:hep-th/0609128];

[91] R. G. Cai and L. M. Cao, Phys. Rev. D **75**, 064008 (2007) [arXiv:gr-qc/0611071];

[92] M. Nozawa and H. Maeda, Class. Quant. Grav. **25**, 055009 (2008) [arXiv:0710.2709 [gr-qc]].

[93] S. Hawking, J. Math. Phys. **9**, 598 (1968).

[94] S. A. Hayward, Phys. Rev. D **49**, 831 (1994) [gr-qc/9303030].

[95] S. A. Hayward, Phys. Rev. D **49**, 6467 (1994).

[96] S. A. Hayward, S. Mukohyama and M. C. Ashworth, Phys. Lett. A **256**, 347 (1999) [arXiv:gr-qc/9810006].

[97] S. A. Hayward, Phys. Rev. Lett. **93**, 251101 (2004) [arXiv:gr-qc/0404077].

[98] S. A. Hayward, Phys. Rev. D **70**, 104027 (2004) [arXiv:gr-qc/0408008].

[99] S. A. Hayward, Phys. Rev. D **74**, 104013 (2006) [arXiv:gr-qc/0609008].

[100] S. A. Hayward, Class. Quant. Grav. **24**, 923 (2007) [arXiv:gr-qc/0611027].

[101] E. Gourgoulhon, Phys. Rev. D **72**, 104007 (2005) [arXiv:gr-qc/0508003].

[102] E. Gourgoulhon and J. L. Jaramillo, Phys. Rev. D **74**, 087502 (2006) [arXiv:gr-qc/0607050].

[103] E. Gourgoulhon and J. L. Jaramillo, New Astron. Rev. **51**, 791 (2008) [arXiv:0803.2944 [astro-ph]].

[104] A. Ashtekar and B. Krishnan, Living Rev. Rel. **7**, 10 (2004) [arXiv:gr-qc/0407042].

[105] A. Ashtekar and A. Magnon, Class. Quant. Grav. **1**, L39 (1984).

[106] A. Ashtekar and S. Das, Class. Quant. Grav. **17**, L17 (2000) [hep-th/9911230].

[107] A. Ashtekar, B. Bonga and A. Kesavan, Class. Quant. Grav. **32**, no. 2, 025004 (2015) [arXiv:1409.3816 [gr-qc]].

[108] H. Bray, S. Hayward, M. Mars and W. Simon, Commun. Math. Phys. **272**, 119 (2007) [arXiv:gr-qc/0603014].

[109] C. C. M. Liu and S. T. Yau, Phys. Rev. Lett. **90**, 231102 (2003) [gr-qc/0303019].

[110] Kijowski, J.: A simple derivation of canonical structure and quasi-local hamiltonians in general relativity. Gen. Relativ. Gravit. **29**, (307-343), 1997.

[111] R. J. Epp, Phys. Rev. D **62**, 124018 (2000) [gr-qc/0003035].

[112] M. T. Wang and S. T. Yau, Phys. Rev. Lett. **102**, 021101 (2009) [arXiv:0804.1174 [gr-qc]].

[113] J. M. Maldacena, Int. J. Theor. Phys. **38**, 1113 (1999) [Adv. Theor. Math. Phys. **2**, 231 (1998)] [hep-th/9711200].

[114] S. S. Gubser, I. R. Klebanov and A. M. Polyakov, Phys. Lett. B **428**, 105 (1998) [hep-th/9802109].

[115] E. Witten, Adv. Theor. Math. Phys. **2**, 253 (1998) [hep-th/9802150].

[116] O. Aharony, S. S. Gubser, J. M. Maldacena, H. Ooguri and Y. Oz, Phys. Rept. **323**, 183 (2000) [hep-th/9905111].

[117] G. 't Hooft, Salamfest 1993:0284-296 [gr-qc/9310026].

[118] L. Susskind, J. Math. Phys. **36**, 6377 (1995) [hep-th/9409089].

[119] V. Balasubramanian and P. Kraus, Commun. Math. Phys. **208**, 413 (1999) [hep-th/9902121].

[120] P. Kraus, F. Larsen and R. Siebelink, Nucl. Phys. B **563**, 259 (1999) [hep-th/9906127].

[121] A. Tautman, "Conervation laws in General relativity", edited by L. Witten, Gravitation: an introduction to current research, John Wiley & Sons, Inc., New York · London, 1962.

[122] J. Norton, Rep. Prog. Phys. **56**, 791458 (1993).

[123] R. M. Wald, Living Rev. Rel. **4**, 6 (2001) [arXiv:gr-qc/9912119].

[124] M. Henneaux and C. Teitelboim, Commun. Math. Phys. **98**, 391 (1985).

[125] S. W. Hawking, C. J. Hunter and M. Taylor, Phys. Rev. D **59**, 064005 (1999) [hep-th/9811056].

[126] G. W. Gibbons, M. J. Perry and C. N. Pope, Class. Quant. Grav. **22**, 1503 (2005) [hep-th/0408217].

Chapter 6

Gravitational Energy and the Gauge Theory Perspective

Chiang-Mei Chen[a,b] [1] and James M. Nester[a,b,c,d] [2]

[a]Department of Physics, National Central University,
Chungli 32001, Taiwan
[b]Center for Mathematical and Theoretical Physics,
National Central University, Chungli 32001, Taiwan
[c]Graduate Institute of Astronomy, National Central University,
Chungli 32001, Taiwan
[d]Leung Center for Cosmology and Particle Astrophysics,
National Taiwan University, Taipei 10617, Taiwan

Abstract: Gravity, and the puzzle regarding its energy, can be understood from a gauge theory perspective. Gravity, i.e., dynamical spacetime geometry, can be considered as a local gauge theory of the symmetry group of Minkowski spacetime: the Poincaré group. The dynamical potentials of the Poincaré gauge theory of gravity are the frame and the metric-compatible connection. The spacetime geometry has in general both curvature and torsion. Einstein's general relativity theory is a special case. Both local gauge freedom and energy are clarified via the Hamiltonian formulation. We have developed a covariant Hamiltonian formulation. The Hamiltonian boundary term gives covariant expressions for the quasi-local energy, momentum and angular momentum. A key feature is the necessity to choose on the boundary a non-dynamic reference. With a best matched reference one gets good quasi-local energy-momentum and angular momentum values.

Dedication: To the memory of Prof. Yi-Shi Duan, who inspired many students and did pioneering work on many subjects—including two that have been our interest: the gauge theory formulation of gravity and identifying good expressions for the energy-momentum and angular momentum of gravitating systems. [43, 44, 45, 46, 47, 48, 49, 53] Prof. Duan

---

[1]Email: cmchen@phy.ncu.edu.tw
[2]Email: nester@phy.ncu.edu.tw

169

did much to encourage attention to these important topics.

6.1 Introduction

The evolution of a generally covariant theory is under-determined. This first became an issue in connection with Einstein's gravity theory, general relativity (GR). One consequence is that gravitational energy has no proper localization. Trying to clarify this fact led ultimately to the gauge theories of physical interactions.

GR with general covariance is the premier gauge theory. The consequences of this, especially regarding gravitational energy and under-determined evolution, were long perplexing. The Hamiltonian approach clarifies these issues. Gravity can be understood as a gauge theory of the local Poincaré symmetries of spacetime.

As noted above, Prof. Duan was much concerned with these issues. Here we present an introduction to our work in this area. It will be noticed that we use many of the same ideas that were used by Prof. Duan. The distinctive features of our approach is that we use the first order Lagrangian and the Hamiltonian formulations, moreover, we always favor a representation in terms of differential forms.

6.2 Some historical background

Dynamical equations obtained from a variational principle formerly had deterministic Cauchy initial value problems, but in GR there appears a differential identity connecting the evolution equations, they were not independent and could not give uniquely determined evolution—this is *the essence of gauge theory*. This type of indeterminism was later found to be best addressed via the Hamiltonian approach. [1]

6.2.1 *Automatic conservation of the source and gauge fields*

In 1916 Einstein showed that local coordinate invariance plus his field equations gives material energy momentum conservation, without using the matter field equations (see Doc. 41 in Vol. 6 of Ref. 2). This is referred to as *automatic conservation of the source* (see section 17.1 in Ref. 25); it uses a Noether second theorem local (gauge) symmetry type of argument to obtain current conservation. Hermann Weyl argued in this way for the electromagnetic current in his papers of 1918 (the name *gauge theory* comes from this work) and 1929,[3] whereas modern field theory generally uses Noether's first theorem for current conservation.

The *essence* of gauge theory is *a local symmetry*, consequently: (i) a differential

---

[3]An English translation of Weyl's seminal papers can be found in Ref. 4.

identity, (ii) under-determined evolution, (iii) restricted type of source coupling, (iv) automatic conservation of the source. Yang–Mills is only one special type. Our gauge approach to gravity does not try to force it into the Yang–Mills mold, but rather simply recognizes the natural local symmetries associated with the spacetime geometry.

6.3 Noether's 1918 contribution

One word well describes 20th century physics: *symmetry*. Most of the theoretical physics ideas involved symmetry—essentially they are applications of Noether's two theorems. [5] The first associates conserved quantities with global symmetries. The second concerns *local symmetries*: it is *the foundation of the modern gauge theories*.

Why did Noether make her investigation? She was a mathematician; her interest was not physics. At the time she was assisting Hilbert and Klein, especially in connection with the puzzling issue of energy in GR. After presenting her two famous theorems she uses them to draw the conclusion that clarifies the situation. [5]

Her result regarding *"the lack of a proper law of energy"* applies not just to Einstein's GR, but *to all geometric theories of gravity*. For gravitating systems there is no well-defined *local* energy-momentum density. The modern view is that energy-momentum is not "local" (i.e., meaningful as a density at a point) but rather *quasi-local*—associated with a closed 2-surface. [6]

6.4 Energy-momentum pseudotensors and the Hamiltonian

The Einstein Lagrangian differs from Hilbert's by a total divergence:

$$2\kappa \mathcal{L}_{\mathrm{E}}(g_{\alpha\beta}, \partial_\mu g_{\alpha\beta}) := -\sqrt{-g}\, g^{\beta\sigma}\Gamma^\alpha{}_{\gamma\mu}\Gamma^\gamma{}_{\beta\nu}\delta^{\mu\nu}_{\alpha\sigma} \equiv \sqrt{-g}\,R - \mathrm{div}. \tag{6.1}$$

From this Einstein constructed the associated canonical energy-momentum density, now known as the *Einstein pseudotensor*:[4]

$$\mathfrak{t}^\mu_{\mathrm{E}\nu} := \delta^\mu_\nu \mathcal{L}_{\mathrm{E}} - \frac{\partial \mathcal{L}_{\mathrm{E}}}{\partial \partial_\mu g_{\alpha\beta}} \partial_\nu g_{\alpha\beta}. \tag{6.2}$$

Using the Einstein equation $\sqrt{-g}\,G^\mu{}_\nu = \kappa \mathfrak{T}^\mu{}_\nu$ one gets a conserved total energy-momentum:

$$\partial_\mu(\mathfrak{T}^\mu{}_\nu + \mathfrak{t}^\mu_{\mathrm{E}\nu}) = 0, \qquad \Longleftrightarrow \qquad \sqrt{-g}\,G^\mu{}_\nu + \kappa \mathfrak{t}^\mu_{\mathrm{E}\nu} = \partial_\lambda \mathfrak{U}^{[\mu\lambda]}{}_\nu. \tag{6.3}$$

A good form for the superpotential $\mathfrak{U}$ was found only much later by Freud in 1939: [7] $\mathfrak{U}^{\mu\lambda}_{\mathrm{F}\ \nu} := -\mathfrak{g}^{\beta\sigma}\Gamma^\alpha{}_{\beta\gamma}\delta^{\mu\lambda\gamma}_{\alpha\sigma\nu}$. Other pseudotensors similarly follow from different super-potentials. They are all inherently coordinate reference frame dependent. Thus

---

[4]It is not a *proper* tensor.

there are two big problems: (1) *which pseudotensor?* (2) *which reference frame?* The Hamiltonian approach, as we shall explain, has answers.

With constant Z^μ, the energy-momentum within a region is

$$-Z^\mu P_\mu(V) := -\int_V Z^\mu(\mathfrak{T}^\nu{}_\mu + \mathfrak{t}^\nu{}_\mu)\sqrt{-g}d^3\Sigma_\nu$$

$$\equiv \int_V \left[Z^\mu\sqrt{-g}\left(\frac{1}{\kappa}G^\nu{}_\mu - T^\nu{}_\mu\right) - \frac{1}{2\kappa}\partial_\lambda\left(Z^\mu\mathfrak{U}^{\nu\lambda}{}_\mu\right)\right]d^3\Sigma_\nu$$

$$\equiv \int_V Z^\mu\mathcal{H}^{\mathrm{GR}}_\mu + \oint_{S=\partial V}\mathcal{B}^{\mathrm{GR}}(Z) \equiv H(Z,V). \tag{6.4}$$

$\mathcal{H}^{\mathrm{GR}}_\mu$ is the well known covariant expression for the *Hamiltonian density*. The *boundary term* 2-surface integral is determined by the superpotential. The value of the pseudotensor/Hamiltonian is thus *quasi-local*, from just the boundary term, since by the initial value constraints the spatial volume integral vanishes.

6.5 The Hamiltonian approach

Noether's work can be combined with the Hamiltonian formulation. In Hamiltonian field theory, *the conserved currents are the generators of the associated symmetry.* For local spacetime "translations" (i.e., infinitesimal diffeomorphisms), the associated current expression (i.e., the energy-momentum density) *is* the Hamiltonian density—the canonical generator of spacetime displacements. Because it can be varied it gives a handle on the conserved current ambiguity. As we will explain, the Hamiltonian variation gives information that tames the ambiguity in the boundary term—namely boundary conditions. In this way problem 1 is under control. Pseudotensor values are values of the Hamiltonian with certain boundary conditions. [8]

The Hamiltonian approach reveals certain aspects of a theory. The constrained Hamiltonian formalism was developed by Dirac [1] and by Bergmann and coworkers. It was applied to GR by Pirani, Schild and Skinner [9] and by Dirac [10]. Later the ADM approach [22] became dominant. For the Poincaré gauge theory of gravity (PG) the Hamiltonian approach was developed by Blagojević and coworkers. [12]

6.6 Gauge and geometry

For a good account of the early history of gauge theory see Ref. 4. Einstein's theory of general relativity (GR) with its principle of *general covariance* was the first recognized gauge theory, the first physical theory where a local gauge symmetry was understood from the beginning as playing a major role. Inspired by GR, Weyl, [13, 14] in his seminal works that developed a gauge theory of electrodynamics, identified the key features of all gauge theories.

In 1916 Einstein showed that local coordinate invariance plus his field equations gives material energy momentum conservation, without using the matter field equations (see Doc. 41 in Vol. 6 of Ref. 2). This is referred to as *automatic conservation of the source* (see section 17.1 in Ref. 25); it uses a Noether second theorem local (gauge) symmetry type of argument to obtain current conservation. Hermann Weyl argued in this way for the electromagnetic current in his papers of 1918 (the name *gauge theory* comes from this work) and 1929,[5] whereas modern field theory generally uses Noether's first theorem for current conservation.

The *essence* of gauge theory is *a local symmetry*, consequently: (i) a differential identity, (ii) under-determined evolution, (iii) restricted type of source coupling, (iv) automatic conservation of the source.

Yang–Mills is only one special type of gauge theory. Our gauge approach to gravity does not try to force it into the Yang–Mills mold, but rather simply recognizes the natural symmetries of spacetime geometry.

Gravity viewed explicitly as a gauge theory was pioneered by Utiyama (1956, 1959), Sciama (1961) and Kibble (1961). For accounts of gravity as a spacetime symmetry gauge theory, see Hayashi & Shirifuji [15], Hehl and coworkers [16, 17, 18, 19], Mielke [20] and Blagojević [21]. A comprehensive reader with summaries, discussions and reprints has recently appeared. [22]

GR can be seen as the original gauge theory: the first physical theory where a local gauge freedom (general covariance) played a key role. Although the electrodynamics potentials with their gauge freedom were known long before GR yet this gauge invariance was not seen as having any important role in connection with the nature of the interaction, the conservation of current, or a differential identity—until the seminal work of Weyl, which post-dated (and was inspired by) GR.

We also should draw attention to the parallel developments of the concept of a connection in geometry by Levi–Civita, Weyl, Schouten, Cartan, Eddington, and others. Riemann–Cartan geometry (with a metric and a metric compatible connection, having both curvature and torsion) is the most appropriate for a dynamic spacetime geometry theory: its local symmetries are just those of the local Poincaré group. The conserved quantities, energy-momentum and angular momentum/center-of-mass momentum are associated with the Minkowski spacetime symmetry, i.e., the Poincaré group.

6.7 Geometry: kinematics and dynamics

For general dynamical geometry (*metric-affine gravity*, MAG), [18] the geometric potentials can be taken as the *metric* $g_{\mu\nu}$, the *co-frame* one-form ϑ^μ and the (a

---

[5] For an English translation of Weyl's papers see Ref. 4.

priori independent) *connection* one-form $\Gamma^{\alpha}{}_{\beta}$. The respective field strengths are

$$Q_{\mu\nu} := -Dg_{\mu\nu} := \quad -dg_{\mu\nu} + \Gamma^{\gamma}{}_{\mu}g_{\gamma\nu} + \Gamma^{\gamma}{}_{\nu}g_{\mu\gamma}, \quad \text{non-metricity one-form} \qquad (6.5)$$

$$T^{\alpha} := \quad D\vartheta^{\alpha} := \qquad d\vartheta^{\alpha} + \Gamma^{\alpha}{}_{\beta} \wedge \vartheta^{\beta}, \quad \text{torsion two-form} \qquad (6.6)$$

$$R^{\alpha}{}_{\beta} := \quad D\Gamma^{\alpha}{}_{\beta} := \qquad d\Gamma^{\alpha}{}_{\beta} + \Gamma^{\alpha}{}_{\gamma} \wedge \Gamma^{\gamma}{}_{\beta}, \quad \text{curvature two-form} \qquad (6.7)$$

which have the respective *Bianchi identities*:

$$DQ_{\mu\nu} \equiv \quad -D^2 g_{\mu\nu} \equiv \quad R_{\mu\nu} + R_{\nu\mu}, \qquad (6.8)$$

$$DT^{\alpha} \equiv \quad D^2\vartheta^{\alpha} \equiv \quad R^{\alpha}{}_{\beta} \wedge \vartheta^{\beta}, \qquad (6.9)$$

$$DR^{\alpha}{}_{\beta} \equiv \quad D^2\Gamma^{\alpha}{}_{\beta} \equiv \quad 0. \qquad (6.10)$$

Second order field equations for dynamical geometry can be obtained by varying the potentials independently in a Lagrangian 4-form:[6]

$$\mathcal{L} = \mathcal{L}(g_{\mu\nu}, \vartheta^{\mu}, \Gamma^{\alpha}{}_{\beta}; Q_{\mu\nu}, T^{\mu}, R^{\alpha}{}_{\beta}). \qquad (6.11)$$

Because of the diffeomorphism and local frame gauge symmetries, the resultant field equations will satisfy the associated differential identities.

An alternative is a *first-order* Lagrangian of the form

$$\mathcal{L}^1 = Dg_{\mu\nu} \wedge \pi^{\mu\nu} + D\vartheta^{\mu} \wedge \tau_{\mu} + D\Gamma^{\alpha}{}_{\beta} \wedge \rho_{\alpha}{}^{\beta} + \Lambda(g, \vartheta, \Gamma; \pi, \tau, \rho), \qquad (6.12)$$

for which independent variations of the potentials and their associated conjugate momentum fields leads to pairs of *first-order* equations. Such a formulation has some advantages. The connection constraints of *metric compatible, symmetic* or *teleparallel* can be easily imposed simply by choosing the potential Λ to be independent of the associated conjugate momentum field; then the field equation obtained from variation with respect to the conjugate momentum leads, respectively, to vanishing non-metricity, torsion or curvature.

Another advantage is that a first-order formulation allows for the construction of a covariant Hamiltonian formulation, as we shall explain below.

The frame can always be restricted to be *orthonormal*, then the metric coefficients are *constants* and one can then eliminate the metric as a dynamical variable; the frame will still have *local Lorentz* gauge freedom. The infinitesimal local symmetry group is then local infinitesimal Lorentz frame gauge transformations plus infinitesimal diffeomorphisms, i.e., local displacements of the spacetime point. This can be recognized as the local Poincaré group, which can be viewed as acting on the local observer and his frame. Thus dynamical geometry gravitational theories are *naturally* local gauge theories of the symmetry group of Minkowski spacetime: *Poincaré gauge theory.*

It is customary to use the term *Poincaré gauge theory of gravity* (PG) to be restricted to the case where the connection is *metric compatible*. The geometry is

---

[6]For a manifestly covariant formulation without any frames or components see Ref. 23.

then Riemann–Cartan, with curvature and torsion. Although one could still use the more general metric-affine formulation with the metric compatible constraint imposed by using the metric's conjugate momentum as a Lagrange multiplier, it is more revealing—and more efficient—to restrict to using an orthonormal frame and drop the metric and its momentum as dynamical variables. In an orthonormal frame with a metric compatible connection both $\Gamma^{\alpha\beta}$ and $R^{\alpha\beta}$ are anti-symmetric (Lorentz Lie algebra valued forms).

The MAG and PG equations and Noether differential identities along with the interaction with source fields are rather lengthy; they have been presented and discussed in detail elsewhere. [18, 24, 25]

6.8 The Poincaré gauge theory of gravity

The standard PG Lagrangian density has a quadratic field strength form:[7]

$$\mathcal{L}_{\text{PG}} \sim \frac{1}{\kappa}\left(\Lambda + \text{curvature} + \text{torsion}^2\right) + \frac{1}{\varrho}\,\text{curvature}^2, \tag{6.13}$$

Varying ϑ, Γ gives quasi-linear 2nd order dynamical equations for the potentials:

$$\kappa^{-1}(\Lambda + \text{curv} + D\,\text{tor} + \text{tor}^2) + \varrho^{-1}\text{curv}^2 = \text{energy-momentum}, \tag{6.14}$$

$$\kappa^{-1}\text{tor} + \varrho^{-1}D\,\text{curv} = \text{spin}. \tag{6.15}$$

In complete detail the general quadratic PG Lagrangian 4-form is [26]

$$\begin{aligned}
\mathcal{L}_{\text{PG}} = \ &\frac{1}{2\kappa}\left(a_0 R\eta + b_0 X\eta - 2\Lambda\eta + \sum_{I=1}^{3} a_I\,^{(I)}T^\alpha \wedge *^{(I)}T_\alpha\right) \\
&+\frac{1}{\kappa}\left(\sigma_1\,^{(1)}T^\alpha \wedge {}^{(1)}T_\alpha + \sigma_2\,^{(2)}T^\alpha \wedge {}^{(3)}T_\alpha\right) \\
&-\frac{1}{2\varrho}\left(\sum_{I=1}^{6} w_I\,^{(I)}R^{\alpha\beta} \wedge *^{(I)}R_{\alpha\beta}\right) \\
&-\frac{1}{2\varrho}\left(\mu_1\,^{(1)}R^{\alpha\beta} \wedge {}^{(1)}R_{\alpha\beta} + \mu_2\,^{(2)}R^{\alpha\beta} \wedge {}^{(4)}R_{\alpha\beta}\right. \\
&\left.\qquad+ \mu_3\,^{(3)}R^{\alpha\beta} \wedge {}^{(6)}R_{\alpha\beta} + \mu_4\,^{(5)}R^{\alpha\beta} \wedge {}^{(5)}R_{\alpha\beta}\right).
\end{aligned} \tag{6.16}$$

Here R is the scalar curvature and X is the pseudoscalar curvature ($X \equiv -\frac{1}{2}R_{\alpha\beta\mu\nu}\eta^{\alpha\beta\mu\nu}$). The torsion has been decomposed into three algebraically irreducible pieces: $T^\alpha = {}^{(1)}T^\alpha + {}^{(2)}T^\alpha + {}^{(3)}T^\alpha$, which are, respectively, a pure tensor (16 components), the trace (vector), and a totally antisymmetric part (dual to an axial vector). Similarly the curvature 2-form has been decomposed into a sum of 6 algebraically irreducible pieces: $R^\alpha{}_\beta = \sum_{I=1}^{6} {}^{(I)}R^\alpha{}_\beta$, namely, in numerical order: weyl, pair-commutator, pseudoscalar, ricci-symmetric, ricci-antisymmetric, and scalar. The respective number of components is (10,9,1,9,6,1). In the above

---

[7] $\kappa := 8\pi G/c^4$ and ϱ^{-1} has the dimensions of action. Λ is the cosmological constant.

Lagrangian the parameters Λ, a_0, a_I, w_I (which multiply even parity 4-forms) are scalars, and the parameters b_0, σ_I, μ_I (which multiply odd parity 4-forms) are pseudoscalars. The general theory has 11 scalar plus 7 pseudoscalar parameters. But they are not all physically independent. They are subject to 1 even parity and 2 odd total differentials, leaving effectively 10 scalar + 5 pseudoscalar = 15 "physical" parameters, as we will briefly explain, referring to Refs. 26,27 for details.

6.8.1 *Topological terms*

Not all of the above parameters are physically independent, since there are 3 topological invariants. Without changing the field equations, one can add to the Lagrangian 4-form (6.16) any multiple of the (odd parity) *Nieh–Yan identity* [28]:

$$T^\alpha \wedge T_\alpha - R_{\alpha\beta} \wedge \vartheta^{\alpha\beta} \equiv d(\vartheta^\alpha \wedge T_\alpha). \tag{6.17}$$

Also one can add a multiple of the (even parity) *Euler 4-form* $R^{\alpha\beta} \wedge R^{\gamma\delta}\eta_{\alpha\beta\gamma\delta}$. Because of the 2nd Bianchi identity (6.10), this makes no contributions to the field equations. Furthermore once can add a multiple of the (odd parity) *Pontryagin* 4-form $R^\alpha{}_\beta \wedge R^\beta{}_\alpha$. Again, thanks to the 2nd Bianchi identity, this has no effect on the field equations. The actual physical equations will only depend on the 15 combinations of the 18 parameters that are invariant under such transformations.

The PG dynamics has been discussed in detail in Ref. 25 including (i) the Lagrangian, both 2nd and 1st order, (ii) the Noether symmetries, conserved currents and differential identities, (iii) the covariant Hamiltonian including the generators of the local Poincaré gauge symmetries, (iv) our preferred Hamiltonian boundary term, (v) the quasi-local energy-momentum and angular momentum/center-of-mass moment obtained therefrom, and (vi) the choice of reference in the boundary term. We will include below a brief report of the Hamiltonian, its boundary term, and the associated quasi-local quantities. The general PG homogeneous and isotropic cosmology has been considered by our group recently. [29]

6.9 The covariant Hamiltonian formulation

Overview: The Hamiltonian for dynamical geometry generates the evolution of a spatial region along a vector field. It includes a boundary term which determines both the value of the Hamiltonian and the boundary conditions. The value gives the quasi-local quantities: energy-momentum, angular-momentum and center-of-mass. The boundary term depends not only on the dynamical variables but also on their reference values; the latter determine the ground state (having vanishing quasi-local quantities). For our preferred boundary term for the PG (including Einstein's GR as a special case) we proposed 4D isometric matching and extremizing the energy to determine the reference metric and connection values.

Although the global *total* energy-momentum is well defined (for spaces with suitable asymptotic regions), for any gravitating system — and hence for all real *physical* systems — the localization of energy-momentum is still an outstanding fundamental problem. [6] Unlike all matter and other interaction fields, the gravitational field itself has *no* proper energy-momentum density. In view of the fact that energy-momentum is conserved, and that sources *exchange* energy-momentum *locally* with the gravitational field, one expects some kind of "local description" of the energy-momentum density of gravity itself. But all attempts at constructing such an expression led only to reference frame dependent quantities, generally referred to as *pseudotensors*. [8] Physically this can be understood as a consequence of *Einstein's equivalence principle*: gravity cannot be detected at a point. As mentioned earlier, Noether showed in her 1918 paper [5, 30] that there is no proper energy density. The energy-momentum of gravity — and thus for all physical systems is inherently *non-local*. The modern idea is *quasi-local*: energy-momentum is associated with a closed surface bounding a region. [31]

From a first order Lagrangian formulation which gives pairs of first order equations for a k-form φ and its conjugate p, we developed a 4D-*covariant* Hamiltonian formalism. [8, 23, 25, 32, 33, 34, 35] The Hamiltonian generates the evolution of a spatial region along a vector field.

The *first order* Lagrangian for a k-form[8] field φ and its *conjugate momentum p* is given by

$$\mathcal{L} = \mathcal{L}(d\varphi; \varphi, p) = d\varphi \wedge p - \Lambda(\varphi, p). \tag{6.18}$$

The variation (with respect to φ and p independently)

$$\delta\mathcal{L} = d(\delta\varphi \wedge p) + \delta\varphi \wedge \frac{\delta\mathcal{L}}{\delta\varphi} + \frac{\delta\mathcal{L}}{\delta p} \wedge \delta p \tag{6.19}$$

gives the equations of motion, with $\varsigma := (-1)^k$,

$$\frac{\delta\mathcal{L}}{\delta p} := d\varphi - \partial_p\Lambda = 0, \qquad \frac{\delta\mathcal{L}}{\delta\varphi} := -\varsigma dp - \partial_\varphi\Lambda = 0. \tag{6.20}$$

Infinitesimal diffeomorphism invariance (in (6.19) replacing δ by the Lie derivative operator on forms: $\pounds_Z = di_Z + i_Z d$), leads to an identity for *any vector* Z:

$$di_Z\mathcal{L} \equiv \pounds_Z\mathcal{L} \equiv d(\pounds_Z\varphi \wedge p) + \pounds_Z\varphi \wedge \frac{\delta\mathcal{L}}{\delta\varphi} + \frac{\delta\mathcal{L}}{\delta p} \wedge \pounds_Z p. \tag{6.21}$$

From this one gets a "translational current" 3-form which is conserved *on shell* (i.e., when the field equations are satisfied):

$$-d\mathcal{H}(Z) \equiv \pounds_Z\varphi \wedge \frac{\delta\mathcal{L}}{\delta\varphi} + \frac{\delta\mathcal{L}}{\delta p} \wedge \pounds_Z p, \tag{6.22}$$

$$\mathcal{H}(Z) := \pounds_Z\varphi \wedge p - i_Z\mathcal{L} \tag{6.23}$$

$$\equiv \varsigma i_Z\varphi \wedge dp + \varsigma d\varphi \wedge i_Z p + i_Z\Lambda + d(i_Z\varphi \wedge p) \tag{6.24}$$

$$=: Z^\mu\mathcal{H}_\mu + d\mathcal{B}(Z). \tag{6.25}$$

---

[8]This can include several fields of various types and grades with their indicies suppressed.

A consequence of the expression (6.25) and (6.22) is

$$d\mathcal{H}(Z) = d[Z^\mu \mathcal{H}_\mu + d\mathcal{B}(Z)] \equiv dZ^\mu \wedge \mathcal{H}_\mu + Z^\mu d\mathcal{H}_\mu, \tag{6.26}$$
$$\implies \quad \mathcal{H}_\mu \text{ vanishes "on shell".}$$

Hence for gravitating systems the Noether translational "charge" — *energy-momentum* — is *quasi-local*, it is given by the integral of the boundary term, $\mathcal{B}(N)$:

$$E(Z, \Sigma) = \int_\Sigma \mathcal{H}(Z) = \oint_{\partial\Sigma} \mathcal{B}(Z). \tag{6.27}$$

But the total differential/boundary term can be completely modified:

$$\mathcal{H}' = \mathcal{H} + d\mathcal{B}' \quad \implies \quad d\mathcal{H} = d\mathcal{H}'. \tag{6.28}$$

This does not change the conservation property (it is an instance of the usual Noether current ambiguity—essentially adding a curl preserves the vanishing divergence property), but such a modification does change the conserved value. The Hamiltonian approach *tames this ambiguity*.

$\mathcal{H}(Z)$ is not merely the Noether translational current, it is also the *generator* of local diffeomorphisms, i.e., the *Hamiltonian* density (3-form) which evolves a spacetime region along the vector field Z:

$$H(Z, \Sigma) = \int_\Sigma \mathcal{H}(Z) = \int_\Sigma Z^\mu \mathcal{H}_\mu + \oint_{S=\partial\Sigma} \mathcal{B}(Z). \tag{6.29}$$

From this perspective one can identify the roles played by its separate pieces: the 3-form $\mathcal{H}_\mu Z^\mu$ generates the infinitesimal displacements along the vector field Z (the Hamiltonian equations); this follows from the easily verified variational identity:

$$\delta\mathcal{H}(N) \equiv -\delta\varphi \wedge \pounds_N p + \pounds_N \varphi \wedge \delta p + d i_N (\delta\varphi \wedge p) - i_N \left(\delta\varphi \wedge \frac{\delta\mathcal{L}}{\delta\varphi} + \frac{\delta\mathcal{L}}{\delta p} \wedge \delta p \right). \tag{6.30}$$

The boundary term, on the other hand, has two roles. (i) The Hamiltonian value—the quasi-local quantities—as can be seen from (6.27), (6.29) are determined only by the surface integral:

$$E(Z, \Sigma) = \int_\Sigma \mathcal{H}(Z) = \int_\Sigma [Z^\mu \mathcal{H}_\mu + d\mathcal{B}(Z)] = \oint_{\partial\Sigma} \mathcal{B}(Z). \tag{6.31}$$

(ii) Our Noether analysis revealed that $\mathcal{B}(N)$ can be adjusted, changing the conserved value to a new value. Fortunately the variational principle contains a (largely overlooked) feature which distinguishes all of these choices. The *boundary variation principle*: the *boundary term* in the *variation* of the Hamiltonian (6.30) shows what is to be held fixed on the boundary—thus it determines the boundary conditions [32, 34]. Hence the *first ambiguity*—which expression?—is clearly under physical control: different Hamiltonian boundary term quasi-local expressions are associated with different types of physical boundary conditions. This is similar to thermodynamics where there are various "energies" (internal, enthalpy, Helmholtz, Gibbs, etc.) which correspond to how the system interacts with the outside through its boundary. The Hamiltonian boundary term should be adjusted to give suitable boundary conditions.

In general (in particular for gravity) it is necessary (in order to guarantee functional differentiability of the Hamiltonian on the phase space with the desired boundary conditions) to adjust the boundary term $\mathcal{B}(N) = i_N \varphi \wedge p$ which is naturally inherited from the Lagrangian (6.18).

We were led to a 2 parameter set of general boundary terms which are linear in $\Delta \varphi := \varphi - \bar{\varphi}$, $\Delta p := p - \bar{p}$, where $\bar{\varphi}, \bar{p}$ are reference values: [32, 34, 35, 36]

$$\mathcal{B}(Z) := i_Z \{ a\varphi + (1-a)\bar{\varphi} \} \wedge \Delta p - \zeta \Delta \varphi \wedge i_Z \{ bp + (1-b)\bar{p} \}. \tag{6.32}$$

The associated variational Hamiltonian boundary term is

$$\delta \mathcal{H}(Z) \sim d \Big[\{ a i_Z \delta \varphi \wedge \Delta p - (1-a) i_Z \Delta \varphi \wedge \delta p \}$$

$$+ \zeta \{ -b \Delta \varphi \wedge i_Z \delta p + (1-b) \delta \varphi \wedge i_Z \Delta p \} \Big]. \tag{6.33}$$

Here the extreme values $a, b = \{(0,0), (0,1), (1,0), (1,1)\}$ represent essentially a choice of Dirichlet (fixed field) or Neumann (fixed momentum) boundary conditions for the space and time parts of the fields separately.

For asymptotically flat spaces the Hamiltonian with any of these boundary term expressions is *well defined*, i.e., the boundary term in its variation vanishes and the quasi-local quantities are well defined—at least on the phase space of fields satisfying Regge-Teitelboim [37] like asymptotic parity/fall-off conditions:

$$\Delta \varphi \approx \mathcal{O}^+(1/r) + \mathcal{O}^-(1/r^2), \qquad \Delta p \approx \mathcal{O}^-(1/r^2) + \mathcal{O}^+(1/r^3). \tag{6.34}$$

Also from (6.30), (6.33) the formalism has natural energy flux expressions. [35]

The Hamiltonian boundary terms determines the values of the quasi-local quantities. For asymptotically flat spaces:

- energy is given by a suitable *timelike* displacement;
- linear momentum is obtained from a *spatial* translation;
- angular momentum from a suitable *rotational* displacement;
- a *boost* will give the center-of-mass moment.

6.10 Application to the PG and GR

A first order Lagrangian for Einstein's (vacuum) gravity theory is

$$\mathcal{L}_{\mathrm{GR}} = \frac{1}{2\kappa} R^\alpha{}_\beta \wedge \eta_\alpha{}^\beta, \tag{6.35}$$

where $\eta^{\alpha\beta} := *(\vartheta^\alpha \wedge \vartheta^\beta)$. Our general formalism with $\varphi \to \Gamma^\alpha{}_\beta$ and $p \to \eta_\alpha{}^\beta$ gives a 2 parameter set of quasi-local expressions for GR:

$$2\kappa \mathcal{B}(Z) := \Delta \Gamma^\alpha{}_\beta \wedge i_Z [a\eta_\alpha{}^\beta + (1-a)\bar{\eta}_\alpha{}^\beta] + [b D_\beta Z^\alpha (1-b) \bar{D}_\beta Z^\alpha] \Delta \eta_\alpha{}^\beta. \tag{6.36}$$

6.10.1 *Preferred Hamiltonian boundary terms*

For GR, we identified a distinguished expression with some desirable properties:

$$\mathcal{B}_\vartheta(Z) := \frac{1}{2\kappa}\left(\Delta\Gamma^\alpha{}_\beta \wedge i_Z\eta_\alpha{}^\beta + \bar{D}_\beta Z^\alpha \Delta\eta_\alpha{}^\beta\right). \tag{6.37}$$

For this expression

$$\delta\mathcal{H}_\vartheta(Z) \simeq d i_Z(\Delta\Gamma^\alpha{}_\beta \wedge \delta\eta_\alpha{}^\beta). \tag{6.38}$$

Hence it corresponds to imposing boundary conditions on a *manifestly covariant object* (the coframe, i.e. essentially the metric—the obvious variable choice). The associated energy flux expression is

$$\pounds_Z\mathcal{H}_\vartheta \simeq d i_Z\left(\Delta\Gamma^\alpha_\beta \wedge \pounds_Z\eta_\alpha{}^\beta\right). \tag{6.39}$$

Like many other choices, for asymptotically flat spaces at spatial infinity, $\mathcal{B}_\vartheta(Z)$ (6.37) gives the standard values for energy-momentum and angular momentum/center-of-mass momentum. [25, 22, 37, 38, 39] Our preferred GR expression has some special virtues, including: (i) at null infinity it gives the Bondi–Trautman energy and Bondi energy flux, [35] (ii) it is covariant, (iii) it can give positive energy, (iv) for a suitable choice of reference it vanishes for Minkowski space.

For the PG our preferred Hamiltonian boundary term is

$$\mathcal{B}_{\mathrm{PG}}(Z) = i_Z\vartheta^\alpha\tau_\alpha + \Delta\Gamma^\alpha{}_\beta \wedge i_Z\rho_\alpha{}^\beta + \bar{D}_\beta Z^\alpha \Delta\rho_\alpha{}^\beta. \tag{6.40}$$

For more details about it please see the works cited above.

6.11 The reference choice

Regarding the *second ambiguity* inherent in our quasi-local energy-momentum expressions: *the choice of reference*. In principle, one could use any reference appropriate to the physical application. In practice one normally wants a very symmetrical reference. Only if the chosen reference is a space of *constant curvature* (positive for de Sitter, negative for anti-de Sitter and zero for Minkowski) will one have 10 reference Killing vector fields that can be used for the vector Z to define 10 quasi-local quantities via (6.37) and (6.40) for GR and the PG, respectively. For a Minkowski reference they are the energy-momentum, angular momentum and center-of-mass moment.

Here we present some details just for the case of GR with a Minkowski reference. One then needs to choose a specific Minkowski space. Recently we proposed (i) *4D isometric matching on the boundary*,[9] and (ii) *energy optimization* as criteria for the

---

[9]The hardest part of 4D isometric matching is the embedding of the 2D surface S into Minkowski space; Yau and coworkers have extensively investigated this. [40]

"best matched" reference on the boundary of the quasi-local region. This proposal has been tested on spherically symmetric and axisymmetric spacetimes. [41]

Essentially one needs a reference geometry in the neighborhood of the boundary of the region. One could view this as embedding a neighborhood of the 2-boundary into Minkowski space. [42] One construction is to choose, in a neighborhood of the desired spacelike boundary 2-surface S, 4 smooth functions $y^i = y^i(x^\mu)$, $i = 0, 1, 2, 3$ with $dy^0 \wedge dy^1 \wedge dy^2 \wedge dy^3 \neq 0$ and use them to define a Minkowski reference:

$$\bar{g} = -(dy^0)^2 + (dy^1)^2 + (dy^2)^2 + (dy^3)^2. \tag{6.41}$$

The reference connection is then

$$\bar{\Gamma}^\alpha{}_\beta = x^\alpha{}_i(\bar{\Gamma}^i{}_j y^j{}_\beta + dy^i{}_\beta) = x^\alpha{}_i dy^i{}_\beta, \tag{6.42}$$

where $dy^i = y^i{}_\alpha dx^\alpha$ and $dx^\alpha = x^\alpha{}_j dy^j$ have been used along with vanishing Minkowski reference connection coefficients, $\bar{\Gamma}^i{}_j = 0$. If Z^μ is a *translational Killing field* of the Minkowski reference, then $\bar{D}Z$ vanishes, and hence so does the 2nd quasi-local term. Our quasi-local expression then takes the simpler form:

$$\mathcal{B}(Z) = Z^k x^\mu{}_k(\Gamma^\alpha{}_\beta - x^\alpha{}_j \, dy^j{}_\beta) \wedge \eta_{\mu\alpha}{}^\beta. \tag{6.43}$$

How we determine the reference choice $y^i{}_\mu$ can be simply explained with the aid of quasi-spherical foliation adapted coordinates t, r, θ, ϕ. Isometric matching on the 2-surface implies

$$g_{AB} = \bar{g}_{AB} = \bar{g}_{ij} y^i_A y^j_B = -y^0_A y^0_B + \delta_{ab} y^a_A y^b_B; \quad a, b = 1, 2, 3; \ A, B = 2, 3 = \theta, \phi, \tag{6.44}$$

where the reference metric on the dynamical space has the components $\bar{g}_{\mu\nu} = \bar{g}_{ij} y^i{}_\mu y^j{}_\nu$. From a classic closed 2-surface into $\mathbb{R}^3$ embedding theorem—as long as one restricts S and $y^0(x^A)$ such that on S

$$g'_{AB} := g_{AB} + y^0_A y^0_B \tag{6.45}$$

is convex—one has a unique isometric embedding. (But, unfortunately, there is no explicit formula.)

6.11.1 *4D isometric matching*

Complete 4D isometric matching on S has 10 constraints: [41, 50, 51, 52]

$$g_{\mu\nu}|_S = \bar{g}_{\mu\nu}|_S = \bar{g}_{ij} y^i{}_\mu y^j{}_\nu|_S. \tag{6.46}$$

(With this condition $\Delta\eta_\alpha{}^\beta$ vanishes, so the 2nd term in our quasi-local expression (6.37) vanishes for all Z.)

There are 12 embedding functions on the constant t, r 2-surface:

$$y^i(\Longrightarrow y^i_\theta, y^i_\phi), \quad y^i_t, \quad y^i_r. \tag{6.47}$$

The 10 constraints split into 3 for the already discussed 2D isometric matching: $g_{\theta\theta}, g_{\theta\varphi}, g_{\varphi\varphi}$ which constrain the 4 y^i; 3 normal bundle algebraic quadratic expressions: g_{tt}, g_{tr}, g_{rr}; and 4 mixed linear algebraic expressions: $g_{t\theta}, g_{t\varphi}, g_{r\theta}, g_{r\varphi}$. The 2D isometric matching can be regarded as a given y^0 uniquely determining y^1, y^2, y^3 on S. The remaining 7 algebraic equations can be regarded as fixing the other 7 embedding variables in terms of y^i and $y^0{}_r$. Thus one can take y^0, $y^0{}_r$ as the embedding control variables. Geometrically $y^0{}_r$ controls a boost in the normal plane.

An alternative approach is to regard 4D isometric matching in terms of orthonormal frames. The reference geometry will have a reference frame of the form $\bar{\vartheta}^i = dy^i$. If there is 4D isometric matching then one can choose the dynamical frame such that it can be Lorentz transformed to match such a reference frame at all points on the boundary:

$$L^i{}_\alpha(p)\vartheta^\alpha(p) = \bar{\vartheta}^i(p), \qquad \forall p \in S. \tag{6.48}$$

Then these 2-forms restricted to S must satisfy the integrability conditions:

$$d(L^i{}_\alpha\vartheta^\alpha)|_S = 0. \tag{6.49}$$

This is 4 2-forms on a 2D space, each 2-form having one component, thus this is 4 restrictions on the 6 parameters of the Lorentz transformation, so again we see that 4D isometric matching has 2 degrees of freedom.

6.11.2 *An optimal choice*

How to fix the remaining 2 degrees of freedom in the reference choice? One can regard the value of the boundary term as a measure of the difference between the dynamical boundary values and the reference boundary values. [41, 50, 51, 52]

For a given S there are 2 related quantities which can be considered: $m^2 = -\bar{g}^{ij}p_i p_j = p_0^2 - p_1^2 - p_2^2 - p_3^2$ and $E(Z, S)$. The *critical points* are distinguished. Consider first finding the critical points of m^2. Technically this is rather complicated: one would be extremizing a linear combination of quadratic quantities, each being an integral over S. It would determine the reference only up to Poincaré transformations. However, once could then use the available Lorentz "gauge" freedom to specialize to the reference in which the linear momentum vanished: $\vec{p} = 0$. In this "center-of-momentum" frame m^2 reduces to p_0^2. But the critical points of p_0^2 are also critical points of p_0. Thus one can arrive at the same reference by considering the much more simple case of the critical points of $E(\partial_{y^0}, S)$.

Based on some physical and practical computational arguments, it is reasonable to expect that one could find a unique solution. Consider being given a set of data from a numerical relativity calculation. Compute the energy given by a large number of reference choices, the critical values will stand out. For our quasi-local values for axisymmetric solutions including Kerr we were able to explicitly find the critical point analytically. [51]

6.12 Summary

For any gravitating system — and hence for all *physical* systems — the *localization* of energy-momentum is an outstanding problem. We have discussed the relation between the covariant Hamiltonian boundary term, the quasi-local quantities and the physical boundary conditions. For gravitating systems, using variables appropriate to a gauge theory perspective, we found certain quasi-local energy-momentum expressions; each is associated with a physically distinct, geometrically clear, boundary condition. We identified certain preferred expressions for the PG and GR. With a 4D isometric *"best matched"* reference, we have a method to determine the Hamiltonian boundary term quasi-locally for locally Poincaré gauge invariant gravity including GR. This in particular gives a way of resolving the ambiguities in determining the quasi-local energy-momentum and angular momentum of classical physical systems.

Acknowledgments

C.M.C. was supported by the Ministry of Science and Technology of the R.O.C. under the grant MOST 106-2112-M-008-010.

Bibliography

[1] P. A. M. Dirac, *Lectures on Quantum Mechanics* (Belfer, Yeshiva Univ., 1964).

[2] The Collected Papers of Albert Einstein: http://einsteinpapers.press.princeton.edu.

[3] C. W. Misner, K. S. Thorne and J. A. Wheeler, *Gravitation* (W. H. Freeman, San Francisco, 1973).

[4] L. O'Raifeartaigh, *The Dawning of Gauge Theory* (Princeton University Press, Princeton, 1997).

[5] Y. Kosmann-Schwarzbach, *The Noether Theorems: Invariance and Conservation Laws in the Twentieth Century* (Springer, New York, 2011).

[6] L. B. Szabados, *Living Rev. Relativ.* **12**, 4 (2009).

[7] Ph. Freud, *Ann. Math.* **40**, 417 (1939).

[8] C.-C. Chang, J. M. Nester and C.-M. Chen, *Phys. Rev. Lett.* **83**, 1897 (1999).

[9] F. Pirani, A. Schild and R. Skinner, *Phys. Rev.* **87**, 452 (1952).

[10] P. A. M. Dirac, *Proc. Roy. Soc.* (London) A **246**, 326 (1958).

[11] R. Arnowitt, S. Deser and C. W. Misner, "The dynamics of general relativity", in *Gravitation: An Introduction to Current Research*, ed. L. Witten (Wiley, New York, 1962), pp. 227–265; reprinted in *Gen. Relativ. Gravit.* **40**, 1997 (2008).

[12] M. Blagojević and I. A. Nikolić, *Phys. Rev.* D **28**, 2455 (1983); I. A. Nikolić, *Phys. Rev.* D **30**, 2508 (1984).

[13] H. Weyl, *Sitzungberichte der Königlich-preussischen Akademie der Wissenschaften zu Berlin* **26**, 465–480 (1918); English translation in O'Raifeartaigh [1997] Ref. [4] pp 24–37.

[14] H. Weyl, "Elektron und Gravitation. I", *Zeit. fur Physik* **56**, 330–352 (1929); English translation in O'Raifeartaigh [1997] Ref. [4] pp 121-144.

[15] K. Hayashi and T. Shirafuji, "Gravity from Poincare Gauge Theory of the Fundamental Particles. *Prog. Theor. Phys.* **64** (1980) 866–882, 883–896, 1435–1452, 2222–2241.

[16] F. W. Hehl, P. von der Heyde, G. D. Kerlik and J. M. Nester, "General Relativity with Spin and Torsion: Foundations and Prospects", *Rev. Mod. Phys.* **48**, 393 (1976).

[17] F. W. Hehl, "Four lectures on Poincaré gauge theory", in *Proc. 6th Course of the Int. School of Cosmology and Gravitation on Spin Torsion and Supergravity*, eds. P. G. Bergmann and V. de Sabbatta (Plenum, New York, 1980).

[18] F. W. Hehl, J. D. McCrea, E. W. Mielke and Y. Ne'eman, "Metric Affine Gauge Theory of Gravity: Field Equations, Noether Identities, World Spinors, and Breaking of Dilation Invariance", *Phys. Rep.* **258**, 1–171 (1995) [gr-qc/9402012].

[19] F. Gronwald and F. W. Hehl, "On the Gauge Aspects of Gravity", in *Proc. 14th Course of the School of Gravitation and Cosmology (Erice)*, eds. P.G. Bergman, V. de Sabata, and H.J. Treder (World Scientific, Singapore, 1996), pp. 148–98.

[20] E. W. Mielke, *Geometrodynamics of Gauge Fields*, (Akademie-Verlag, Berlin, 1987).

[21] M. Blagojević, *Gravitation and Gauge Symmetries* (Institute of Physics, Bristol, 2002).

[22] M. Blagojević and F. W. Hehl, *Gauge Theories of Gravitation* (Imperial College Press, London, 2013).

[23] J. M. Nester, "A Manifestly Covariant Hamiltonian Formalism for Dynamical Geometry", *Prog. Theor. Phys. Suppl.* **172**, 30–39 (2008).

[24] J. M. Nester and C.-M. Chen, "Gravity: a gauge theory perspective", *Int. J. Mod. Phys. D* **25** (2016) no. 13, 1645002; in *Everything about gravity: 2nd LeCosPa International Symposium* NTU, Taipei, Taiwan, 14–18 December 2015, ed. Pisin Chen (World Scientific, Singapore, 2017) pp 8—18.

[25] C.-M. Chen, J. M. Nester, and R.-S. Tung, *Int. J. Mod. Phys.* **24**, 1530026 (2015); in *One Hundred Years of General Relativity: From Genesis and Empirical Foundations to Gravitational Waves, Cosmology and Quantum Gravity* Vol 1., ed. W.-T. Ni (World Scientific, Singapore, 2017) pp I-187—I-252.

[26] P. Baekler, F. W. Hehl, and J. M. Nester, *Phys. Rev. D* **83**, 024011 (2011).

[27] P. Baekler and F. W. Hehl, *Class. Quantum Grav.* **28**, 215017 (2011).

[28] H. T. Nieh and M. L. Yan, J. Math. Phys. **23**, 373 (1982).

[29] F. H. Ho, H. Chen, J. M. Nester and H. J. Yo, "General Poincaré Gauge Theory Cosmology", *Chin. J. Phys.* **53** (2015) 110109.

[30] D. E. Rowe, The Göttingen response to general relativity and Emmy Noether's theorems, in *The Symbolic Universe: Geometry and Physics 1890–1930*, ed. J. Gray (Oxford University Press, Oxford, 1999), pp. 189–233.

[31] R. Penrose, "Quasi-local Mass and Angular Momentum in General Relativity", *Proc. R. Soc. London* A **381**, 53–63 (1982).

[32] C.-M. Chen and J. M. Nester, *Class. Quantum Grav.* **16**, 1279 (1999).

[33] C.-M. Chen and J. M. Nester, *Grav. Cosmol.* **6**, 257 (2000).

[34] C.-M. Chen, J. M. Nester and R.-S. Tung, *Phys. Lett.* A **203**, 5 (1995).

[35] C.-M. Chen, J. M. Nester and R.-S. Tung, *Phys. Rev. D* **72**, 104020 (2005).

[36] L. L. So, "A Modification of the Chen-Nester quasilocal expressions", *Int. J. Mod. Phys. D* **16** (2007) 875 [gr-qc/0605149].

[37] T. Regge and C. Teitelboim, "Role of Surface Integrals in the Hamiltonian Formulation of General Relativity", *Ann. Phys.* (N.Y.) **88**, 286–318 (1974).

[38] L. B. Szabados, "On the Roots of the Poincare Structure of Asymptotically Flat Space-times", *Class. Quantum Grav.* **20**, 2627–2662 (2003) [gr-qc/0302033].

[39] R. Beig and N. Ó Murchadha, "The Poincare Group as the Symmetry Group of Canonical General Relativity", *Ann. Phys.* (N.Y.) **174**, 463–498 (1987).

[40] M.-T. Wang and S.-T. Yau, *Phys. Rev. Lett.* **102** (2009) 021101; *Commun. Math. Phys.* **288**, 919 (2009).

[41] G. Sun, C.-M. Chen, J.-L. Liu and J. M. Nester, *Chinese J. Phys.* **53**, 110107 (2015).

[42] M.-F. Wu, C.-M. Chen, J.-L. Liu and J. M. Nester, *Phys. Rev. D* **84**, 084047 (2011); *Gen. Relat. Gravit.* **44**, 2401 (2012).

[43] Y. S. Duan, "Gauge Theories of Gravitation", *Commun. Theor. Phys.* **4** (1985) 661.

[44] Y. S. Duan, J. C. Liu, and X. G. Dong, "Generally Covariant Energy-Momentum Conservation Law in General Spacetime" *Gen. Relativ. Gravit.* **20** (1988) 485.

[45] Y. S. Duan and S. S. Feng, "Generally Covariant Conservation Law of Angular Momentum in General Relativity", *Commun. Theor. Phys.* **25** (1996) 99–104.

[46] Y. S. Duan and Y. Jiang, "Can torsion play a role in angular momentum conservation law?", *Gen. Rel. Grav.* **31** (1999) 3.

[47] S. S. Feng and Y. S. Duan, "Conservative quantities and their algebra in the selfdual gravity", *Gen. Rel. Grav.* **27** (1995) 887.

[48] S. X. Feng and Y. S. Duan, "About the energy of the universe" *Chin. Phys. Lett.* **13** (1996) 409.

[49] S. S. Feng and Y. S. Duan, "Generally covariant conservative energy momentum for gravitational anyons", *Class. Quant. Grav.* **16** (1999) 3237.

[50] J. M. Nester, C.-M. Chen, J.-L. Liu, and G. Sun, in *Relativity and Gravitation: 100 Years after Einstein in Prague*, Eds. J. Bičák and T. Ledvinka, (Springer, 2014); [arXiv:1210.6148 [gr-qc]].

[51] G. Sun, C.-M. Chen, J.-L. Liu and J. M. Nester, *Chinese J. Phys.* **52**, 111–125 (2014).

[52] G. Sun, C.-M. Chen, J.-L. Liu and J. M. Nester, "Covariant Hamiltonian boundary term: Reference and quasi-local quantities", in *Everything about gravity: 2nd LeCosPa International Symposium* NTU, Taipei, Taiwan, 14–18 December 2015, ed. Pisin Chen (World Scientific, Singapore, 2017) pp 620–626.

[53] Y. X. Liu, Z. H. Zhao, J. Yang and Y. S. Duan, "The Total energy-momentum of the universe in teleparallel gravity", arXiv:0706.3245 [gr-qc].

Chapter 7

Energy-Momentum in General Relativity

Xiaoning Wu[1] and Xiao Zhang[2]

Institute of Mathematics, Academy of Mathematics and Systems Science,
Chinese Academy of Sciences, Beijing 100190, China

Abstract: We briefly review of the definitions of the total energy, the total linear momentum and the angular momentum of gravitational field when the cosmological constant is zero. In particular, we show pseudo-tensor's definition of the energy and the momentum given by Prof. Duan in 1963 agree with the ADM total energy-momentum and the Bondi energy-momentum at spatial and null infinity respectively. We also review the relevant energy-momentum inequalities. Finally, we provide a short review of the positive energy theorem and the peeling property of the Newmann–Penrose quantities when the cosmological constant is positive.

Keywords: General relativity; total energy-momentum; positive energy theorem; peeling property

7.1 Energy-momentum of gravitational systems

7.1.1 *The total conserved quantities for matter fields*

It is well known that conserved quantities, such as energy and momentum, are very important both in physics and in mathematics. For ordinary matter fields, people usually use the energy-momentum tensor T_{ab} to describe the distribution of energy and momentum. Such tensor comes from the variation of Lagrangian function of the matter field. For simplicity, we only consider Lagrangian which contains the physical field ϕ and its first order derivative, i.e. $L_m = L_m(\phi, \nabla\phi, g_{ab})$, ϕ is a (p, q)

---

[1] Email: wuxn@amss.ac.cn
[2] Email: xzhang@amss.ac.cn

type tensor field. The space-time metric g_{ab} in the Lagrangian is only served as a geometric back ground, which means the physical field ϕ is living on a fixed space-time and the back reaction of the matter field to the space-time is neglected. Since the Lagrangian is a scalar function on space-time, it should be natural to require that it is diffeomorphism invariant, i.e. $L_m(\psi_t^*[g_{ab}], \psi_t^*[\phi], \psi_t^*[\nabla\phi]) = \psi_t^*[L_m(g_{ab}, \phi, \nabla\phi)]$ holds for any diffeomorphism map ψ_t^*. Given the Lagrangian L_m, the energy-momentum tensor can be calculated as following [1]:

$$T_{ab}(\phi) := \frac{2}{\sqrt{|g|}} \frac{\delta L_m}{\delta g^{ab}}$$

$$= 2\frac{\partial L_m}{\partial g^{ab}} - g_{ab}L_m + \frac{1}{2}\nabla^e(\sigma_{abe} + \sigma_{bae} - \sigma_{aeb} - \sigma_{bea} - \sigma_{eab} - \sigma_{eba}). \quad (7.1)$$

where σ_{eab} is called canonical spin tensor and is defined as

$$\sigma_b^{ea} := (-1)\left(\frac{\partial L_m}{\partial \nabla_e \phi_{d\cdots}^{c\cdots}} \Delta_{bd\cdots h\cdots}^{ac\cdots g\cdots} \phi_{g\cdots}^{h\cdots}\right). \quad (7.2)$$

This term appears because of the deformation of the covariant derivative ∇. $\Delta_{bd\cdots h\cdots}^{ac\cdots g\cdots}$ is the $(p+q+1, p+q+1)$ type invariant tensor, which is built from the Kronecker deltas. Some simple calculation will show that the diffeomorphism invariance of L_m ensures that T_{ab} is divergence-free. If O is a physical observer, he can measure the energy density ρ and energy flux j^a at each point of space-time, which can expressed as

$$\rho := T(e_0, e_0),$$
$$j_i := T(e_0, e_i), \quad i = 1, 2, 3, \quad (7.3)$$

where e_0 is the 4-velocity of this observer and e_i are normal basis chosen by him. If Σ is a Cauchy surface, one may naively think the total energy of field ϕ on Σ should be

$$E = \int_\Sigma \rho. \quad (7.4)$$

Unfortunately, such simple idea is NOT correct. A simple observation is above defined total energy E may not be conserved although the energy-momentum tensor is divergence-free. One can understand such problem in Minkowski case. If the background space-time is Minkowski and O are inertial observers, it is easy to show that the energy defined by (7.4) is conserved. If O are not inertial observer, the conservativeness of E is broken. The key difference between the two kinds of observers is that the 4-velocity of initial observers is a Killing vector field. In fact, it is quite difficult to define a conserved total energy for general observers. For stationary space-time, a suitable definition of the total energy on Σ is

$$E = \int_\Sigma T(t, n)dV, \quad (7.5)$$

where t^a is the time-like Killing vector and n is the normal vector of Σ. Using the property of Killing vector field, it is easy to show that such energy is conserved.

For non-stationary case, such conserved total energy does not exist. A physical explanation for this phenomena is that the non-stationary back ground gravitational filed will inject energy to the matter filed which makes the total energy of ϕ non-constant.

Similarly, one can also define the linear momentum P_i measured by an observer O as

$$P_i := T(e_0, e_i). \tag{7.6}$$

If Σ is still Cauchy surface, a space-like Killing vector k will give a conserved quantity P_k as

$$P_k := \int_\Sigma T(k, n)dV. \tag{7.7}$$

We notice that existence of a conserved quantity is always associated with the existence of a Killing vector. This is nothing new but result of the well-known Noether theorem. This result also has a more direct way to see. If ξ is a Killing vector, it will generate a diffeomorphism map ψ_t^* which satisfies $\psi_t^*[g_{ab}] = g_{ab}$ and $[\psi_t^*, \nabla] = 0$. The diffeomorphism invariance imply

$$\psi_t^*[L_m] = L_m(g_{ab}, \psi_t^*[\phi], \nabla\psi_t^*[\phi]). \tag{7.8}$$

Taking the t derivative, one gets

$$\xi^\mu \nabla_\mu L_m = \left(\frac{\partial L_m}{\partial \phi} - \nabla_a \frac{\partial L_m}{\partial \nabla_a \phi}\right) L_\xi \phi + \nabla_a \left(\frac{\partial L_m}{\partial \nabla_a \phi} L_\xi \phi\right),$$

$$\Rightarrow \nabla_a \left[\left(\frac{\partial L_m}{\partial \nabla_a \phi} L_\xi \phi\right) - \xi^a L_m\right] = \left(\frac{\partial L_m}{\partial \phi} - \nabla_a \frac{\partial L_m}{\partial \nabla_a \phi}\right) L_\xi \phi. \tag{7.9}$$

If ϕ is a solution of Euler-Lagrangian equation, $J^a[\xi, \phi] := \left(\frac{\partial L_m}{\partial \nabla_a \phi} L_\xi \phi\right) - \xi^a L_m$ is a conserver current. On any Cauchy surface, there is a conserved quantity $Q[\xi, \phi]$ as

$$Q[\xi, \phi] := \int_\Sigma J \cdot d\Sigma. \tag{7.10}$$

One can prove [1] that $E = Q[t, \phi]$ and $P_k = Q[k, \phi]$. Furthermore, the diffeomorphism invariance of L_m also implies another equation which is called Noether identity. Consider a smooth vector field K^a, the diffeomorphism invariance of the action implies [1, 2, 3]

$$\left(\frac{\partial L_m}{\partial \phi} - \nabla_a \frac{\partial L_m}{\partial \nabla_a \phi}\right) L_K \phi + \frac{1}{2} T_{ab} L_K g^{ab} + \nabla_e C^e = 0, \tag{7.11}$$

$$C^e[K, \phi] = \theta^{ec} K_c + \left(\sigma^{e[ab]} + \sigma^{a[be]} + \sigma^{b[ae]}\right) \nabla_a K_b,$$

$$\theta^{ab} = -L_m g^{ab} + \frac{\partial L_m}{\partial \nabla_a \phi} \nabla_b \phi.$$

where θ^{ab} is called *canonical energy-momentum tensor* and $C^a[K]$ is called *canonical Noether current*. The difference between θ_{ab} and T_{ab} is some divergence term.

Unlike T_{ab}, θ_{ab} contains some "unphysical" information. Different choice of gauge, coordinates or boundary condition may give different θ_{ab}. This is also the reason why people can construct many different type energy definition for gravitational field [1, 12]. Starting from this equation and using the fact that ϕ is a solution and K is Killing vector, one can get the conserved current J^a again.

7.1.2　*The total conserved quantities for gravity*

In last section, we discussed the total energy and momentum of a matter field. The Noether's principle is a nice tool to define conserved quantities. For gravity case, things become more complex. For matter case, the space-time (M, g_{ab}) is only served as a background and does not involve the dynamics. The non-dynamical background structure enables us to do variation for dynamical variables ϕ and for background g_{ab} separately and corresponding results are the dynamical equation and the energy-momentum tensor. If we consider gravity, ϕ and g_{ab} are all dynamical variable so we have no natural energy-momentum tensor for gravitational field. This fact also can be seen as the result of equivalent principle. Roughly speaking, Eq.(7.1) shows that the energy-momentum tensor of matter field will depend on the first derivative of the field variables in quadratic form, but the existence of Riemann normal coordinates implies such term will always vanish. This fact implies that one need some "additional structure" to introduce the conserved quantities for gravity.

The first "additional structure" is Killing vector. As we have noticed in last section, the existence of Killing vector is helpful for finding conserved quantities. With the help of Killing vector, one can introduce conserved quantities following the idea of Noether theorem.

Consider the Einstein-Hilbert action

$$I_{EH} = \frac{1}{16\pi} \int_M R\sqrt{-g}dx^4. \tag{7.12}$$

Obviously, $L_{EH} = R\sqrt{-g}$ is diffeomorphism invariant. Suppose there exists a Killing vector field ξ on space-time, ψ_t^* is the diffeomorphism map generated by ξ,

$$\begin{aligned}
0 &= \xi^a \nabla_a(R\sqrt{-g}) \\
&= [g^{ac}\delta R_{ab} + G_{ab}\delta g^{ab}]\sqrt{-g} \\
&= \nabla^a[\nabla^b L_\xi g_{ab} - g^{bc}\nabla_a L_\xi g_{bc}] \\
&= \nabla^a(R_{ab}\xi^b).
\end{aligned} \tag{7.13}$$

So we get a conserved current $J_a = R_{ab}\xi^b$. Using the Killing equation, such current can be rewritten as

$$\begin{aligned}
J &= \nabla^b\nabla_a\xi_b - \nabla^b\nabla_b\xi_a \\
&= \nabla^b[\delta_a^e\delta_b^f - \delta_b^e\delta_a^f]\nabla_e\xi_f \\
&= \epsilon_{abcd}\nabla^b\epsilon^{cdef}\nabla_e\xi_f \\
&= *d*d\xi.
\end{aligned} \tag{7.14}$$

Using Stokes' theorem, the conserved charge Q is

$$Q[\xi] = \frac{1}{8\pi} \int_{\Sigma} d * d\xi = \frac{1}{8\pi} \int_{\partial\Sigma} *d\xi. \tag{7.15}$$

This is the famous Komar integral which is found by Komar in 1959 [4]. If ξ is a time-like Killing vector, $Q[\xi]$ is the famous Komar mass. If ξ is axial-symmetric Killing vector, $Q[\xi]$ is Komar angular-momentum.

Second possible "additional structure" is coordinates, or in other words replacing the the Levi–Civita covariant derivative by the ordinary partial derivative. In a fixed local coordinates, since the gravitational Lagrangian is diffeomorphism invariant, one can use Eq.(7.11) to get the canonical energy-momentum tensor. Since such tensor depends on local coordinates, it is not covariant under the coordinate transformation. This is the famous pseudotensors method. In the early time of pseudotensors method, people get the pseudotensors just by rewriting the Einstein equation, for example see [5, 6, 7]. It is found that the Einstein equation can be rewritten into following form under suitable choice of coordinates,

$$16\pi(\sqrt{-g}T_b^a + t_b^a) = \partial_a U_b^{ac}, \tag{7.16}$$

where T_{ab} is the energy momentum tensor of matter field, U_b^{ac} is called *superpotential* and satisfies $U_b^{ac} = U_b^{[ac]}$. Using the symmetric property of U_b^{ac}, it is easy to see that

$$\partial_a(\sqrt{-g}T_b^a + t_b^a) = 0. \tag{7.17}$$

Since T_{ab} is divergence-free, above equation implies

$$\partial_a t_b^a = 0. \tag{7.18}$$

t_b^a is called *the energy-momentum pseudotensor* of gravitational filed. Weinberg gave a more direct way to see why people call t_b^a energy-momentum pseudotensor [7]. He introduces a "Minkowski-like" coordinates in asymptotic flat space-time and reexpresses the metric as $g_{ab} = \eta_{ab} + h_{ab}$. If h_{ab} is small enough, the Einstein equation becomes

$$R_{ab}^{(1)} - \frac{1}{2}R^{(1)}\eta_{ab} = 8\pi(T_{ab} + t_{ab}), \tag{7.19}$$

where $R_{ab}^{(1)} - \frac{1}{2}R^{(1)}\eta_{ab}$ is the linear term of h_{ab} in Einstein tensor. In this equation, it is quite clear that t_{ab} is contributed by the curved geometry and can be seen as the energy-momentum pseudotensor. Since the energy-momentum pseudotensor is related with different coordinate choice, there are many different kinds of energy-momentum pseudotensor which were found by different physical motivations, such as [5, 6, 7, 8, 9, 10].

To understand more about the energy-momentum pseudotensor, analysis of last section is very helpful. One can understand the energy-momentum pseudotensor in terms of the Noether's principle. Starting from the standard Einstein-Hilbert Lagrangian $L_{EH} = \frac{1}{16\pi} R\sqrt{-g}$, one can rediscover Moeller's energy-momentum pseudotensor by calculating the canonical energy-momentum tensor θ_{ab} [11]. The reason why there exist so many energy-momentum pseudotensor is that there is another "additional structure" which has not been discussed. That is the boundary term of the Lagrangian. It is well-known that a total divergent term in Lagrangian will not contribute to the dynamical equation, but such term will contribute to the energy-momentum pseudotensor. Adding different boundary term, one can get different energy-momentum pseudotensor by calculating associated canonical energy-momentum tensor θ_{ab} [12].

With the help of energy-momentum pseudotensor, one can define the total energy and linear momentum of gravitational field as

$$P^\mu(\Sigma) := \int_\Sigma (T^\mu_\nu + t^\mu_\nu) n^\nu dx^3 = \frac{1}{16\pi} \int_\Sigma \partial_\alpha U^{\mu\alpha}_0 dx^3 = \frac{1}{16\pi} \int_{\partial\Sigma} U^{\mu\alpha}_0 dS_\alpha. \quad (7.20)$$

The energy-momentum pseudotensor is a quite useful tool for people to study gravitational physics. With help of it, people explained the energy loss of binary system, which was the first indirect evidence for the existence of gravitational wave [13]. People also used this method got the correct answer for the energy translation of the tidal friction processes, which was gravitational energy being translated into thermal energy caused by the tidal force of planet's elliptic orbits. Such phenomena was first observed on Io, a moon of Jupiter [14].

Although the energy-momentum pseudotensor is very useful in many cases, the coordinates dependence is still a disadvantage. To avoid this disadvantage, many research work has been done. One way to solve this problem is to use tetrad instead of metric to rewrite the Lagrangian. It is well-known that the moving frame method is a very powerful method in differential geometry. Compare with the metric g_{ab}, the tetrad $\{e_{(\alpha)}\}$ has more freedom. Such advantage has been noticed since the beginning of 1960's. In 1961, Moeller began to use "absolute parallelism" to reconsider pseudotensor problem [15]. Newman, Penrose and their colleagues developed techniques so called "Newman-Penrose formulism", which was using null tetrad to consider the gravitational radiation [17].

7.1.3 *Duan's energy-momentum*

Almost the same time, unlike Moeller's absolute parallelism, Prof. Duan used ordinary orthonormal tetrad to study this problem [18]. This is a more natural choice.

Using orthonormal tetrad $\{e_{(\alpha)}\}$, one can rewrite the Einstein-Hilbert Lagrangian as [19]

$$L = \frac{1}{16\pi} \left[(\nabla_\nu e^\nu_{(\alpha)})(\nabla_\mu e^\mu_{(\alpha)}) - (\nabla_\mu e^\nu_{(\alpha)})(\nabla_\nu e^\mu_{(\alpha)}) \right], \quad (7.21)$$

where $e^{\mu}_{(\alpha)}$ is the tetrad, (α) is the tetrad index and μ, ν are coordinate index. Based on the idea of Noether theorem, considering the variation of above Lagrangian caused by diffeomorphism map, Prof. Duan and his colleagues gave the expression of the total energy and momentum as [19, 20]:

$$P_{(\alpha)} = \int_S \sqrt{g}\, V^{0j}_{(\alpha)} dS_j, \qquad \texttt{greek letter} = 0, 1, 2, 3 \tag{7.22}$$

$$V^{\mu\nu}_{(\alpha)} = \frac{1}{8\pi}\left[e^{\mu}_{(\beta)} e^{\nu}_{(\gamma)} \eta_{(\alpha\beta\gamma)} + (e^{\mu}_{(\alpha)} e^{\nu}_{(\beta)} - e^{\nu}_{(\alpha)} e^{\mu}_{(\beta)}) \eta_{(\beta)} \right], \tag{7.23}$$

$$\eta_{(\alpha\beta\gamma)} = \frac{1}{2}\left\{ \frac{\partial e_{\sigma(\alpha)}}{\partial x^{\lambda}} \left[e^{\lambda}_{(\beta)} e^{\sigma}_{(\gamma)} - e^{\lambda}_{(\gamma)} e^{\sigma}_{(\beta)} \right] + \frac{\partial e_{\sigma(\beta)}}{\partial x^{\lambda}} \left[e^{\lambda}_{(\alpha)} e^{\sigma}_{(\gamma)} - e^{\lambda}_{(\gamma)} e^{\sigma}_{(\alpha)} \right] \right.$$
$$\left. + \frac{\partial e_{\sigma(\gamma)}}{\partial x^{\lambda}} \left[e^{\lambda}_{(\beta)} e^{\sigma}_{(\alpha)} - e^{\lambda}_{(\alpha)} e^{\sigma}_{(\beta)} \right] \right\}, \tag{7.24}$$

$$\eta_{(\alpha)} = \eta_{(\beta\alpha\beta)} = \frac{1}{\sqrt{g}} \partial_{\mu} \left[\sqrt{g} e^{\mu}_{(\alpha)} \right], \tag{7.25}$$

where $P_{(0)}$ is the total energy and $P_{(i)}$ are total momentum components.

7.1.4 *ADM total energy and linear momentum*

Let's remember Weinberg's work [7] again. He tries to introduce some "Minkwoski coordinates" on space-time such that there is a "Minkowksi metric" η_{ab} as a back ground. If the space-time is asymptotic flat, his idea makes sense. So the asymptotic flat condition also can been as a kind of "additional structure". For Asymptotic flat space-time (M, g_{ab}), if Σ is an asymptotic flat Cauchy surface of (M, g_{ab}), it should satisfies following conditions :

(i) C is a compact subset of Σ, $\Sigma \backslash C$ is diffeomorphic to $R^3 \backslash B_R$ and $\{x^i\}$ is the Euclidean coordinates on this region.

(ii) In the region $\Sigma \backslash C$, the induced metric h_{ij} and the extrinsic curvature K_{ij} should satisfy following asymptotic conditions:

$$h_{ij} = \delta_{ij} + a_{ij}, \ a_{ij} \sim (r^{-\alpha}), \ \partial_k h_{ij} \sim O(r^{-\alpha-1}), \ \partial_k \partial_l h_{ij} \sim O(r^{-\alpha-2}),$$

$$K_{ij} \sim O(r^{-\alpha-1}), \qquad \partial_k K_{ij} \sim O(r^{-\alpha-2}), \quad \frac{1}{2} < \alpha \leq 1$$

$$i, j, k, l = 1, 2, 3, \qquad r^2 = \sum_{i=1}^{3} (x^i)^2. \tag{7.26}$$

With such asymptotic flat condition, Arnowitt, Deser and Misner [22, 23, 24] introduce the ADM energy and momentum as the total energy and momentum of gravitational field on Σ

$$E_{ADM} := \frac{1}{16\pi} \int_{S_\infty} (\partial_j h_{ij} - \partial_j h_{jj}) dS^i, \tag{7.27}$$

$$P_i^{ADM} := \frac{1}{8\pi} \int_{S_\infty} (K_{ij} - h_{ij} tr K) dS^j. \tag{7.28}$$

This is a well accepted definition of the total energy and momentum on Σ.

7.1.5 *Bondi's energy-momentum*

It was a serious problem to describe gravitational waves at the early age of general relativity. At the beginning, people just considered the linearized Einstein equation in Minkowski space and found that equation was very similar to Maxwell equation. Analog to electromagnetic wave, people called the wave-like solutions as gravitational waves. But how to describe the concept of gravitational wave in non-linear case was a quite difficult problem. That question was answered by Bondi and his colleagues in 1960' [27,28]. They gave famous Bondi–Sachs' radiating metrics, which are wave-like, vacuum solutions of the Einstein field equations, to describe the non-linear effect of gravitational waves. Such metrics take the following asymptotical forms

$$ds^2_{BS} = -\left(1 - \frac{2M}{r}\right)du^2 - 2dudr + 2ldud\theta + 2\bar{l}\sin\theta dud\psi$$

$$+ r^2\left[(1 + \frac{2c}{r})d\theta^2 + \frac{4d}{r}\sin\theta d\theta d\psi + (1 - \frac{2c}{r})\sin^2\theta d\psi^2\right] \tag{7.29}$$

$$+ \text{ lower order terms,}$$

where u is retarded coordinate (u-slices are null hypersurfaces), $r = x^1$, θ and ψ are polar coordinates, M, c, d are smooth functions of u, θ, ψ defined on $\mathbb{R} \times S^2$ with regularity condition $\int_0^{2\pi} c(u,\theta,\psi)d\psi = 0$ for $\theta = 0,\pi$ and all u, $l = c_{,\theta} + 2c\cot\theta + d_{,\psi}\csc\theta$, $\bar{l} = d_{,\theta} + 2d\cot\theta - c_{,\psi}\csc\theta$. With this metric, they also studied how gravitational waves carry away energy and momentum. Denote $n^0 = 1$, $n^1 = \sin\theta\cos\psi$, $n^2 = \sin\theta\sin\psi$, $n^3 = \cos\theta$. At null infinity, the Bondi energy-momentum on slice $\{u = u_0\}$ are defined as

$$m_\nu(u_0) = \frac{1}{4\pi}\int_{S^2} M(u_0,\theta,\psi)n^\nu dS \tag{7.30}$$

for $\nu = 0,1,2,3$. Consider the time derivative of Bondi energy, one has

$$\frac{d}{du}m_0(u) = -\frac{1}{4\pi}\int_{S^2} (c_{,u})^2 + (d_{,u})^2 \leq 0. \tag{7.31}$$

This is called Bondi energy-loss formula. The non-increasing property of Bondi energy indicates that the energy can be carried away by gravitational waves and the Bondi energy-momentum can be viewed as the total energy-momentum measured after the loss due to the gravitational radiation up to that time. Let $|m(u)| = \sqrt{m_1^2(u) + m_2^2(u) + m_3^2(u)}$. If $|m(u)| \neq 0$, Huang, Yau and Zhang proved the more general energy-loss formula [16]

$$\frac{d}{du}\left(m_0(u) - |m(u)|\right) = -\frac{1}{4\pi}\int_{S^2}\left[(c_{,u})^2 + (d_{,u})^2\right]\left(1 - \frac{m_i n^i}{|m|}\right) \leq 0. \tag{7.32}$$

It is an interesting question whether the total angular momentum can be detected near or at null infinity. As it is still open to find smooth Bondi–Sachs' coordinates for Kerr spacetime, it is unclear what rotation means for gravitational systems traveling in the speed of light. In 1962, Newman and Unti introduced another coordinates, which is slightly different to Bondi–Sachs' coordinates, to describe asymptotic flat space-time [29]. The Newman–Unti coordinates for Kerr space-time has been worked out in [30, 31]. Wu and Bai also proved that Kerr metric is the unique stationary, axial-symmetric, asymptotic flat space-time under some algebraic conditions on Weyl curvature [30].

7.2 Equivalence

In this section, we show that the energy-momentum defined by Prof. Duan are the same as the ADM energy-momentum, Bondi energy-momentum . Let's consider the ADM case first. With the asymptotic condition (7.26), we need to choose a suitable tetrad $e^i_{(\alpha)}$ and a suitable coordinate to do the calculation. For coordinates, we choose the standard "3+1" coordinates [25] as

$$ds^2 = -(N^2 - |\beta|^2)dt^2 + 2\beta_i dt dx^i + h_{ij} dx^i dx^j, \tag{7.33}$$

where

$$N = 1 + f, \quad f \sim O(r^{-1}), \quad \partial_k N \sim O(r^{-1-\varepsilon}), \quad \partial_t N \sim O(r^{-1-\varepsilon}),$$
$$\beta_i \sim O(r^{-1}), \quad \partial_k \beta_i \sim O(r^{-1-\varepsilon}), \quad \varepsilon > 0$$
$$h_{ij} \text{ is asymptotic flat.} \tag{7.34}$$

Under such coordinates, one can choose tetrad as

$$e^a_{(0)} = n^a = \frac{1}{N}\frac{\partial}{\partial t} - \frac{\beta^i}{N}\frac{\partial}{\partial x^i},$$
$$e^{(0)}_a = n_a = -N dt,$$
$$e^a_{(k)} = \partial_k - \frac{1}{2}a_{kj}\partial_j + O(a^2),$$
$$e^{(k)}_a = dx^k + \frac{1}{2}a_{kj}dx^j + \beta^k dt + O(a^2). \tag{7.35}$$

With such tetrad and coordinates, one can simplifies Eq .(7.22) as

$$E := \frac{1}{8\pi}\int_{S_\infty} \sqrt{g} \left[e^0_{(\beta)}e^j_{(\gamma)}\eta_{(0\beta\gamma)} + (e^0_{(0)}e^j_{(\beta)} - e^j_{(0)}e^0_{(\beta)})\eta_{(\beta)} \right] dS_j$$
$$= \frac{1}{8\pi}\int_{S_\infty} N\sqrt{h} \left[e^0_{(0)}e^j_{(\gamma)}\eta_{(00\gamma)} + e^0_{(0)}e^j_{(\beta)}\eta_{(\beta)} - e^j_{(0)}e^0_{(0)}\eta_{(0)} \right] dS_j. \tag{7.36}$$

By definition,

$$\eta_{00\gamma} = \frac{1}{2}\left\{ \frac{\partial e_{\sigma(0)}}{\partial x^\lambda}\left[e^\lambda_{(0)}e^\sigma_{(\gamma)} - e^\lambda_{(\gamma)}e^\sigma_{(0)} \right] + \frac{\partial e_{\sigma(0)}}{\partial x^\lambda}\left[e^\lambda_{(0)}e^\sigma_{(\gamma)} - e^\lambda_{(\gamma)}e^\sigma_{(0)} \right] \right.$$
$$\left. + \frac{\partial e_{\sigma(\gamma)}}{\partial x^\lambda}\left[e^\lambda_{(0)}e^\sigma_{(0)} - e^\lambda_{(0)}e^\sigma_{(0)} \right] \right\}$$
$$= -\frac{\partial e^{(0)}_\sigma}{\partial x^\lambda}\left[e^\lambda_{(0)}e^\sigma_{(\gamma)} - e^\lambda_{(\gamma)}e^\sigma_{(0)} \right]. \tag{7.37}$$

Therefore

$$
\begin{aligned}
e^0_{(0)}e^j_{(\gamma)}\eta_{(00\gamma)} &= -e^0_{(0)}e^j_{(\gamma)}\frac{\partial e^{(0)}_\sigma}{\partial x^\lambda}\left[e^\lambda_{(0)}e^\sigma_{(\gamma)} - e^\lambda_{(\gamma)}e^\sigma_{(0)}\right]\\
&= -e^0_{(0)}g^{j\sigma}e^\lambda_{(0)}\frac{\partial e^{(0)}_\sigma}{\partial x^\lambda} + e^0_{(0)}g^{j\lambda}e^\sigma_{(0)}\frac{\partial e^{(0)}_\sigma}{\partial x^\lambda}\\
&= \partial_j\frac{1}{N} + o(r^{-2}).
\end{aligned}
\tag{7.38}
$$

The second term is $e^0_{(0)}e^j_{(\beta)}\eta_{(\beta)}$. Using the asymptotic flat condition of tetrad (7.35), this term becomes

$$
\begin{aligned}
e^0_{(0)}e^j_{(\beta)}\eta_{(\beta)} &= e^0_{(0)}e^j_{(\beta)}\left[e^\lambda_{(\beta)}e^\sigma_{(\alpha)} - e^\lambda_{(\alpha)}e^\sigma_{(\beta)}\right]\frac{\partial e_{\sigma(\alpha)}}{\partial x^\lambda}\\
&= e^0_{(0)}g^{j\lambda}e^\sigma_{(\alpha)}\frac{\partial e_{\sigma(\alpha)}}{\partial x^\lambda} - e^0_{(0)}g^{j\sigma}e^\lambda_{(\alpha)}\frac{\partial e_{\sigma(\alpha)}}{\partial x^\lambda}\\
&= \frac{1}{N}e^\sigma_{(\alpha)}\frac{\partial e_{\sigma(\alpha)}}{\partial x^j} - \frac{1}{N}e^\lambda_{(\alpha)}\frac{\partial e_{j(\alpha)}}{\partial x^\lambda} + o(r^{-2})\\
&= \frac{1}{N}e^\sigma_{(0)}\frac{\partial e_{\sigma(0)}}{\partial x^j} + \frac{1}{N}e^\sigma_{(k)}\frac{\partial e_{\sigma(k)}}{\partial x^j} - \frac{1}{N}e^\lambda_{(0)}\frac{\partial e_{j(0)}}{\partial x^\lambda} - \frac{1}{N}e^\lambda_{(k)}\frac{\partial e_{j(k)}}{\partial x^\lambda} + o(r^{-2})\\
&= -\partial_j\frac{1}{N} + \frac{1}{2N}\partial_j h_{ii} - \frac{1}{2N}\partial_i h_{ij} + o(r^{-2}).
\end{aligned}
\tag{7.39}
$$

The third term is $-e^j_{(0)}e^0_{(0)}\eta_{(0)}$, detail calculation shows it only contributes $o(r^{-2})$ terms. Summarize above results and back to the definition of energy (7.22), it is easy to see that Prof. Duan's definition of E agrees with the ADM energy (7.27) for asymptotic flat case.

Now let's consider the total linear momentum $P_{(i)}$. By definition (7.22),

$$
\begin{aligned}
P_{(i)} &= \int_{S_\infty} \sqrt{g}\, V^{0j}_{(i)} dS_j\\
&= \frac{1}{8\pi}\int_{S_\infty} \sqrt{g}\left[e^0_{(\beta)}e^j_{(\gamma)}\eta_{(i\beta\gamma)} + (e^0_{(i)}e^j_{(\beta)} - e^j_{(i)}e^0_{(\beta)})\eta_{(\beta)}\right]dS_j\\
&= \frac{1}{8\pi}\int_{S_\infty} N\sqrt{h}\left[\frac{1}{N}e^j_{(0)}\eta_{(i00)} + \frac{1}{N}e^j_{(k)}\eta_{(i0k)} - \frac{1}{N}e^j_{(i)}\eta_{(0)}\right]dS_j.
\end{aligned}
\tag{7.40}
$$

The value of the first term is vanishes since $\eta_{(i00)} = 0$. The second term is

$$
\frac{1}{N}e^j_{(k)}\eta_{(i0k)} = \frac{1}{N^2}e^j_{(k)}\frac{1}{2}\left[\frac{\partial e^{(i)}_m}{\partial x^0}e^m_{(k)} - \frac{\partial e^{(i)}_0}{\partial x^l}e^l_{(k)} + \frac{\partial e^{(k)}_m}{\partial x^0}e^m_{(i)} - \frac{\partial e^{(k)}_0}{\partial x^l}e^l_{(i)}\right] + o(r^{-2}).
\tag{7.41}
$$

Based on the choice of tetrad (7.35), above equation becomes

$$
\begin{aligned}
\frac{1}{N}e^j_{(k)}\eta_{(i0k)} &= \frac{1}{2N^2}\delta^j_k\left[\frac{1}{2}\partial_0 a_{mi}\delta^m_k - \partial_l\beta^i\delta^l_k + \frac{1}{2}\partial_0 a_{mk}\delta^m_i - \partial_l\beta^k\delta^l_i\right] + o(r^{-2})\\
&= \frac{1}{2}\left[\partial_0 h_{ij} - \partial_i\beta_j - \partial_j\beta_i\right] + o(r^{-2}).
\end{aligned}
\tag{7.42}
$$

Remember $K_{ij} = \frac{1}{2}L_n h_{ij}$, with the help of asymptotic condition for metric, we get

$$K_{ij} = \frac{1}{2}L_n h_{ij} = \frac{1}{2}\left(\frac{1}{N}\partial_t h_{ij} - \frac{\beta^k}{N}\partial_k h_{ij} - h_{ik}\partial_j\frac{\beta^k}{N} - h_{jk}\partial_i\frac{\beta^k}{N}\right)$$

$$= \frac{1}{2}[\partial_t h_{ij} - \partial_i\beta_j - \partial_j\beta_i] + o(r^{-2}). \tag{7.43}$$

Then

$$\frac{1}{N}e^j_{(k)}\eta_{(i0k)} = K_{ij} + o(r^{-2}). \tag{7.44}$$

Similarly, the third term is

$$-\frac{1}{N}e^j_{(i)}\eta_0 = -\left[\frac{1}{N}e^j_{(i)}e^\lambda_{(0)}e^\sigma_{(\alpha)}\frac{\partial e_{\sigma(\alpha)}}{\partial x^\lambda}\right] + \left[\frac{1}{N}e^j_{(i)}e^\lambda_{(\alpha)}e^\sigma_{(0)}\frac{\partial e_{\sigma(\alpha)}}{\partial x^\lambda}\right]$$

$$= -\delta^j_i\left[\partial_0 N + \frac{1}{2}\delta^m_k\partial_0 h_{mk}\right] + \delta^j_i\left[\partial_0 N + \partial_k\beta^k\right] + o(r^{-2})$$

$$= -\delta^j_i trK + o(r^{-2}). \tag{7.45}$$

Back to Eq.(7.40), it is clear that Prof. Duan's total momentum also agrees with the ADM linear momentum in the asymptotic flat case.

The ADM energy characterize the total energy of an asymptotic flat Cauchy surface Σ. If one wants to consider gravitational radiation, The Bondi energy is need. The Bondi–Sachs metrics are [27, 28]

$$ds^2 = (e^{2\beta}\frac{V}{r} - r^2 h_{\mu\nu}U^\mu U^\nu)du^2 + 2e^{2\beta}dudr$$

$$+ 2r^2 h_{\mu\nu}U^\nu dudx^\mu - r^2 h_{\mu\nu}dx^\mu dx^\nu, \tag{7.46}$$

where $\mu, \nu = 2, 3$, $U^2 = U$, $U^3 = W\csc\theta$,

$$(h_{\mu\nu}) = \begin{pmatrix} e^{2\gamma}\cosh 2\delta & \sinh 2\delta\sin\theta \\ \sinh 2\delta\sin\theta & e^{-2\gamma}\cosh 2\delta\sin^2\theta \end{pmatrix}, \tag{7.47}$$

asymptotic behavior:

$$\gamma = \frac{c(u,\theta,\phi)}{r} + (C(u,\theta,\phi) - \frac{1}{6}c^3 - \frac{3}{2}cd^2)\frac{1}{r^3} + \cdots,$$

$$\delta = \frac{d(u,\theta,\phi)}{r} + (D(u,\theta,\phi) - \frac{1}{6}d^3 + \frac{1}{2}c^2 d)\frac{1}{r^3} + \cdots,$$

$$\beta = -\frac{c^2 + d^2}{4r^2} + \cdots, \quad W = -(\frac{\partial d}{\partial\theta} + 2d\cot\theta - \frac{\partial c}{\partial\phi}\csc\theta)\frac{1}{r^2} + \cdots,$$

$$U = -(\frac{\partial c}{\partial\theta} + 2c\cot\theta + \frac{\partial d}{\partial\phi}\csc\theta)\frac{1}{r^2} + \cdots,$$

$$V = r - 2M(u,\theta,\phi) + \frac{V_1(u,\theta,\phi)}{r} + \cdots. \tag{7.48}$$

In the Bondi coordinates, one can choose tetrad as [52]

$$e^{\mu}{}_{(0)} = (1, \frac{1}{2}(e^{-2\beta} - \frac{V}{r}), U, W \csc \theta),$$

$$e^{\mu}{}_{(1)} = (-1, \frac{1}{2}(e^{-2\beta} + \frac{V}{r}), -U, -W \csc \theta),$$

$$e^{\mu}{}_{(2)} = (0, 0, \frac{e^{-\gamma}}{r\sqrt{\cosh 2\delta}}, 0),$$

$$e^{\mu}{}_{(3)} = (0, 0, -\frac{e^{-\gamma} \sinh 2\delta}{r\sqrt{\cosh 2\delta}}, \frac{e^{\gamma} \sqrt{\cosh 2\delta}}{r \sin \theta}). \tag{7.49}$$

With the choice of coordinates and tetrad, the definition of energy (7.22) becomes

$$\begin{aligned}
P_{(0)} &= \frac{1}{8\pi} \int_{S_\infty} \sqrt{g} \left[e^0_{(\beta)} e^j_{(\gamma)} \eta_{(0\beta\gamma)} + (e^0_{(0)} e^j_{(\beta)} - e^j_{(0)} e^0_{(\beta)}) \eta_{(\beta)} \right] dS_j \\
&= \frac{1}{8\pi} \int_{S_\infty} \sqrt{g} \left[e^j_{(\gamma)} \eta_{(00\gamma)} - e^j_{(\gamma)} \eta_{(01\gamma)} + (e^j_{(\beta)} - e^j_{(0)} e^0_{(\beta)}) \eta_{(\beta)} \right] dS_j \\
&= \frac{1}{8\pi} \int_{S_\infty} \sqrt{g} \left[e^r_{(1)} \eta_{001} + e^r_{(0)} \eta_{001} + e^r_{(0)} \eta_{(0)} + e^r_{(1)} \eta_{(1)} \right] dS. \tag{7.50}
\end{aligned}$$

By definition, straight calculation will show

$$\eta_{001} = (-2M - c\frac{\partial c}{\partial u} - d\frac{\partial d}{\partial u}) \frac{1}{r^2} + \mathcal{O}(r^{-3}),$$

$$\begin{aligned}
\eta_0 &= (2 + \frac{\partial M}{\partial u}) \frac{1}{r} + \Big[ ]\cot\theta \csc\theta \frac{\partial d}{\partial \phi} - \csc^2\theta \frac{\partial^2 c}{\partial \phi^2} + 2\cot\theta \frac{\partial c}{\partial \theta} \\
&\quad + 2\csc\theta \frac{\partial^2 d}{\partial \theta \partial \phi} + \frac{\partial^2}{\partial \theta^2} + c(-2\csc^2\theta + \frac{9}{2}\frac{\partial c}{\partial u}) - \frac{11}{2} d\frac{\partial d}{\partial u} \\
&\quad + 12d \sin\theta \frac{\partial d}{\partial u} - 4d \sin^2\theta \frac{\partial d}{\partial u} - \frac{1}{2}\frac{\partial V_1}{\partial u} \Big] \frac{1}{r^2} + \mathcal{O}(r^{-3}),
\end{aligned}$$

$$\eta_1 = -\frac{2}{r} - (c\frac{\partial c}{\partial u} + d\frac{\partial d}{\partial u}) \frac{1}{r^2} + \mathcal{O}(r^{-3}). \tag{7.51}$$

But the covariant derivatives are used in the definition. Using the asymptotic behaviors of 3-index symbols [27, 28], we finally obtain

$$P_{(0)} = \frac{1}{4\pi} \lim_{r\to\infty} \int_{S_\infty} \frac{M(u,\theta)}{r^2} dS = \frac{1}{4\pi} \int_{S^2} M(u,\theta) dS. \tag{7.52}$$

So Prof. Duan's energy definition (7.22) agrees with the Bondi energy at null infinity. Similar straight calculation also can show Prof. Duan's momentum definition also agrees with the Bondi momentum.

7.3   Positivity of the total energy

It was conjectured that the total energy is nonnegative both at spatial infinity and at null infinity for isolated physical systems satisfying the dominant energy condition.

For the recent progress of the topics, we refer to [32] for details. The positive energy conjecture for the ADM total energy-momentum at spatial infinity was first proved by Schoen and Yau [33,34,35], later by Witten using spinors [36,37]. More precisely, they proved

The positive energy theorem: Let (Σ, h, K) be asymptotically flat, with possibly a finite number of black holes. Suppose the dominant energy condition holds, then
(i) $E \geq \sqrt{P_1^2 + P_2^2 + P_3^2}$ for each end;
(ii) That $E = 0$ for some end implies M has only one end, and space-time is flat along Σ.

The total mass is $\sqrt{E^2 - P_1^2 - P_2^2 - P_3^2}$ which is a Lorentzian invariant. When the positive energy theorem holds, the total mass is well-defined. In 1999, Zhang proved the positive energy theorem for *generalized asymptotically flat initial data set* (Σ, h, p) where p is general 2-tensor which is not necessary symmetric [38]. Geometrically, the second fundamental form p is nonsymmetric when spacetimes equip with affine connections with torsion. In this case matter translates, meanwhile, it rotates. The idea using connection with torsion was initially duo to E. Cartan [39,40,41,42]. In [38], Zhang defined the following generalized linear momentum counting both translation and rotation

$$\bar{P}_k = \frac{1}{8\pi} \int_{S_\infty} (p_{ki} - h_{ki} tr_h(p)) * dx^i, \tag{7.53}$$

and proved the following theorem.

The generalized positive energy theorem: Let (Σ, h, p) be generalized asymptotically flat, with a finite number of black holes. Suppose the generalized dominant energy condition

$$\frac{1}{2}(R + (tr_h(p))^2 - |p|^2) \geq \max\{|\omega|, |\omega + \chi|\}, \tag{7.54}$$

holds, $\omega_j = \nabla^i p_{ji} - \nabla_j tr_h(p)$, $\chi_j = 2\nabla^i(p_{ij} - p_{ji})$, then
(i) $E \geq \sqrt{\bar{P}_1^2 + \bar{P}_2^2 + \bar{P}_3^2}$ for each end;
(ii) That $E = 0$ for some end implies M has only one end, and

$$R_{ijkl} + p_{ik}p_{jl} - p_{il}p_{jk} = 0, \quad \nabla_i p_{jk} - \nabla_j p_{ik} = 0, \quad \nabla^i(p_{ij} - p_{ji}) = 0. \tag{7.55}$$

In 1974, Regge–Teitelboim defined the total angular momentum for asymptotically flat initial data sets [43],

$$J_k(x_0) = \frac{1}{8\pi} \int_{S_\infty} \epsilon_{kuv}(x^u - x_0^u)\pi^v_i * dx^i, \quad \pi^v_i = K^v_i - h^v_i tr_K(h). \tag{7.56}$$

In general, the integrand is $O(\frac{1}{r})$ which may not be integrable. This ambiguity resolution requires stronger "Regge–Teitelboim" conditions on ends

$$h(x) - h(-x) = O(r^{-3}), \quad \pi(x) + \pi(-x) = O(r^{-3}). \tag{7.57}$$

The integrand in Regge–Teitelboim's definition is not tensor, it can not relate the local density to the total angular momentum. To resolve this difficulty, Zhang defined trace free, non-symmetric tensor of local angular momentum density [38]

$$\tilde{h}^z_{ij} = \frac{1}{2}\epsilon_i{}^{uv}\left(\nabla_u\rho_z^2\right)\left(K_{vj} - h_{vj}tr_h(K)\right), \tag{7.58}$$

where ρ_z is the distance function w.r.t some $z \in M$. If $(\Sigma, g, \tilde{h}^z_{ij})$ is generalized asymptotically flat, Zhang defined the total angular momentum [38]

$$J_k = \frac{1}{8\pi}\int_{S_\infty}\tilde{h}^z_{ki} * dx^i. \tag{7.59}$$

J is also a geometric quantity which is independent on the choice of coordinates $\{x^i\}$ if $|\nabla\tilde{h}^z|$ is integrable. In Kerr spacetime, $K_{ij} = O(\frac{1}{r^4})$, so the total angular momentum is well-defined and it was found $J = (0, 0, ma)$ [44]. By taking $p_{ij} = C\tilde{h}^z_{ij}$ for certain constant $C > 0$, Zhang proved the Kerr constraint $E \geq C|J|$ under the generalized dominant energy condition [38]. It is worth pointed out that the dominant energy condition does not yield to the Kerr constraint. In [45], Huang, Schoen and Wang showed that it is possible to perturb arbitrary vacuum asymptotically flat initial data sets to new vacuum ones having exactly the same total energy, but with the arbitrary large total angular momentum.

At null infinity, it was conjectured that the Bondi energy is nonnegative. The proofs of this conjecture were claimed by Schoen–Yau using geometric analysis methods, as well as by physicists using Witten's positive energy arguments (see [46] and references therein). However, extra conditions are required when two methods are worked out rigorously and completely [16]. In particular, by using the positive energy theorem near null infinity [47], Huang, Yau and Zhang proved

Positivity of Bondi energy: Suppose there exists u_0 in vacuum Bondi's radiating spacetime such that $c(u_0) = d(u_0) = 0$.
(i) $m_0(u_0) \geq |m(u_0)|$, and the Bondi energy-momentum loss formula gives $m_0(u) \geq |m(u)|$ for all $u \leq u_0$;
(ii) If $m_0(u_0) = |m(u_0)|$ and there is $u_1 < u_0$ such that $m_0(u_1) = |m(u_1)|$, then $c(u_0) = d(u_0) = 0$ on $[u_1, u_0]$. Thus the spacetime is flat in the region $(u_1, u_0]$.

It ensures that the Bondi mass $\sqrt{m_0^2 - m_1^2 - m_2^2 - m_3^2}$ is well-defined.

7.4   Positive cosmological constant

Based on the experimental dates, e.g., Planck 2015[3], the actual universe's metric is asymptotic to the FLRW metric with $k = 0$

$$\tilde{g}_{FLRW} = -dt^2 + e^{2Ht}g_\delta, \quad g_\delta = (dx^1)^2 + (dx^2)^2 + (dx^3)^2, \tag{7.60}$$

---

[3]Planck 2015 results I. Overview of Scientific Results. ArXiv: 1502.01582; XIII. Cosmological Parameters. ArXiv: 1502.01589.

where $H > 0$ is the Hubble constant, $3H^2 = \rho_m + \Lambda_c$, $\rho_m \cong 0.3156 \times 3H^2$ is the matter density containing dark matter, $\Lambda_c \cong 0.6844 \times 3H^2$ is the real value of cosmological constant representing dark energy. Denote $\Lambda = 3H^2$, $\lambda = H^{-1}$. The initial data set is $(\mathbb{R}^3, \check{g} = e^{2Ht}g_\delta, \check{K} = H\check{g})$.

To define the total energy-momentum from the Hamiltonian point of view, an asymptotically de Sitter initial data set (M, g, K) should satisfy

$$g - \check{g} = e^{2Ht_0}a = O(\frac{1}{r}), \qquad K - \check{K} = e^{Ht_0}b = O(\frac{1}{r^2}) \tag{7.61}$$

for constant t_0 on ends. Then the ten Killing vectors $U_{\alpha\beta}$ of $\mathbb{R}^{4,1}$ and a and b are used to define the ADM-like total energy-momentum [48]. But there is no energy-momentum inequalities for them in general. Alternatively, there is different approach to define the total energy-momentum from the initial data set point of view. An initial data set (M, g, K) is *$\mathcal{P}$-asymptotically de Sitter* if $g = \mathcal{P}^2\bar{g}$, $h = \mathcal{P}\bar{h}$ for certain constant $\mathcal{P} > 0$, and $(M, \bar{g}, \bar{h})$ is asymptotically flat. Let $\bar{E}$, $\bar{P}_k$ and $\bar{J}_k(z)$ be the total energy, the total linear momentum and the total angular momentum of the end M_l for $(M, \bar{g}, \bar{h})$ respectively. The corresponding quantities for $\mathcal{P}$-asymptotically de Sitter initial data set (M, g, K) are

$$E = \mathcal{P}\bar{E}, \quad P_k = \mathcal{P}^2\bar{P}_k, \quad J_k(z) = \mathcal{P}^2\bar{J}_k(z), \tag{7.62}$$

and one of the positive energy theorem proved by Luo, Xie and Zhang [49] is as follows.

The positive energy theorem: Let (M, g, K) be a $\mathcal{P}$-asymptotically de Sitter initial data set in spacetime $\mathbf{L}^{3,1}$ satisfying the Einstein field equations

$$\mathbf{R}_{\alpha\beta} - \frac{\mathbf{R}}{2}\mathbf{g}_{\alpha\beta} + \Lambda\mathbf{g}_{\alpha\beta} = T_{\alpha\beta}.$$

Suppose $tr_g(K) \leq \sqrt{3\Lambda}$. If $\mathbf{L}^{3,1}$ satisfies the dominant energy condition, then
(i) $E \geq \sqrt{P_1^2 + P_2^2 + P_3^2}$ for any end;
(ii) That $E = 0$ for some end implies

$$(M, g, K) \equiv \left(\mathbb{R}^3, \mathcal{P}^2 g_\delta, \sqrt{\frac{\Lambda}{3}}\mathcal{P}^2 g_\delta\right) \tag{7.63}$$

and the spacetime $\mathbf{L}^{3,1}$ is de Sitter along M.

It ensures that the total mass $\sqrt{E^2 - P_1^2 - P_2^2 - P_3^2}$ is well-defined.

The two definitions of total energy are the same both in the Hamiltonian formulation and in the initial data set formulation. But the definitions of the total linear momentum are completely different, while it is finite, defined in terms of h, the total linear momentum is infinite in general, defined in terms of $K - \check{K}$. So, in the Hamiltonian formulation, there should not be any energy-momentum inequality, but $E \geq 0$ if $tr_g(K) \leq \sqrt{3\Lambda}$ in this case.

In 2012, Liang and Zhang constructed counterexamples of the above positive energy theorem while the condition $tr_g(K) \leq \sqrt{3\Lambda}$ is violated [50].

By suitable coordinate transformation, the FLRW metric can be transferred into the retarded coordinates which are used to study the nonlinear theory of gravitational waves. In general, if spacetimes have a family of non-intersecting null hypersurfaces given by the level sets of smooth function u, gravitational waves can also be described as the Bondi–Sachs metrics even if the cosmological constant is nonzero. For the positive cosmological constant, u is only continuous with discontinuous derivatives across the cosmological horizon, so the above metrics are valid only inside and outside the cosmological horizon.

The theory of gravitational waves for positive cosmological constant has been studied extensively in recent years. In [51], a detail asymptotic analysis of Bondi–Sachs metrics was provided by assuming Sommerfeld's radiation condition, which is natural in numerical simulations,

$$\gamma = \frac{c}{r} + \left(-\frac{1}{6}c^3 - \frac{3}{2}d^2c + C \right)\frac{1}{r^3} + O\left(\frac{1}{r^4}\right),$$

$$\delta = \frac{d}{r} + \left(-\frac{1}{6}d^3 + \frac{1}{2}c^2d + D \right)\frac{1}{r^3} + O\left(\frac{1}{r^4}\right), \tag{7.64}$$

with the regularity condition $\int_0^{2\pi} c(u,\theta,\psi)d\psi = 0$ for $\theta = 0, \pi$ and for all u. The key point is that, for the positive cosmological constant, u can be only continuous with discontinuous derivatives across the cosmological horizon, so the above expansions are valid only near and inside the cosmological horizon as well as near infinity outside the cosmological horizon. For the negative cosmological constant, the series expansions are taken near $r = \infty$, the peeling property is not affected even if the coefficients involve Λ. But for the positive cosmological constant, the series expansions are taken both near and inside the cosmological horizons where $r \sim \sqrt{3\Lambda^{-1}}$ is finite, and near $r = \infty$ outside the cosmological horizon. The peeling property can be affected if the coefficients involve Λ near the cosmological horizon. So suitable series expansions are required.

From [51, 52], we can find the following asymptotic expansions for the vacuum field equations

$$\beta = B - \frac{c^2 + d^2}{4r^2} + O\left(\frac{1}{r^3}\right),$$

$$W = X + 2e^{2B}B_{,\phi}\csc\theta\frac{1}{r} + O\left(\frac{1}{r^2}\right),$$

$$U = Y + 2e^{2B}B_{,\theta}\frac{1}{r} + O\left(\frac{1}{r^2}\right),$$

$$V = -\frac{e^{2B}\Lambda}{3}r^3 + (\cot\theta Y + \csc\theta X_{,\phi} + Y_{,\theta})r^2 + e^{2B}\Big(4B_{,\phi}^2\csc^2\theta$$

$$+2B_{,\phi\phi}\csc^2\theta + 2B_{,\theta}\cot\theta + 4B_{,\theta}^2 + 2B_{,\theta\theta} + 1\Big)r - 2M + O\left(\frac{1}{r}\right), \tag{7.65}$$

where $X = \Lambda a\sin\theta$, $Y = \Lambda b\sin\theta$, and a, b satisfy

$$a_{,\phi} - \sin\theta b_{,\theta} = -\frac{2}{3}e^{2B}c, \qquad b_{,\phi} + \sin\theta a_{,\theta} = \frac{2}{3}e^{2B}d. \tag{7.66}$$

Thus a, b can be determined uniquely by B, c, d up to some functions $\rho(u)$, $\sigma(u)$. Moreover, we can choose B, a, b which are independent on Λ.

Let Ψ_k, $k = 0, \ldots, 4$, be the Newmann–Penrose quantities. It was proved in [52] that, under Sommerfeld's radiation condition together with the nontrivial B, X, Y, the following peeling property holds

$$\Psi_k = -\left[(\Psi_k^{5-k})^0 + O(\Lambda) \right] \frac{1}{r^{5-k}} + O\left(\frac{1}{r^{6-k}} \right), \tag{7.67}$$

where coefficients $\left(\Psi_k^{5-k} \right)^0$ are given by B, a, b and other Λ-independent functions appeared in series expansions of γ, δ, β, W, U and V. The cosmological constant affects the experimental data only in a scale of Λ which can be ignored.

Note that $\left(\Psi_4^1 \right)^0 \neq 0$ when B is nontrivial, $a = a(u)$, $b = b(u)$ are functions of u which give $c = d = 0$. This indicates that there exist gravitational waves without Bondi news, which may be referred as B-gravitational waves. In [52], some nonstationary vacuum Bondi–Sachs metrics are constructed with $\gamma = \delta = 0$. Hence $c = d = 0$.

$$\begin{aligned}
g = -\Bigg[&-(C + \cos\theta)^2 \frac{\Lambda}{3} r^2 - \sin^2\theta \left(\frac{C'}{C^2 - 1} \right)^2 r^2 - 2(C\cos\theta + 1)\frac{C'}{C^2 - 1} r \\
&+ C^2 - 1 - \frac{2m(C^2 - 1)^{\frac{3}{2}}}{r(C + \cos\theta)} \Bigg] du^2 - 2(C + \cos\theta)\,dudr \\
&+ 2r\sin\theta \left(1 + \frac{C'}{C^2 - 1} r \right) dud\theta + r^2 \left(d\theta^2 + \sin^2\theta d\phi^2 \right).
\end{aligned} \tag{7.68}$$

The above metrics have black holes when $m \neq 0$. Moreover, the Newmann–Penrose quantity Ψ_k satisfy

$$\Psi_0 = \Psi_1 = 0, \qquad\qquad \Psi_2 = -\frac{m(C^2 - 1)^{\frac{3}{2}}}{r^3(C + \cos\theta)^3},$$

$$\Psi_3 = \frac{3m\sin\theta(C^2 - 1)^{\frac{3}{2}}}{\sqrt{2}r^3(C + \cos\theta)^4}, \quad \Psi_4 = -\frac{3m\sin^2\theta(C^2 - 1)^{\frac{3}{2}}}{r^3(C + \cos\theta)^5}. \tag{7.69}$$

They fall faster than usual, which may be missed in the experimental data.

We refer to [53] for an alternative boundary condition in the axi-symmetric case, without assuming Sommerfeld's radiation condition, but deforming 2-sphere with

$$\gamma = \Lambda f(u, \theta) + \frac{c(u, \theta)}{r} + O\left(\frac{1}{r^3} \right) \tag{7.70}$$

and taking $B = X = Y = 0$. The vacuum field equations give

$$f(u, \theta) = f(-\infty, \theta) + \frac{1}{3} \int_{-\infty}^u c(s, \theta)ds. \tag{7.71}$$

Physically, $f(-\infty, \theta)$ exists. Thus, in order that $f(u, \theta)$ exists for any $u < \infty$ and for $u \to +\infty$, Bondi news must satisfy $\int_{-\infty}^u c(s, \theta)ds < \infty$ for any $u < \infty$ and for $u = +\infty$. This boundary condition is rather restricted which actually excludes gravitational waves with $\int_{-\infty}^{u} c(s,\theta)ds = \infty$ or $\int_{u}^{+\infty} c(s,\theta)ds = \infty$ for some $u < \infty$ or for $u = +\infty$. In a series of papers [54, 55, 56, 57], asymptotics with $\Lambda > 0$ was discussed in framework of conformal compactification, and the linearization theory as well as the quadrupole formula were derived. Some relevant works on the linearization theory can also be found in [58, 59, 60]. The papers [61, 62] discussed the asymptotic vacuum and the electromagnetism Newman–Penrose equations as well as Bondi mass for $\Lambda \neq 0$. The boundary condition in [61, 62] is essentially equivalent to that given in [53].

The peeling property for $\Lambda \neq 0$ was proved previously by Penrose in framework of conformal compactification [63], and by Saw without conformal compactification [61, 62]. It was pointed out in [52] that the induced metric on conformal boundary J^+ are

$$\frac{e^{2B}\Lambda}{3}du^2 + 2Y\,dud\theta + 2X\,dud\phi + d\theta^2 + \sin^2\theta d\phi^2. \tag{7.72}$$

So the boundary condition given in [51, 52] does not seem to consist with Penrose's framework. This new boundary condition is natural in three aspects that it consists with nontrivial Bondi news, gives rise to the peeling property and features B-gravitational waves without Bondi news.

7.5　Acknowledgement

This work is partially supported by the National Science Foundation of China (grants 11571345, 11575286) and the project of mathematics and interdisciplinary sciences of Chinese Academy of Sciences.

Bibliography

[1] Laszlo B. Szabados, Living Rev. Relativity **12** (2009) 4. [Online Article]: http://www.livingreviews.org/lrr-2009-4

[2] V. Iyer and R. M. Wald, Phys. Rev. D **50** (1994) 846.

[3] V. Iyer and R. M. Wald, Phys. Rev. D **52** (1995)4430.

[4] A. Komar, Phys. Rev. **113**(1959)934.

[5] A, Einstein, Berlin. Ber. **778**(1915) 154.

[6] L. D. Landau and E. M. Lifshitz, *The Classical Theory of Fields*, 2nd edn. (Addison-Wesley, Reading, MA, 1962).

[7] S. Weinberg, *Gravitation and Cosmology*, 1972 (New York: Wiley).

[8] A. Papapetrou, Proc. Roy. Irish Acad. (Sect. A) **52A** (1948) 11

[9] R. C. Tolman, *Relativity Thermodynamics and Cosmology*, 1950.

[10] C. Moeller, Ann. Phys. (N.Y.) **4** (1958) 347.

[11] L. B. Szabados, Canonical pseudotensors, Sparling's form and Noether currents, KFKI Report 1991- 29/B, (KFKI Research Institute for Particle and Nuclear Physics (RMKI), Budapest, 1991). Online version (accessed 29 January 2004): http://www.rmki.kfki.hu/ lbszab/doc/sparl11.pdf.

[12] Chiang-Mei Chen, James M. Nester and Roh-Suan Tung, International Journal of Modern Physics D **24**, No. 11 (2015) 1530026

[13] R. A. Hulse and J. H. Taylor, Astrophys. J. **195** (1975) L51–L53.

[14] S. J. Peale, P. Cassen and R. T. Reynolds, Science **203** (1979) 892.

[15] C. Moeller, Mat.-Fys. Skr. K. Danske Vid. Selsk. **1**(10) (1961) 1–50.

[16] W-L. Huang, S.T. Yau, X. Zhang, Positivity of the Bondi mass in Bondi's radiating spacetimes. Rend. Lincei Mat. Appl. **17**, 335–349 (2006).

[17] E. T. Newman and R. Penrose, J. Math. Phys. **3** (1963) 896.

[18] Yi-Shi Duan and Jing-Ye Zhang, Acta Physica Sinca **18** (1962) 211.

[19] Yi-Shi Duan and Jing-Ye Zhang, Acta Physica Sinca **19** (1963) 689.

[20] Yi-Shi Duan and You-Tang Wang, Scentia Sinca (Series A) XXVI (1983) 961.

[21] Yi-Shi Duan and Shi-Xiang Feng, Acta Physica Sinca **44**(1995) 1373.

[22] R. Arnowitt, S. Deser, C. Misner, The dynamics of general relativity. pp.227–264, in *Gravitation: an introduction to current research*, L. Witten, ed. Wiley, New York, 1962.

[23] C. B. Liang, *Introduction for Differential Geometry and General Relativity*, Beijing Normal University Press, 2001

[24] R. M. Wald, *General Relativity*, The University of Chicago Press, 1984.

[25] C. W. Misner, K. S. Thorne and J. A. Wheeler, *Gravitation*, W. H. Freeman and Compeny, 1973.

[26] H. Bondi, Nature, **186**(1960)535.

[27] H. Bondi, M. G. J. van der Burg and A. W. K. Metzner, Proc. R. Soc. London, Ser. A, **269**, 21–52 (1962).

[28] R. K. Sachs, Waves in asymptotically flat space-time, Proc. R. Soc. London A **270**(103) (1962).

[29] E. T. Newman and T. W. J. Unti, J. Math. Phys. **3** (1962) 891.

[30] X. Wu and S. Bai, Phys. Rev. D **78** (2008) 124009.

[31] S. Bai, Z. Cao, X. Gong, X. Wu and Y. K. Lau, Phys. Rev. D **75** (2007) 044003.

[32] X. Zhang, International Journal of Modern Physics A, **30**, Nos. 28–29 (2015) 1545018 (20 pages)

[33] R. Schoen, S.T. Yau, On the proof of the positive mass conjecture in general relativity. Commun. Math. Phys. **65**, 45–76 (1979).

[34] R. Schoen, S.T. Yau, The energy and the linear momentum of spacetimes in general relativity. Commun. Math. Phys. **79**, 47–51 (1981).

[35] R. Schoen, S.T. Yau, Proof of the positive mass theorem. II. Commun. Math. Phys. **79**, 231–260 (1981).

[36] E. Witten, A new proof of the positive energy theorem. Commun. Math. Phys. **80**, 381–402 (1981).

[37] T. Parker, C. Taubes, On Witten's proof of the positive energy theorem. Commun. Math. Phys. **84**, 223–238 (1982).

[38] X. Zhang, Angular momentum and positive mass theorem. Commun. Math. Phys. **206**, 137–155 (1999).

[39] E. Cartan, Sur une généralisation de la notion de courbure de Riemann et les espaces à torsion. C. R. Acad. Sci. (Paris) **174**, 593–595 (1922).

[40] E. Cartan, Sur les variétés à connexion affine et la théorie de la relativité généralisée. Part I: Ann. Éc. Norm. **40**, 325–412 (1923).

[41] E. Cartan, Sur les variétés à connexion affine et la théorie de la relativité généralisée. Part II: Ann. Éc. Norm. **41**, 1–25 (1924).

[42] E. Cartan, Sur les variétés à connexion affine et la théorie de la relativité généralisée. Part III: Ann. Éc. Norm. **42**, 17–88 (1925).

[43] T. Regge, C. Teitelboim, Role of surface integrals in the Hamiltonian formulation of general relativity. Ann. Phys. **88**, 286–318 (1974).

[44] X. Zhang, Remarks on the total angular momentum in general relativity. Commun. Theor. Phys. **39**, 521–524 (2003).

[45] L-H. Huang, R. Schoen, M-T. Wang, Specifying angular momentum and center of mass for vacuum initial data sets. Commun. Math. Phys. **306**, 785–803 (2011).

[46] P. Chruściel, J. Jezierski, S. Leski, The Trautman-Bondi mass of initial data sets. Adv. Theor. Math. Phys. **8**, 83 (2004).

[47] X. Zhang, A definition of total energy-momentua and the positive mass theorem on asymptotically hyperbolic 3-manifolds I. Commun. Math. Phys. **249**, 529–548 (2004).

[48] D. Kastor, J. Traschen, A positive energy theorem for asymptotically de Sitter spacetimes. Class. Quantum Gravity **19**, 5901–5920 (2002).

[49] M. Luo, N. Xie, X. Zhang, Positive mass theorems for asymptotically de Sitter spacetimes. Nucl. Phys. B **825**, 98–118 (2010).

[50] Z. Liang, X. Zhang, Spacelike hypersurfaces with negative total energy in de Sitter spacetime. J. Math. Phys. **53**, 022502 (2012).

[51] H. Ge, M. Luo, Q. Su, D. Wang, X. Zhang, Bondi–Sachs metrics and photon rockets, Gen. Relativ. Gravit. **43**, 2729 (2011).

[52] F. Xie, X. Zhang, Peeling property of Bondi–Sachs metrics for nonzero cosmological constant, arXiv:1704.06015[gr-qc].

[53] X. He, Z. Cao, New Bondi-type outgoing boundary condition for the Einstein equations with cosmological constant, Int. J. Mod. Phys. D **24**, 1550081 (2015).

[54] A. Ashtekar, B. Bonga, A. Kesavan, Asymptotics with a positive cosmological constant: I. Basic framework, Class. Quantum Grav. **32**, 025004 (2015).

[55] A. Ashtekar, B. Bonga, A. Kesavan, Asymptotics with a positive cosmological constant. II. Linear fields on de Sitter spacetime, Phys. Rev. D **92**, 044011 (2015).

[56] A. Ashtekar, B. Bonga, A. Kesavan, Asymptotics with a positive cosmological constant. III. The quadrupole formula, Phys. Rev. D **92**, 104032 (2015).

[57] A. Ashtekar, B. Bonga, A. Kesavan, Gravitational Waves from Isolated Systems: Surprising Consequences of a Positive Cosmological Constant, Phys. Rev. Lett. **116**, 051101 (2016).

[58] N. T. Bishop, Gravitational waves in a de Sitter universe, Phys. Rev. D **93**, 044025 (2016).

[59] G. Date, S. J. Hoque, Gravitational Waves from Compact Sources in de Sitter Background, Phys. Rev. D **94**, 064039 (2016).

[60] G. Date, S. J. Hoque, Cosmological Horizon and the Quadrupole Formula in de Sitter Background, arXiv:1612.09511v2 [gr-qc] 5 Apr 2017.

[61] V-L. Saw, Mass-loss of an isolated gravitating system due to energy carried away by gravitational waves with a cosmological constant, Phys. Rev. D **94**, 104004 (2016).

[62] V-L. Saw, Behaviour of asymptotically electro-Λ spacetimes, Phys. Rev. D **95**, 084038 (2017)

[63] R. Penrose, Zero rest mass fields including gravitation: asymptotic behavior, Proc. R. Soc. A **284**, 159 (1965).

Chapter 8

Introduction to Extra Dimensions and Thick Braneworlds

Yu-Xiao Liu [1]

Institute of Theoretical Physics, Lanzhou University,
Lanzhou 730000, China

Abstract: In this review, we give a brief introduction on the aspects of some extra dimension models and the five-dimensional thick brane models in extended theories of gravity. First, we briefly introduce the Kaluza–Klein theory, the domain wall model, the large extra dimension model, and the warped extra dimension models. Then some thick brane solutions in extended theories of gravity are reviewed. Finally, localization of bulk matter fields on thick branes is discussed.

Keywords: Extra dimensions, braneworlds, extended theories of gravity, localization

8.1 Introduction

The concept of extra dimensions has been proposed for more than one hundred years. In 1914, a Finnish physicist Gunnar Nordström first introduced an extra spatial dimension to unify the electromagnetic and gravitational fields [1, 2]. It is known that Nordström's work is not successful because it appeared before Einstein's general relativity. A few years later, a German mathematics teacher Theodor Kaluza put forward a five-dimensional theory that tries to unify Einstein's general relativity and Maxwell's electromagnetism [3]. Subsequently, in 1926 the Swedish physicist Oskar Klein suggested that the extra spatial dimension should be "compactified": it is curled up on a circle with a microscopically small radius so that it cannot be directly observed in everyday physics [4, 5]. This theory is referred to Kaluza–Klein (KK) theory. Since then extra dimensions have aroused intense interest and study from physicists. Specifically, KK theory is usually regarded as an important

---

[1] Email: liyx@lzu.edu.cn

predecessor to string theory, which attempts to address a number of fundamental problems of physics.

However, the major breakthrough of the research along phenomenological lines occurred at the end of the 20th century. In 1982, Keiichi Akama presented a picture that we live in a dynamically localized 3-brane in higher-dimensional space-time, which is also called "braneworld" in modern terminology [6]. In 1983, Valery Rubakov and Mikhail Shaposhnikov proposed an extra dimension model, i.e., the domain wall model [7, 8], which assumes that our observable universe is a domain wall in five-dimensional space-time. The most remarkable characteristic of the two models is that the extra dimensions are non-compact and infinite [6, 7], which is also the seed of the subsequent thick brane models with curved extra dimensions. In 1990, Ignatios Antoniadis examined the possibility of the existence of a large internal dimension at relatively low energies of the order of a few TeV [9]. Such a dimension is a general prediction of perturbative string theories and this scenario is consistent with perturbative unification up to the Planck scale.

But what really triggered the revolution of the extra dimension theory is the work done by Nima Arkani-Hamed, Savas Dimopoulos, and Georgi Dvali (ADD) in 1998 [10], which has provided an important solution to the gauge hierarchy problem. Since the extra dimensions in the ADD model are large (compared to the Planck scale) and compact (similar to KK theory), now it has been a paradigm of large extra dimension models. Ignatios Antoniadis, Nima Arkani-Hamed, Savas Dimopoulos, and Georgi Dvali (AADD) [11] gave the first string realization of low scale gravity and braneworld models, and pointed out the motivation of TeV strings from the stabilization of mass hierarchy and the graviton emission in the bulk.

One year later, Lisa Randall and Raman Sundrum (RS) proposed that it is also possible to solve the gauge hierarchy problem by using a non-factorizable warped geometry [12]. This model is also called RS-1 model and has been a paradigm of warped extra dimension models now. One of the basic assumptions of the two models is that the Standard Model particles are trapped on a three-dimensional hypersurface or brane, while gravity propagates in the bulk. This type of model is also known as braneworld model. It is worth mentioning that Merab Gogberashvili also considered a similar scenario [13, 14, 15]. After the ADD model and RS-1 model, the study of extra dimensions enters a new epoch and some of the extended extra dimension models are also well known, such as the RS-2 model [16], the Gregory–Rubakov–Sibiryakov (GRS) model [17], the Dvali–Gabadadze–Porrati (DGP) model [18], the thick brane models [19, 20, 21], the universal extra dimension model [22], etc. It should be noted that the universal extra dimension model [22] is a particular case of the proposal of TeV extra dimensions in the Standard Model [9].

Here, we list some review references and books. References [24, 23, 25, 26, 27, 28] are very suitable for beginners. References [29, 30, 31, 32, 33, 34, 35] provide very detailed introductions to extra dimension theories, including phenomenological research. There are also some books [36, 37], which may be of great help to the readers who want to do some related research in this direction. In this review, we mainly focus on thick brane models. Some review papers for some thick brane models can be found in Refs. [32, 38, 39].

8.2 Some extra dimension theories

In this section, we will give a brief review of some extra dimension theories, including the KK theory, the domain wall with a non-compact extra dimension, the braneworld with large extra dimensions, and the braneworld with a warped extra dimension.

In this review, we use capital Latin letters (such as M, N, $\cdots$) and Greek letters (such as μ, ν,...) to represent higher-dimensional and four-dimensional indices, respectively. The coordinate of the five-dimensional space-time is denoted by $x^M = (x^\mu, y)$ with x^μ and y the usual four-dimensional and the extra-dimensional coordinates, respectively. A five-dimensional quantity is described by a "sharp hat" (in this section). For example, $\hat{R}$ indicates the scalar curvature of the higher-dimensional space-time.

8.2.1 *KK theory*

First of all, let us review KK theory [3, 4]. It is the first unified field theory of Maxwell's electromagnetism theory and Einstein's general relativity built with the idea of an extra spatial dimension beyond the usual four of space and time. The three-dimensional space is homogeneous and infinite and the fourth spatial dimension y is a compact circle with a radius R_{ED} (see Fig. 8.1). So, this model was also known as "cylinder world" [40] in the early days. This theory is a purely classical extension of general relativity to five dimensions. Therefore, it assumes that there is only gravity in the five-dimensional space-time and the four-dimensional electromagnetism and gravity can be obtained by dimensional reduction. KK theory is viewed as an important precursor to string theory.

KK theory is described by the five-dimensional Einstein-Hilbert action

$$S_{\mathrm{KK}} = \frac{1}{2\kappa_5^2} \int d^4x \, dy \, \sqrt{-\hat{g}} \, \hat{R} \tag{8.1}$$

and the following metric ansatz

$$\widehat{ds}^2 = \hat{g}_{MN} dx^M dx^N = e^{2\alpha\phi} g_{\mu\nu} dx^\mu dx^\nu + e^{-4\alpha\phi}(dy + A_\mu dx^\mu)^2, \tag{8.2}$$

where the five-dimensional gravitational constant κ_5 is related to the five-dimensional Newton constant $G_N^{(5)}$ and the five-dimensional mass scale M_* as

$$\kappa_5^2 = 8\pi G_N^{(5)} = \frac{1}{M_*^3}, \tag{8.3}$$

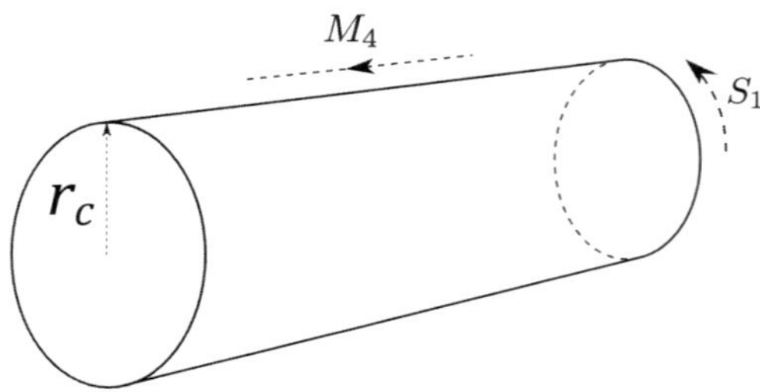

Fig. 8.1   The basic picture of KK theory with topology $M_4 \times S^1$ [41]. The radius of the extra dimension $r_c = R_{\mathrm{ED}}$.

α is a parameter, and all these components $g_{\mu\nu}$, A_μ, ϕ are functions of x^μ only (the so-called cylinder condition). The components of the five-dimensional metric $\hat{g}_{MN}$ are

$$\hat{g}_{\mu\nu} = e^{2\alpha\phi} g_{\mu\nu} + e^{-4\alpha\phi} A_\mu A_\nu, \tag{8.4}$$

$$\hat{g}_{\mu 5} = e^{-4\alpha\phi} A_\mu, \tag{8.5}$$

$$\hat{g}_{55} = e^{-4\alpha\phi}. \tag{8.6}$$

Note that, among the 15 components of $\hat{g}_{MN}$, 10 components are identified with the four-dimensional metric $g_{\mu\nu}$, four components with the electromagnetic vector potential A_μ, and one component with a scalar field called "radion" or "dilaton". Substituting the above metric into (8.1) and integrating the extra dimension y yields the following four-dimensional effective action

$$S_{\mathrm{KK}} = \int d^4 x \sqrt{-g} \left(\frac{1}{2\kappa_4^2} R - \frac{1}{2} g^{\mu\nu} \, \nabla_\mu \phi \, \nabla_\nu \phi - \frac{1}{4} e^{6\alpha\phi} F_{\mu\nu} F^{\mu\nu} \right), \tag{8.7}$$

where $\kappa_4^2 = \sqrt{8\pi G_N} = 1/M_{\mathrm{Pl}}$ with M_{Pl} the four-dimensional Plack mass, R is the four-dimensional scalar curvature defined by the metric $g_{\mu\nu}$, ϕ is a dilaton field, and $F_{\mu\nu} = \partial_\mu A_\nu - \partial_\nu A_\mu$ is the four-dimensional field strength of the vector field $A_\mu(x^\lambda)$. Thus, one obtains a four-dimensional scalar-vector-tensor theory (8.7) from a five-dimensional pure gravity. This theory only contains gravity and electromagnetic fields when ϕ is a constant:

$$S_{\mathrm{KK}} = \int d^4 x \sqrt{-g} \left(\frac{1}{2\kappa_4^2} R - \frac{1}{4} F_{\mu\nu} F^{\mu\nu} \right). \tag{8.8}$$

Thus, Einstein's general relativity and Maxwell's electromagnetic theory in four-dimensional space-time can be unified in the five-dimensional KK theory. The relation between the fundamental Planck scale M_* and the four-dimensional effective one M_{Pl} is

$$M_{\mathrm{Pl}}^2 = (2\pi R_{\mathrm{ED}}) M_*^3. \tag{8.9}$$

Here, R_{ED} is the radius of the extra dimension. It is easy to see that when the radius of the extra dimension R_{ED} is large, one can get a four-dimensional effective Planck scale M_{Pl} from a small fundamental scale M_*. This characteristic inspired the later large extra dimension model that tries to solve the hierarchy problem. In 1926, in order to explain the cylinder condition, Oskar Klein gave this classical theory a quantum interpretation by introducing the hypothesis that the fifth dimension is curled up and microscopic [4, 5]. He also calculated a scale for the fifth dimension based on the quantization of charge.

As an early theory of extra dimensions, KK theory is not a completely self-consistent theory. Now let us consider the following translation in the fifth coordinate

$$x^\mu \to x'^\mu = x^\mu, \qquad y \to y' = y + \kappa\xi(x), \tag{8.10}$$

which leads to the gauge transformation of the electromagnetic vector potential A_μ:

$$A_\mu(x) \to A'_\mu(x) = A_\mu(x) + \kappa\partial_\mu\xi(x). \tag{8.11}$$

The above coordinate translation (8.10) also leads to the gauge transformation of each KK mode $\Phi^{(n)}(x)$ of a bulk scalar $\Phi(x, y) = \sum_{n=0}^{\infty} \Phi^{(n)}(x)e^{iny/R_{ED}}$:

$$\Phi^{(n)}(x) \to \Phi'^n(x) = \Phi^{(n)}(x)e^{in\kappa\xi(x)/R_{ED}}, \tag{8.12}$$

which indicates that each scalar KK mode has charge

$$Q_n = n\frac{\kappa}{R_{\mathrm{ED}}} = ne, \tag{8.13}$$

with the charge quanta e given by

$$e = \frac{\kappa}{R_{\mathrm{ED}}} = \frac{\sqrt{16\pi G_N}}{R_{\mathrm{ED}}} = \sqrt{4\pi\alpha} = \sqrt{4\pi/137}. \tag{8.14}$$

Thus, one will obtain a tiny scale of the extra dimension:

$$R_{\mathrm{ED}} \sim 10^{-33}\mathrm{m}, \tag{8.15}$$

which is not much larger than the Plank length $l_{\mathrm{Pl}} \sim 10^{-35}m$. Such a tiny scale means that detecting the extra dimension is almost hopeless. On the other hand, in KK theory, the mass spectrum of KK modes of a bulk field with mass M_0 is given by

$$m_n = \sqrt{M_0^2 + \frac{n^2}{R_{\mathrm{ED}}}} \simeq n \times 10^{17}\mathrm{GeV}, \tag{8.16}$$

where the electroweak scale parameter M_0 is neglected. So, the masses of the massive KK modes of a bulk field will be much larger than the order of TeV. It is difficult to detect such heavy KK particles in the present and future experiments. Therefore, only the zero modes of the bulk fields are observable. However, it is not acceptable that the charges of the KK modes of the bulk fields must satisfy $Q_n = ne$, i.e., all zero modes that describe the observed particles are neutral. Therefore, the predictions of KK theory about four-dimensional particles are completely inconsistent with experiments. This is the main reason why KK theory was not taken seriously for nearly half a century.

More details and related issues about KK theory can be found in the book [36] or Refs. [42, 43, 44, 45] and the references therein.

8.2.2 *Non-compact extra dimension: domain wall model*

In 1982, Akama presented a picture that we live in a dynamically localized 3-brane in a higher-dimensional space-time [6]. As an example, it was considered that our four-dimensional space-time is localized on a 3-brane by the dynamics of the Nielsen–Olesen-type vortex in six-dimensional space-time. At low energies, matters and gravity are trapped in the 3-brane.

In 1983, Rubakov and Shaposhnikov proposed a domain wall model in a five-dimensional Minkowski space-time [7, 8], which is completely different from KK theory. In this model, our four-dimensional universe is restricted to a domain wall formed by a bulk scalar field, and the extra dimension is non-compact and infinitely large (see Fig. 2). There is an effective potential well around the domain wall, which can trap the lower energy KK modes of a bulk fermion field, i.e., the four-dimensional fermions, on the domain wall. So in general, the Standard Model particles are localized on the domain wall due to the potential well and we can only feel a three-dimensional space. Only when the energy of a particle is higher than the edges of the potential well, can one detect the effects of the extra dimension.

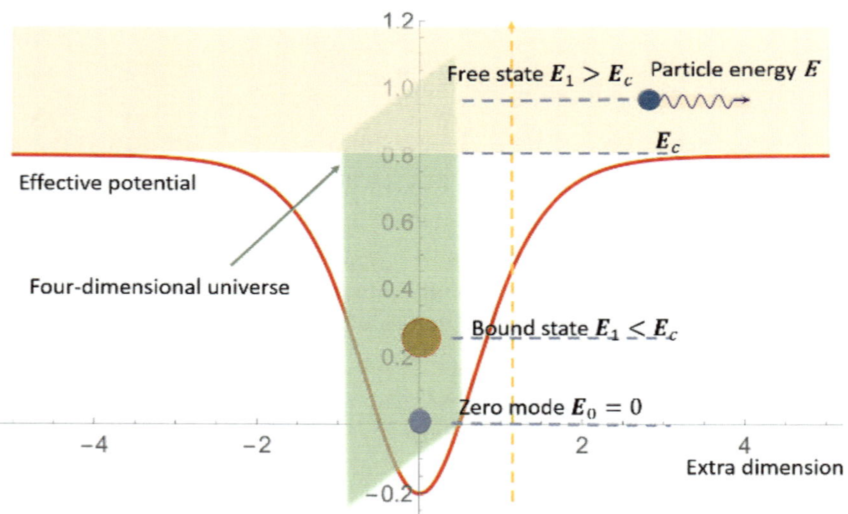

Fig. 8.2   The basic picture of the Rubakov–Shaposhnikov domain wall model [39].

In the original domain wall model, Rubakov and Shaposhnikov considered the following ϕ^4 model of a scalar field in a five-dimensional Minkowski space-time [7]:

$$L_{\mathrm{DW}} = -\frac{1}{2}\eta^{MN}\partial_M\phi\partial_N\phi - \frac{k^2}{2v^2}\left(\phi^2 - v^2\right)^2, \tag{8.17}$$

where v and k are positive parameters. The ϕ^4 model usually gives a double-well potential and the minima of the potential are located at $\phi = \pm v$. The model has the following static domain wall solution:

$$\phi(y) = v\tanh(ky). \tag{8.18}$$

The energy density of the system along extra dimension with respect to a static observer $u^M = (1,0,0,0,0)$ is

$$\rho(y) = T_{MN}u^M u^N = \frac{1}{2}\eta^{MN}\partial_M\phi\partial_N\phi + \frac{k^2}{2v^2}\left(\phi^2 - v^2\right)^2 = k^2v^2\text{sech}^4(ky). \tag{8.19}$$

The shapes of the scalar field (8.18) and energy density (8.19) are shown in Fig. 8.3.

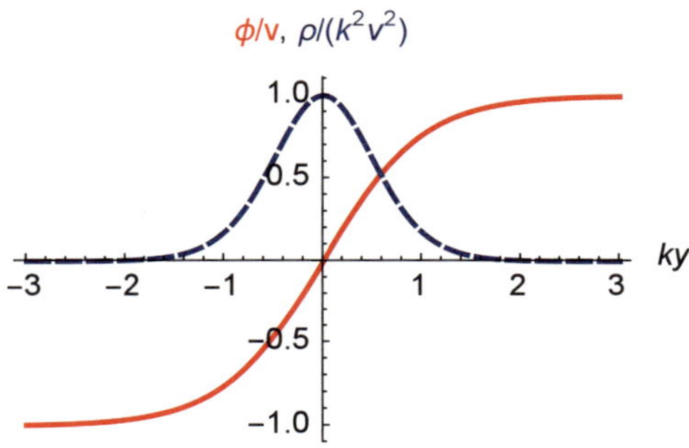

Fig. 8.3 The scalar field (8.18) (red line) and energy density (8.19) (blue dashed line) for the Rubakov–Shaposhnikov domain wall.

Next, we show that the zero mode of a bulk Dirac fermion Ψ coupling with the background scalar ϕ can be localized on the domain wall even though the extra dimension is infinite. Suppose that there is a Yukawa coupling between a five-dimensional fermion field Ψ and the background scalar field ϕ:

$$L_\Psi = \bar{\Psi}\gamma^M\partial_M\Psi - \eta\bar{\Psi}\phi\Psi, \tag{8.20}$$

where η is the coupling parameter. The equation of motion is given by $(\gamma^M\partial_M - \eta\phi)\Psi = 0$. Then with the KK decomposition

$$\Psi(x,y) = \sum_n \Psi_n(x,y) = \sum_n \left[\psi_{Ln}(x)f_{Ln}(y) + \psi_{Rn}(x)f_{Rn}(y)\right], \tag{8.21}$$

where $\psi_{Ln} = -\gamma^5\psi_{Ln}$ and $\psi_{Rn} = \gamma^5\psi_{Rn}$ are the left- and right-chiral components of the Dirac fermion field, respectively, one can obtain the four-dimensional Dirac equations for $\psi_{Ln,Rn}(x)$:

$$\begin{aligned}(\gamma^\mu\partial_\mu - m_n)\psi_{Ln}(x) = 0, \\ (\gamma^\mu\partial_\mu - m_n)\psi_{Rn}(x) = 0,\end{aligned} \tag{8.22}$$

and the equations of motion for the KK modes $f_{Ln,Rn}(y)$:

$$[-\partial_y^2 + V_L(y)]f_{Ln}(y) = m_n^2 f_{Ln}(y), \tag{8.23}$$

$$[-\partial_y^2 + V_R(y)]f_{Rn}(y) = m_n^2 f_{Rn}(y), \tag{8.24}$$

where m_n is the mass of the four-dimensional fermion and the effective potentials are given by

$$V_{L,R}(y) = \eta^2 \phi^2(y) \mp \eta \partial_z \phi(y) = \eta^2 v^2 \left(\tanh^2(ky) \mp \frac{k}{\eta v} \mathrm{sech}^2(ky) \right). \qquad (8.25)$$

The shapes of the effective potentials are plotted in Fig. 8.4. It can be seen that whether there is a potential well for V_R with $\eta > 0$ is determined by the ratio $k/(\eta v)$. When $k/(\eta v) < 1$, there is a potential well, which may trap some bound massive KK fermions.

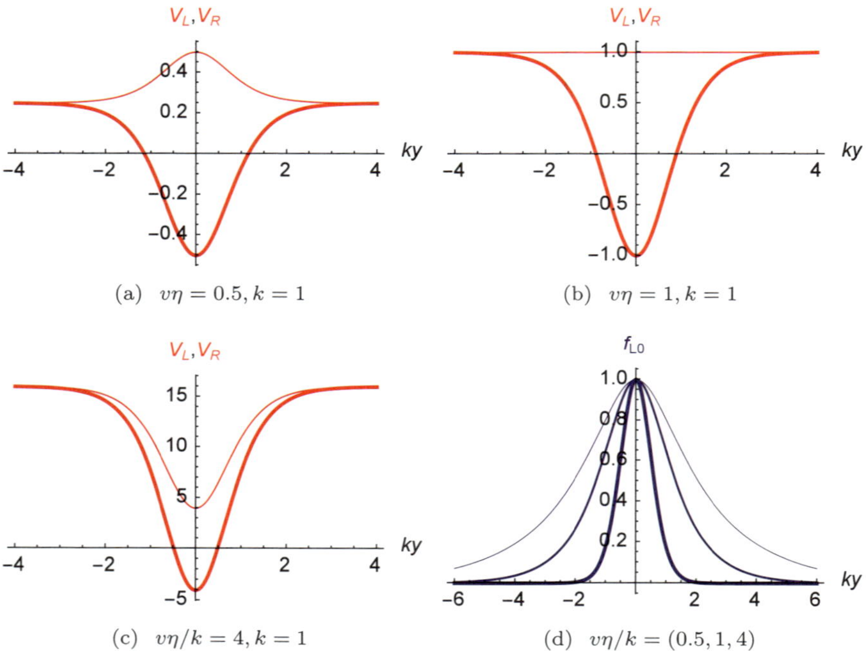

(a) $v\eta = 0.5, k = 1$

(b) $v\eta = 1, k = 1$

(c) $v\eta/k = 4, k = 1$

(d) $v\eta/k = (0.5, 1, 4)$

Fig. 8.4   The effective potentials (8.25) (thick lines for V_L and thin lines for V_R) and the non-normalized left-chiral fermion zero mode (8.26) (thickness of lines increases with $v\eta/k$) for the Rubakov–Shaposhnikov domain wall.

One can derive the zero modes of the left- and right-chiral fermion fields:

$$f_{L0,R0}(y) \propto \exp\left(\mp \eta \int \phi(y) dy \right) = \cosh(ky)^{\mp \eta v/k}. \qquad (8.26)$$

So, when $\eta > 0$, the left-chiral fermion field could be localized on the domain wall (to localize the right-chiral Fermion field one needs $\eta < 0$), which is the most prominent feature of the model. The above localized left-chiral fermion zero mode is plotted in Fig. 8.4(d). It is clear that, even though the extra dimension is infinite, the zero mode of the left-chiral fermion can be localized on the domain wall through the

Yukawa coupling with the background scalar field, while the right-chiral one cannot. Therefore, the fermion zero modes, i.e., the massless four-dimensional fermions localized on the domain wall, can be used to mimic our matters. They propagate with the speed of light along the domain wall, but do not move along the extra dimension. Therefore, domain wall is also called braneworld. In the real world, fermions have masses. So, in realistic theories fermion zero modes should acquire small masses by some mechanism. For the case of weak coupling $|\eta| < k/v$, there is no bound massive fermion KK modes. However, there may be one or more bound massive KK modes on the wall if the coupling is large enough ($\eta \gg k/v$). Besides, there is a continuous part of the spectrum starting at $m = \eta v$. These continuous states correspond to five-dimensional fermions that can escape to $|y| = \infty$.

For the localization of gauge fields and more details about this domain wall model, one can refer to Refs. [7, 46, 47].

It should be noted that the idea of Akama, Rubakov and Shaposhnikov is important because it provides a way basically distinct from the "compactification" to hide the extra dimensions. However, this domain wall model has a fatal weakness. Since the extra dimension is flat and infinitely large, the zero mode of gravity cannot be localized on the domain wall. Obviously, if extra dimensions are infinite and flat, i.e., for a D-dimensional Minkowski space-time, the gravitational force between any two static massive particles would be $F \sim 1/r^{D-2}$ instead of the inverse square law. Because of this shortcoming, this flat domain wall model is not taken seriously for a long time. After lRandall and Sundrum proposed the RS-2 model, the Rubakov–Shaposhnikov domain wall model was reconsidered in a warped five-dimensional space-time [19, 20, 21] (the jargon for this kind of model is thick brane model), which will be introduced in Sections 8.3 and 8.4.

8.2.3 *Large extra dimensions: ADD braneworld model*

After the Akama brane model and the Rubakov–Shaposhnikov domain wall model, the phenomenological lines of extra dimensions almost have no significant development for a long time, except for compactifications at the electroweak scale (see, e.g., Refs. [48, 9, 49]). Until 1998, Arkani-Hamed, Dimopoulos, and Dvali suggested ingeniously that the gauge hierarchy problem can be addressed by the large extra dimension model (also called as ADD braneworld model) [10]. Then extra dimension theories re-attracted attentions of theoretical physicists.

Before introducing the ADD model, it is necessary to describe what the hierarchy problem is. It is usually expressed as the huge discrepancy between the gravitation and electroweak interactions. In quantum field theory, it has another expression, i.e., why the Higgs boson mass is so much lighter than the Planck scale. These two expressions are equivalent and we will briefly introduce the latter one. In the Standard Model, the physical mass μ and bare mass μ_0 of the Higgs boson satisfy the following relationship:

$$\mu^2 = \mu_0^2 + \delta\mu_0^2. \tag{8.27}$$

Here, $\delta\mu_0^2 \sim \Lambda^2$ is the loop correction of the bare mass and Λ is a truncation parameter. According to the effective field theory, Λ should be the energy scale up to which the Standard Model is valid. It is known that the physical mass of the Higgs boson is $\mu \sim 10^2 \text{GeV}$. Assuming that the new physics appears at the Planck scale $M_{\text{Pl}} \sim 10^{19}\text{GeV}$, Eq. (8.27) cannot be satisfied unless there is an unnatural fine-tuning between the bare mass and the loop correction. The essential reason why bare mass of Higgs boson requires such a high-precision adjustment is because of the huge hierarchy between the weak scale $M_{\text{EW}} \approx 246\text{GeV}$ and the Planck scale $M_{\text{Pl}} \sim 10^{19}\text{GeV}$. If the new physics appears at the low-energy scale (such as 1TeV), there is no serious fine-tuning problem. Next, we will see that the ADD model does solve the hierarchy problem by assuming that the new physics in the bulk appears at $M_* \sim 1\text{TeV}$.

In the ADD model, the space-time is assumed to be $(4+d)$-dimensional and the corresponding action is given by [10]

$$S_{\text{ADD}} = \frac{M_*^{d+2}}{2} \int d^{4+d}x \sqrt{-\hat{g}}\hat{R}. \tag{8.28}$$

If these extra spatial dimensions have the same radius R_{ED}, then one can obtain the following relationship

$$M_{\text{Pl}}^2 = M_*^{d+2}(2\pi R_{\text{ED}})^d, \tag{8.29}$$

whose physical meaning and calculation are similar to Eq. (8.9). In the bulk space, the fundamental scale of gravity is no longer the Plank mass M_{Pl} but M_*. To avoid the emergence of hierarchy, one assumes $M_* \sim 1\text{TeV}$. Then the radius of the extra dimensions reads as

$$R_{\text{ED}} = \frac{1}{2\pi M_*}\left(\frac{M_{\text{Pl}}}{\hat{M}}\right)^{2/d} \sim \frac{1}{2\pi}10^{32/d}\text{TeV}^{-1} \sim 10^{-17} \times 10^{32/d}\text{cm}. \tag{8.30}$$

For $d = 1$, the radius of the extra dimension should be as large as 10^{13}m in order to address hierarchy problem. Obviously, it is against the tests of the gravitational inverse-square law [50, 51, 52], which constrain the radius of the extra dimensions to be less than sub-millimeter. Therefore, according to the present gravity experiments, the number of the extra dimensions in the ADD model should be more than two. Of course, if the sizes of these extra dimensions are different, the result would be complex.

More importantly, if one assumes that other fields live in the bulk, then the radius of the extra dimensions should be much smaller (according to the recent experiments it should be less than 10^{-18}m or more) in order not to violate the experiments at Large Hadron Collider (LHC). Even at the end of the last century, according to the nuclear-related researches, it can be deduced that the radius of the extra dimensions should usually be much less than 1μm. For this reason, ADD proposed another supposition inspired by string theory: except the gravitational field, all the Standard Model particles are bounded on a four-dimensional hypersurface or brane by an unknown natural mechanism. This hypothesis is the main difference between the ADD model and the KK theory.

It should be noted that although the ADD model can eliminate the hierarchy between the weak scale and the Planck scale based on the assumptions that the extra dimensions are large (as compared to the Planck length) and the Standard Model particles are localized on a brane, the ratio between the fundamental scale M_* and the scale corresponding to the size of the extra dimensions, R_{ED}, is not acceptable. From Eq. (8.30), we have

$$\frac{M_*}{1/R_{\mathrm{ED}}} \sim \frac{1}{2\pi}\left(\frac{M_{\mathrm{Pl}}}{M_*}\right)^{2/d} = \frac{1}{2\pi}10^{32/d}. \tag{8.31}$$

If one requires that $1/R_{\mathrm{ED}}$ and M_* are in the same order of magnitude, the number of extra dimensions should be 32 or so. Naturally, a question arises: why there are so many extra dimensions? Therefore, the ADD model does not really solve the hierarchy problem.

Other issues in the ADD model (including the difference between the ADD model and the KK theory, the features of the KK states, how the KK states interact with the fields on the brane, etc.) are discussed in detail in Refs. [53, 54, 23]

8.2.4 *Warped extra dimension: RS braneworld models*

The common feature of the KK theory, the domain wall model, and the ADD model is that extra dimensions are flat. They have solved some problems but left some new problems. In this section, we will see that two new extra dimension models in curved space-time give different physical pictures for our world.

8.2.4.1 *RS-1 model*

Inspired by the ADD model, in 1999 Randall and Sundrum proposed a braneworld model with a warped extra dimension to address the hierarchy problem, which is now called the RS-1 model [16]. The basic assumptions of the model are listed as follows:

- There is only one extra spatial dimension, which is compactified on an S^1/Z_2 orbifold with a radius R_{ED} ($y \in [-\pi R_{\mathrm{ED}}, \pi R_{\mathrm{ED}}]$).
- There are two branes at the fixed points $y = 0$ (called the hidden brane or Planck brane or UV brane) and $y = \pi R_{\mathrm{ED}}$ (called the visible brane or TeV brane or IR brane) in the bulk. And all the Standard Model particles are bounded on the visible brane.

- The form of the five-dimensional metric is supposed to be

$$ds^2 = e^{2A(y)}\eta_{\mu\nu}dx^\mu dx^\nu + dy^2 = e^{2A(y)}\eta_{\mu\nu}dx^\mu dx^\nu + R_{\mathrm{ED}}^2 d\phi^2, \qquad (8.32)$$

where the warp factor $A(y)$ is a function of the extra dimension $y = R_{\mathrm{ED}}\phi$ only.

- The bulk is a five-dimensional anti-de Sitter (AdS) space-time, i.e., there is only a negative cosmological constant in the bulk.

Therefore, the total action of the RS-1 model consists of three parts [16]:

$$S_{\mathrm{RS\text{-}1}} = S_{\mathrm{gravity}} + S_{\mathrm{vis}} + S_{\mathrm{hid}}$$

$$= \int d^4x \int dy \sqrt{-\hat{g}} \left(\frac{M_*^3}{2}\hat{R} - \Lambda \right)$$

$$+ \int d^4x \sqrt{g_{\mathrm{vis}}}(L_{\mathrm{vis}} - V_{\mathrm{vis}}) + \int dx^4 \sqrt{g_{\mathrm{hid}}}(L_{\mathrm{hid}} - V_{\mathrm{hid}}). \qquad (8.33)$$

Through a series of calculations and simplifications, one can obtain the four-dimensional effective action of the RS-1 model:

$$S_{\mathrm{eff}} = \frac{M_{\mathrm{Pl}}^2}{2} \int d^4x \sqrt{-g}R. \qquad (8.34)$$

Here, R is the scalar curvature defined by the four-dimensional metric $g_{\mu\nu}$ and M_{Pl} is similar to Eq (8.29):

$$M_{\mathrm{Pl}}^2 = \frac{M_*^3}{k}(1 - e^{-2k\pi R_{\mathrm{ED}}}), \qquad (8.35)$$

where the parameter k has the dimension of mass. To avoid new hierarchy problem, one requires that the parameter k satisfies $k/M_* \sim 1$. When the value of kR_{ED} becomes large, the fundamental scale M_* and the Plank scale M_{Pl} will be the same order. Note that the RS-1 model assumes that the fundamental scale M_* is still equivalent to the Planck scale, which is very different from the ADD model. But how does one get the four-dimensional TeV scale for the weak interaction?

The RS-1 model assumes that the Higgs boson is bounded on the visible brane and the four-dimensional effective mass of the Higgs boson is given by

$$m_{\mathrm{H}} = \sqrt{\lambda}\, e^{-kR_{\mathrm{ED}}\pi}v_0 = e^{-kR_{\mathrm{ED}}\pi}\hat{m}_{\mathrm{H}}, \qquad (8.36)$$

where λ is a dimensionless parameter, and v_0 is the vacuum expected value of the Higgs field in the five-dimensional space-time. The mass of the Higgs boson in the five-dimensional space-time is $\hat{m}_{\mathrm{H}} = \sqrt{\lambda}v_0$.

To eliminate the hierarchy, the RS-1 model requires that the fundamental parameters M_*, k, and v_0 are all at the order of M_{Pl}. Although the fundamental mass of the Higgs boson $\hat{m}_H$ in the five-dimensional space-time is the truncation scale M_{Pl}, the effective physical mass on the four-dimensional brane could be "red-shifted" to the order of TeV as long as the radius of the extra dimension satisfies $R_{\mathrm{ED}} \sim 10/k$. Noted that R_{ED}^{-1} is also a fundamental parameter. Since the exponential function is introduced into Eq. (8.36), the ratio of k to R_{ED}^{-1} does not need to be too large to "red-shift" $\hat{m}_{\mathrm{H}}$ to TeV. This is why we usually say that the RS-1 model solves the hierarchy problem without introducing new hierarchy.

8.2.4.2 *RS-2 model*

As mentioned earlier, the fundamental scale of the five-dimensional space-time and the effective Plank scale in the KK theory and ADD model satisfy Eq. (8.29), which requires that the radius of extra dimensions is finite. However, because of the warped space-time, it can be seen from Eq. (8.35) that the scale of the extra dimension may be infinite if one forgets the hierarchy problem. Enlightened by the RS-1 model, Randall and Sundrum provided another braneworld model (called as the RS-2 model) [12]) to solve the puzzle left by the domain wall model: the localization of gravity on the brane with an infinite co-dimension. The focus of the RS-2 model is mainly on how to restore the four-dimensional gravity on a thin brane when the extra dimension is infinite. Roughly, compared to the RS-1 model, the RS-2 model has made the following changes:

- We live on the Plank brane at $y = 0$ (the Standard Model particles are bounded on this brane).
- The TeV brane located at $y = \pi R_{\text{ED}}$ is moved to infinity, i.e., $R_{\text{ED}} \to \infty$. So the KK spectrum in the RS-2 model is continuous.

The metric of the RS-2 model can be obtained by taking the limit $R_{\text{ED}} \to \infty$ in the metric (8.32). Similarly, the KK spectrum in the RS-2 model can also be reduced from that of the RS-1 model [12]:

$$m_n \approx \left(n + \frac{1}{4}\right)\pi k e^{-\pi k R_{\text{ED}}}. \tag{8.37}$$

Obviously, since $R_{\text{ED}} \to \infty$, the KK spectrum is continuous. In general, the Newton's gravitational potential between two static massive particles on the brane is contributed by all the KK gravitons. In the RS-2 model, due to the presence of the continuous massive KK states, it is necessary to consider how these KK states affect the Newton's gravitational potential. Randall and Sundrum showed that the Newton's gravitational potential of two static particles with masses m_1 and m_2 and distance $|\vec{x}|$ on the brane has the following form [12]:

$$\begin{aligned}
V(|\vec{x}|) &\sim \frac{m_1 m_2}{|\vec{x}|} + \int_0^\infty \frac{dm}{k}\, \frac{m_1 m_2 e^{-m|\vec{x}|}}{|\vec{x}|}\, \frac{m}{k} \\
&\sim \frac{m_1 m_2}{|\vec{x}|}\left(1 + \frac{1}{|\vec{x}|^2 k^2}\right).
\end{aligned} \tag{8.38}$$

Here, the first term is contributed from the zero mode of graviton, which corresponds to the standard Newton's gravitational potential. The second one is the contribution of all the massive KK modes, and it is the correction to the Newton's gravitational potential. As the distance $|\vec{x}|$ increases, the correction term decays quickly. The effect of the extra dimension appears at the scale of the Planck length. So Randall and Sundrum proved that even if there is an infinite extra dimension, as long as it warps in some way, one can still get an effective four-dimensional Newtonian gravity.

8.3   Solutions of thick brane models in extended theories of gravity

In this section, we introduce some thick brane models. It is known from the RS-2 model that, if the extra dimension is warped, it is possible to realize the localization of the matter fields and gravitational fields on a domain wall or thin brane with an infinite extra dimension. In the RS-2 model, the thickness of the brane is neglected and so the brane is called as thin brane. However, a brane without thickness is idealistic and a real braneworld should have a thickness. Furthermore, it may be hard to find thin brane solutions in some higher-order derivative gravity theories, such as the $f(R)$ theory. It is easy to guess that a brane could be dynamically generated by some background fields, such as one or more scalar fields. Therefore, by combining the domain wall model and the RS-2 model, theoretical physicists investigated the so-called thick braneworld models, where the brane solutions are smooth.

In literature, most five-dimensional thick branes are generated by one or more scalar fields with kink-like and/or bump-like configurations [55,56,57,58,60,61,62,59, 63,64], but a few brane models are based on vector fields or spinor fields [65,66,67]. Higher-dimensional thick branes were also considered [68, 69]. They are smooth generalizations of the RS-2 model. Note that these fields should not be thought of as the matter fields that are related with those in the standard model. They are the matter fields generating a brane. There are also some brane models without matter fields [70, 72, 71]. These branes are embedded in higher-dimensional space-times which are not necessarily AdS far from the branes. In this review, we mainly consider scalar-field-generated thick branes embedded in AdS space-time and these smooth branes appear as domain walls interpolating between various vacua of the scalar fields. Unlike the thin RS-2 model, such thick brane solutions do not have any matter fields living on the brane [73]. In fact, all matter fields are assumed as bulk fields in thick brane models and one needs to investigate the localization of these bulk fields on the branes, which is the subject of the next section. In this section, we mainly consider constructions and solutions of thick brane models in extended theories of gravity. The system is described by the action

$$S = \int d^5x \sqrt{-g} \left[\frac{1}{2\kappa_5^2} \mathcal{L}_{\text{G}} + \mathcal{L}_{\text{M}}(g_{MN}, \phi^I, \nabla_M \phi^I) \right], \tag{8.39}$$

where the five-dimensional gravitational constant κ_5 is related to the five-dimensional Newton constant $G_N^{(5)}$ and the five-dimensional Planck mass scale M_* as

$$\kappa_5^2 = 8\pi G_N^{(5)} = \frac{1}{M_*^3}. \tag{8.40}$$

Sometimes one sets $\kappa_5 = 1$ for convenience. $\mathcal{L}_{\text{G}}$ is the Lagrangian of gravity, and $\mathcal{L}_{\text{M}}(g_{MN}, \phi^I)$ is the Lagrangian of the matter fields that generate the thick brane. Note that, a five-dimensional quantity is no longer described by a "sharp hat" from now on for convenience. For the simplest case of general relativity and a canonical scalar field, we have

$$\mathcal{L}_{\mathrm{G}} = R, \tag{8.41}$$

$$\mathcal{L}_{\mathrm{M}} = -\frac{1}{2}g^{MN}\partial_M\phi\partial_N\phi - V(\phi), \tag{8.42}$$

for which the energy-momentum tensor is

$$T_{MN} = \partial_M\phi\partial_N\phi - g_{MN}\left(\frac{1}{2}\partial^P\phi\partial_P\phi + V(\phi)\right). \tag{8.43}$$

In this review, we only consider static branes. The five-dimensional line-element which preserves four-dimensional Poincaré invariance is assumed as

$$ds^2 = g_{MN}dx^M dx^N = \mathrm{e}^{2A}ds^2_{\mathrm{brane}} + dy^2, \tag{8.44}$$

where

$$ds^2_{\mathrm{brane}} = \tilde{g}_{\mu\nu}(x^\lambda)dx^\mu dx^\nu \tag{8.45}$$

describes the geometry of the brane. Usually, we are concerned with three typical branes:

$$ds^2_{\mathrm{brane}} = \begin{cases} \eta_{\mu\nu}dx^\mu dx^\nu & \text{flat brane} \\[2mm] \mathrm{e}^{2Hx_3}(-dt^2 + dx_1^2 + dx_2^2) + dx_3^2 & \text{AdS brane} \\[2mm] -dt^2 + e^{2Ht}dx^i dx^i & \text{dS brane} \end{cases} \tag{8.46}$$

For a static brane, the warp factor A and scalar fields ϕ^I are functions of the extra dimensional coordinate y or z only, and the non-vanishing components of the energy-momentum tensor (8.43) are

$$T_{\mu\nu} = -g_{\mu\nu}\left(\frac{1}{2}g^{55}(\partial_5\phi)^2 + V(\phi)\right), \tag{8.47}$$

$$T_{55} = \frac{1}{2}(\partial_5\phi)^2 - g_{55}V(\phi). \tag{8.48}$$

One can make a coordinate transformation $dz = \mathrm{e}^{-A}dy$ and rewrite the five-dimensional line-element as

$$ds^2 = \mathrm{e}^{2A}(ds^2_{\mathrm{brane}} + dz^2), \tag{8.49}$$

which is very useful in the derivation of the perturbation equations of gravity and localization equations of various bulk matter fields in the current and next sections. The dynamical field equations read as

$$R_{MN} - \frac{1}{2}R\,g_{MN} = \kappa_5^2 T_{MN}, \tag{8.50}$$

$$\Box^{(5)}\phi = V_\phi, \tag{8.51}$$

whose non-vanishing component equations in the (x^μ, y) and (x^μ, z) coordinates are

$$3\left(\varepsilon H^2 e^{-2A} - A'' - 2A'^2\right) = \kappa_5^2 \left(\frac{1}{2}\phi'^2 + V\right), \tag{8.52}$$

$$6\left(-\varepsilon H^2 e^{-2A} + A'^2\right) = \kappa_5^2 \left(\frac{1}{2}\phi'^2 - V\right), \tag{8.53}$$

$$4A'\phi' + \phi'' = V_\phi, \tag{8.54}$$

and

$$3\left(\varepsilon H^2 - \partial_z^2 A - \partial_z A^2\right) = \kappa_5^2 \left(\frac{1}{2}(\partial_z\phi)^2 + e^{2A(z)}V\right), \tag{8.55}$$

$$6\left(-\varepsilon H^2 + \partial_z A^2\right) = \kappa_5^2 \left(\frac{1}{2}(\partial_z\phi)^2 - e^{2A(z)}V\right), \tag{8.56}$$

$$e^{-2A}\left(3\partial_z A \partial_z\phi + \partial_z^2\phi\right) = V_\phi, \tag{8.57}$$

respectively. Here, $\Box^{(5)} = g^{MN}\nabla_M\nabla_N$ is the five-dimensional d'Alembert operator, the primes denotes the derivatives with respect to the extra dimensional coordinate y, and $\varepsilon = 1,\ -1$, and 0 for de Sitter, AdS, and flat brane solutions, respectively. Note that only two of the above equations (8.52)–(8.54) (or (8.55)–(8.57)) are independent.

Before going to extended theories of gravity, we introduce some brane solutions in GR. The first example of a flat brane was given in Ref. [73]:

$$e^{2A(y)} = \operatorname{sech}^{\frac{4v^2}{9}}(ky)e^{\frac{v^2}{9}\operatorname{sech}^2(ky)}, \tag{8.58}$$

$$\phi(y) = \frac{v}{\kappa_5}\tanh(ky), \tag{8.59}$$

$$V(\phi) = \frac{k^2}{54\kappa_5^2 v^2}\left[27v^4 - 18v^2\left(2v^2 + 3\right)\kappa_5^2\phi^2 + 3\left(8v^2 + 9\right)\kappa_5^4\phi^4 - 4\kappa_5^6\phi^6\right]. \tag{8.60}$$

A de Sitter thick brane solution in a five-dimensional space-time for the potential

$$V(\phi) = \frac{1 + 3\alpha}{2\alpha\kappa_5^2}3H^2\left(\cos\left(\frac{\kappa_5\phi}{v}\right)\right)^{2(1-\alpha)} \tag{8.61}$$

was found in Ref. [74]:

$$e^{2A(z)} = \operatorname{sech}^{2\alpha}\left(\frac{Hz}{\alpha}\right), \tag{8.62}$$

$$\phi(z) = \frac{v}{\kappa_5}\arcsin\left(\tanh\left(\frac{Hz}{\alpha}\right)\right), \tag{8.63}$$

where $v = \sqrt{3\alpha(1-\alpha)}$, $0 < \alpha < 1$, $H > 0$. The domain wall configuration with warped geometry is dynamically generates by the soliton scalar. At last, we list the AdS brane solution found in Ref. [57]:

$$V(\phi) = \frac{3k^2}{2\kappa_5^2}\left(4 - v^2\cosh^2\left(\frac{\kappa_5\phi(y)}{v}\right)\right), \tag{8.64}$$

$$\phi(y) = \frac{v}{\kappa_5}\operatorname{arcsinh}(\tan(ky)), \tag{8.65}$$

$$e^{2A(y)} = \frac{3H^2}{k^2\left(v^2 - 3\right)}\cos^2(ky). \tag{8.66}$$

It is obvious to see that the thick brane is bounded in the interval $y \in \left(-\frac{\pi}{2k}, \frac{\pi}{2k}\right)$.

Gauge-invariant fluctuations of the branes were analyzed in Ref. [75], where scalar, vector and tensor modes of the geometry were classified according to four-dimensional Lorentz transformations. It was shown that the tensor zero mode is localized on the brane, which ensures the four-dimensional Newton's law of gravitation, while the scalar and vector fluctuations have no normalizable zero mode and hence are not localized on the brane. In Ref. [76], Herrera-Aguilar et al. considered the mass hierarchy problem and the corrections to Newton's law in thick branes with Poincaré symmetry both in the presence of a mass gap in the graviton spectrum and without it.

There is a special class of theories with noncanonical fields, namely the K-field theory, which was first proposed to drive the inflation with generic initial conditions [77, 78]. This kind of theory introduces a general noncanonical Lagrangian $\mathcal{L} \equiv \mathcal{L}_{\mathrm{M}}(X, \phi)$, where

$$X = -\frac{1}{2} g^{MN} \partial_M \phi \partial_N \phi. \tag{8.67}$$

For example, $\mathcal{L} = \mathcal{F}(X) - V(\phi)$ is a simple one. The flat thick braneworld models generated by K field were considered in Ref. [79], and the solutions with perturbative procedure in Ref. [80]. In Ref. [79], Christoph Adam et al. chose a (non-standard) kinetic term such that the resulting kink in the extra dimension is a compacton, both with and without gravitational backreaction. This is slightly different from the thick branes of Refs. 113-115 which are not compactons. Compactons have some peculiar features (e.g. linear fluctuations are restricted to within the compacton, and the spectrum of perturbations is purely discrete). It was shown that, even with gravity included, the brane solutions remain compactons and are linearly stable [81]. The exact solutions with specific model $\mathcal{L} = X - \alpha X^2 - V(\phi)$ were given in Ref. [82]. The system of a flat thick braneworld consists of the equations

$$-3\partial_y^2 A = \kappa_5^2 \mathcal{L}_X (\partial_y \phi)^2, \tag{8.68a}$$

$$6(\partial_y A)^2 = \kappa_5^2 (\mathcal{L} + \mathcal{L}_X (\partial_y \phi)^2), \tag{8.68b}$$

and

$$(\partial_y^2 \phi)(\mathcal{L}_X + 2X\mathcal{L}_{XX}) + L_\phi - 2X\mathcal{L}_{X\phi} = -4\mathcal{L}_X(\partial_y\phi)(\partial_y A). \tag{8.69}$$

The work [80] developed a first-order formalism to solve the K-brane system by assuming

$$\partial_y A = -\frac{1}{3} W(\phi). \tag{8.70}$$

Using this assumption Eqs. (8.68) can be written as

$$W_\phi = \kappa_5^2 \mathcal{L}_X (\partial_y \phi)^2, \tag{8.71a}$$

$$\frac{2}{3} W^2 = \kappa_5^2 (\mathcal{L} + \mathcal{L}_X (\partial_y \phi)^2). \tag{8.71b}$$

For small parameter α, one can get the analytic solutions. Reference [82] gives another approach which is able to get the exact solutions. The strategy is to assume

$$\partial_y A = -\frac{1}{3}\left[W(\phi) + \alpha Y(\phi)\right],\tag{8.72}$$

$$\partial_y \phi = W_\phi,\tag{8.73}$$

where $Y(\phi)$ satisfies $Y_\phi = W_\phi^3$. One can choose the superpotential as

$$W = k\phi_0^2 \sin\left(\frac{\phi}{\phi_0}\right)\tag{8.74}$$

to get the Sine-Gordon solution:

$$\phi = \phi_0 \arcsin\left(\tanh(ky)\right),\tag{8.75}$$

$$A = -\left(1 + \frac{2}{3}k^2\alpha\phi_0^2\right)\ln(\cosh(ky)) - \frac{1}{6}k^2\alpha\phi_0^2 + \frac{1}{6}k^2\alpha\phi_0^2\mathrm{sech}^2(ky),\tag{8.76}$$

$$V = -\frac{k^2\phi_0^2}{18}\left(6 + 5k^2\alpha\phi_0^2 + k^2\alpha\phi_0^2\cos\left(\frac{2\phi}{\phi_0}\right)\right)^2\sin^2\left(\frac{\phi}{\phi_0}\right)$$

$$+\frac{1}{2}k^2\phi_0^2\cos^2\left(\frac{\phi}{\phi_0}\right) + \frac{3}{4}k^4\alpha\phi_0^4\cos^2\left(\frac{\phi}{\phi_0}\right).\tag{8.77}$$

The analysis of tensor and full linear perturbations can be found in Refs. [80, 83]. The tensor mode is similar to the standard case, while the scalar mode has significant difference. The work [83] shows that the scalar perturbation mode cannot be canonically normalized in the conformally flat coordinate z, and needs another coordinate transformation. The scalar zero mode cannot be localized on the brane, provided that $\mathcal{L}_X > 0$ and $1 + 2\frac{\mathcal{L}_{XX}X}{\mathcal{L}_X} > 0$.

In Ref. [84], a de Sitter tachyon thick braneworld was considered with the following action:

$$S = \int d^5x\sqrt{-g}\left[\frac{1}{2\kappa_5^2}R - \Lambda_5 - V(\phi)\sqrt{1 + g^{MN}\partial_M\phi\partial_N\phi}\right],\tag{8.78}$$

where ϕ is a tachyonic bulk scalar field. It was shown that the four-dimensional gravity is localized on the brane, and it is separated by a continuum of massive KK modes by a mass gap. The corrections to Newton's law in this model decay exponentially. The stability of the de Sitter tachyon braneworld under the scalar sector of fluctuations for vanishing and negative bulk cosmological constant was also investigated [85]. Corrections to Coulomb's law and fermion field localization were computed in Ref. [86].

Next we will introduce the thick brane models in extended theories of gravity.

8.3.1   *Metric $f(R)$ theory*

Among the large amount of proposals of extended theories of gravity, the $f(R)$ theory [87] has received growing interests due to its unique advantage: it is the simplest modification with higher-derivative curvature invariants, which are required by renormalization. In addition, some other theories with curvature invariants like $R_{MN}R^{MN}$ and $R_{MNPQ}R^{MNPQ}$ (except the Guass-Bonnet term) would inevitably lead to Ostrogradski instability [88]. This makes the $f(R)$ theory most likely the only tensor theory of gravity that allows higher derivatives.

Now let us review the five-dimensional thick brane model in the metric $f(R)$ theory coupled with a canonical scalar field with the Lagrangian (8.42). The action is given by

$$S_{\text{met}} = \int d^5x \sqrt{-g} \left[\frac{1}{2\kappa_5^2} f(R) - \frac{1}{2} g^{MN} \partial_M \phi \partial_N \phi - V(\phi) \right]. \tag{8.79}$$

The gravitational field equations read as

$$f_R R_{MN} - \frac{1}{2} f \, g_{MN} - \left(\nabla_M \nabla_N - g_{MN} \Box^{(5)} \right) f_R = \kappa_5^2 T_{MN}, \tag{8.80}$$

$$\Box^{(5)} \phi \equiv g^{MN} \nabla_M \nabla_N \phi = V_\phi, \tag{8.81}$$

where f_R and V_ϕ are defined as $f_R \equiv \frac{df(R)}{dR}$ and $V_\phi \equiv \frac{dV(\phi)}{d\phi}$. We only consider flat branes generated by a canonical scalar field, for which the line-element is given by

$$ds^2 = g_{MN} dx^M dx^N = e^{2A(y)} \eta_{\mu\nu} dx^\mu dx^\nu + dy^2$$
$$= e^{2A(z)} \left(\eta_{\mu\nu} dx^\mu dx^\nu + dz^2 \right). \tag{8.82}$$

Then the field equations (8.80) and (8.81) read as

$$f + 2f_R \left(4A'^2 + A'' \right) - 6f_R' A' - 2f_R'' = \kappa_5^2 (\phi'^2 + 2V), \tag{8.83}$$

$$-8f_R \left(A'' + A'^2 \right) + 8f_R' A' - f = \kappa_5^2 (\phi'^2 - 2V), \tag{8.84}$$

$$4A'\phi' + \phi'' = V_\phi, \tag{8.85}$$

where the primes represent derivatives with respect to the coordinate y. This is a system with fourth-order derivatives on the metric. In general, it would be extremely hard to solve these fourth-order non-linear differential equations analytically. However, it is widely believed that the $f(R)$ theory is equivalent to the Brans–Dicke theory with the Brans–Dicke parameter $\omega_0 = 0$. Hence, it would be more comfortable to operate in the Brans–Dicke theory, which contains only up to second-order derivatives. This is actually the strategy used in Ref. [89]. Some thick brane solutions in the higher-order frame were studied in Refs. [90,91]. However, they are not perfect since the solution in Ref. [90] has singularity while the solution in Ref. [91] is numerical.

The first exact solution in higher-order frame was given in Ref. [92], where the specific model with

$$f(R) = R + \gamma R^2 \tag{8.86}$$

was considered. The solution of equations (8.83)-(8.85) comes from the observation that only two of these equations are independent because of the conservation of the energy-momentum tensor [93]. This implies that one can solve the equations by giving one of the three functions, namely, the warp factor $e^{A(y)}$, the scalar field $\phi(y)$, and the scalar potential $V(\phi)$. The prior choice is to assume the solution of $e^{A(y)}$ since the fourth-order derivatives only act on $A(y)$. With such choice, one only needs to solve the second-order field equations. To get an asymptotically AdS_5 geometry, the warp factor can be assumed as [92]

$$e^{A(y)} = \operatorname{sech}(ky).$$ (8.87)

Then the scalar field and scalar potential can be solved as [92]

$$\phi(y) = v \tanh(ky),$$ (8.88)

$$V(\phi) = \lambda^{(5)}(\phi^2 - v^2)^2 + \frac{\Lambda_5}{2\kappa_5^2},$$ (8.89)

where the parameters are related by

$$\lambda^{(5)} = \frac{29}{98}\kappa_5^2 k^2, \quad v = \sqrt{\frac{3}{29}}\frac{7}{\kappa_5}, \quad \Lambda_5 = -\frac{318}{29}k^2, \quad \gamma = \frac{3}{232k^2}.$$ (8.90)

This is a ϕ^4 type potential with the vacua located at $\phi = \pm v$. It can be seen that the spatial boundaries $y = \pm\infty$ are mapped to the minima of the scalar potential. For the same $f(R)$ (8.86), a general solution with the warp factor $e^{A(y)} = \operatorname{sech}^B(ky)$ was constructed in Ref. [94].

Bazeia et al. [94] considered the generalised model with some other polynomial and nonpolynomial potential solutions. They expanded the scalar field as $\phi(y) = \phi_0(y) + \alpha\phi_\alpha(y)$, and then wrote the scalar potential and warp function as $V(\phi) = V_0(\phi) + \alpha V_\alpha(\phi)$ and $A(y) = A_0(y) + \alpha A_\alpha(y)$, with α a small parameter. To the first order of α, the solution is [94]

$$\phi(y) = v \tanh(ky) + \alpha \sum_{n=0}^{3} C_{\phi n} \tanh^{2n+1}(ky),$$ (8.91)

$$V(\phi(y)) = \sum_{n=0}^{3} C_{Vn} \tanh^{2n}(ky) + \alpha \sum_{n=0}^{6} C_{Vn+4} \tanh^{2n}(ky),$$ (8.92)

$$A(y) = C_{A1}\left(\operatorname{sech}^2(ky) + 4\ln\operatorname{sech}(ky)\right) + \alpha \sum_{n=2}^{4} C_{An}\operatorname{sech}^{2n}(ky),$$ (8.93)

where $C_{\phi n}$, C_{Vn}, and C_{An} are some coefficients related to the parameters α, γ, and κ_5. They also considered the case of multiple scalar fields in Ref. [95], where the first-order formalism method was developed.

There are also some pure geometric thick $f(R)$-branes without any scalar field. Zhong and Liu [71] gave the solutions for triangular $f(R)$ and polynomial $f(R)$. Here we list the solution for the latter [71]:

$$f(R) = \Lambda_5 + c_1 R - \frac{c_2}{k^2}R^2 + \frac{c_3}{k^4}R^3,$$ (8.94)

$$e^{A(y)} = \cosh^{-20}(ky),$$ (8.95)

where Λ_5 is the five-dimensional cosmological constant, and c_n are dimensionless constants. Furthermore, a class of exact solutions in D dimensions were found by Lü et al. [72]. One of solutions of the warp factor is

$$e^{A(y)} = \left[e^{bDy}\cosh^2(ky)\right]^{\frac{2}{\alpha D}}. \tag{8.96}$$

The corresponding $f(R)$ has a complex form.

The stability of the tensor perturbation of general background was analyzed in Ref. [96]. The perturbed metric is

$$ds^2 = e^{2A(z)}\left[(\eta_{\mu\nu} + h_{\mu\nu})dx^\mu dx^\nu + dz^2\right], \tag{8.97}$$

where $h_{\mu\nu}$ is a transverse-traceless tensor, namely, $\eta^{\mu\nu}h_{\mu\nu} = 0 = \partial_\mu h^\mu_{\ \nu}$. By making the decomposition

$$h_{\mu\nu}(x^\rho, z) = (a^{-3/2}f_R^{-1/2})\epsilon_{\mu\nu}(x^\rho)\psi(z), \tag{8.98}$$

where $a \equiv e^{2A}$, one can derive that the KK mode $\psi(z)$ of the tensor perturbation satisfies the following Schrödinger-like equation [96]:

$$\left[-\partial_z^2 + W(z)\right]\psi(z) = m^2\psi(z), \tag{8.99}$$

where the effective potential is given by

$$W(z) = \frac{3}{4}\frac{(\partial_z a)^2}{a^2} + \frac{3}{2}\frac{\partial_z^2 a}{a} + \frac{3}{2}\frac{\partial_z a \partial_z f_R}{a f_R} - \frac{1}{4}\frac{(\partial_z f_R)^2}{f_R^2} + \frac{1}{2}\frac{\partial_z^2 f_R}{f_R}. \tag{8.100}$$

One can check that this equation can be factorized as

$$\mathcal{K}\mathcal{K}^\dagger \psi(z) = m^2\psi(z) \tag{8.101}$$

with

$$\mathcal{K} = \partial_z + \frac{3}{2}\frac{\partial_z a}{a} + \frac{1}{2}\frac{\partial_z f_R}{f_R}, \tag{8.102}$$

$$\mathcal{K}^\dagger = -\partial_z + \frac{3}{2}\frac{\partial_z a}{a} + \frac{1}{2}\frac{\partial_z f_R}{f_R}, \tag{8.103}$$

which ensures that there is no graviton mode with $m^2 < 0$. The graviton zero mode (the four-dimensional massless graviton) can be solved as

$$\psi_0 \propto (a^3 f_R)^{1/2}. \tag{8.104}$$

Note that to make sure ψ_0 is real, f_R should be positive. This also avoids the graviton ghost. The recovering of four-dimensional gravity on the brane requires the normalization of graviton zero mode, namely

$$\int_{-\infty}^{+\infty} (\psi_0)^2 dz < \infty. \tag{8.105}$$

For the solution given in Ref. [92], this condition can certainly be satisfied. Thus the four-dimensional gravity can be obtained. Besides the bound graviton zero mode, there are continuous unbound massive graviton KK modes. They will have a contribution to Newton's law of gravitation at short distance. The structure of other $f(R)$-brane models given in Ref. [94, 97] was analyzed in Ref. [98], where the effective potential for the graviton KK modes may have a singular structure and there is a series of graviton resonant modes.

There is a problem that should be mentioned here. The above analysis involved the tensor mode only. This is not complete since the full perturbations contain tensor, vector, and scalar modes. In the original RS-1 model [16], the fluctuation of the extra dimension radius gives a scalar mode (radion). If the extra dimension is not stabilized, the radion will be massless, which will contribute a long range fifth force. This is undoubtedly unacceptable. If the Goldberger–Wise mechanism [99] is introduced, the extra dimension radius can be stabilized and the radion will becomes massive.

In thick braneworld scenario constructed with a background scalar field, the radion-like scalar mode has a continuous mass spectrum. But there is still a massless radion-like scalar mode. So the recovering of four-dimensional gravity implies that the scalar zero mode should not be localized. For general relativity coupled with a scalar field, the scalar zero mode is not localized.

However, the situation is completely different for the case of the $f(R)$ gravity. The tensor and vector modes are similar to the case of general relativity while the scalar mode is very different. In the higher-order frame, the dynamical equation of the scalar mode would be fourth order, which implies that there are actually two scalar degrees of freedom. Recall that the $f(R)$ theory is equivalent to the Brans–Dicke theory. Based on this fact, the scalar perturbations of the $f(R)$ theory can be investigated in the frame work of scalar-tensor theory, and we will review this part in section 8.3.4.

8.3.2 *Palatini $f(\mathcal{R})$ theory*

It is well known that there are two different formalisms in $f(R)$ theories of gravity, namely, the metric formalism and the Palatini formalism [87]. In Palatini formalism, metric and connection are two independent fundamental variables. In general relativity these two formalisms are completely equivalent, but usually they will lead to different predictions in modified theories of gravity. Different from the metric $f(R)$ theory, the Palatini $f(\mathcal{R})$ theory will lead to a second-order system. In this subsection, we consider thick brane models in the Palatini $f(R)$ theory. The action is given by

$$S_{\text{Pal}} = \int d^D x \sqrt{-g} \left[\frac{1}{2\kappa_D^2} f(\mathcal{R}(g, \Gamma)) - \frac{1}{2} g^{MN} \partial_M \phi \partial_N \phi - V(\phi) \right], \quad (8.106)$$

and the metric is also described by (8.82). Variation with respect to the metric g and the connection Γ yields two equations of motion:

$$f_{\mathcal{R}}\mathcal{R}_{MN} - \frac{1}{2}f\,g_{MN} = \kappa_D^2 T_{MN},\tag{8.107}$$

$$\tilde{\nabla}_A\left(\sqrt{-g}f_{\mathcal{R}}g^{MN}\right) = 0,\tag{8.108}$$

where $\tilde{\nabla}_A$ is compatible with the independent connection Γ. For the case of $f(\mathcal{R}) = \mathcal{R}$, the theory is equivalent to general relativity. It would be convenient to define an auxiliary metric q_{MN} by

$$\sqrt{-q}\,q^{MN} \equiv \sqrt{-g}f_{\mathcal{R}}g^{MN}.\tag{8.109}$$

Now $\Gamma^P{}_{MN}$ and $\mathcal{R}_{MN}(\Gamma)$ can be viewed as the connection and Ricci tensor constructed from the auxiliary metric q_{MN}, respectively. One can eliminate the independent connection from the field equations by using the relation (8.109), and obtain the following equations concerned with the metric only:

$$G_{MN} = \frac{\kappa_D^2 T_{MN}}{f_{\mathcal{R}}} - \frac{1}{2}g_{MN}\left(\mathcal{R} - \frac{f}{f_{\mathcal{R}}}\right) + \frac{1}{f_{\mathcal{R}}}\left(\nabla_M\nabla_N - g_{MN}\nabla_A\nabla^A\right)f_{\mathcal{R}}$$

$$- \frac{D-1}{(D-2)f_{\mathcal{R}}^2}\left(\nabla_M f_{\mathcal{R}}\nabla_N f_{\mathcal{R}} - \frac{1}{2}g_{MN}\nabla_A f_{\mathcal{R}}\nabla^A f_{\mathcal{R}}\right).\tag{8.110}$$

In addition, contracting Eq. (8.107) with g^{MN}, one gets an algebraic equation of $\mathcal{R}$ and T:

$$f_{\mathcal{R}}\mathcal{R} - \frac{D}{2}f = \kappa_D^2 T.\tag{8.111}$$

This implies that $f(\mathcal{R})$ is just an algebraic expression of T. From this point of view, we can see that Eq. (8.110) clearly tells that the Palatini $f(\mathcal{R})$ theory modifies the matter sector of Einstein equations. Note that there are derivatives on $f(\mathcal{R})$ and $f_{\mathcal{R}}(\mathcal{R})$ and thus T. This structure would lead to surface singularity of stars [100], and also give higher derivatives on the matter fields. This is a significant difference with the metric $f(R)$ theory, which has higher derivatives on the metric. For more details about the Palatini $f(\mathcal{R})$ theory, see Refs. [87, 101].

The thick braneworld model in five-dimensional space-time with a scalar field in the Palatini $f(\mathcal{R})$ theory was first considered in Ref. [102]. The authors considered the flat braneworld model and introduced two methods to solve the system. In the first method, they assumed the following relations

$$\frac{d\phi}{dy} = b\cos(b\phi),\tag{8.112}$$

$$f_{\mathcal{R}} = \left(1 + ab^2\cos^2(b\phi)\right)^{-3/4},\tag{8.113}$$

from which one can get $\phi(y)$ and $f(\mathcal{R}(y))$, but it is hard to get the expression of $f(\mathcal{R})$. The solution for $\phi(y)$ and $A(y)$ is [102]

$$\phi(y) = \frac{1}{b}\arcsin(\tanh(b^2 y)),\tag{8.114}$$

$$A(y) = -A_0 + \ln(\mathcal{U}) + \frac{2\kappa_5^2}{27b^2}\mathcal{U}^3 - \frac{\kappa_5^2}{3b^2}\left(1 + \frac{2ab^2}{3}\right)\left[\operatorname{arctanh}\left(\frac{1}{\mathcal{U}}\right) + \arctan(\mathcal{U})\right],$$

$$\tag{8.115}$$

where $\mathcal{U}(y) = \left(1 + ab^2\text{sech}^2(b^2y)\right)^{1/4}$. They also used the perturbative method to consider the model $f(\mathcal{R}) = \mathcal{R} + \epsilon\mathcal{R}^n$ with ϵ a small parameter. To the first order of ϵ, the analytic solutions were obtained.

The complete solutions were first obtained in Ref. [103], and the tensor perturbation was also investigated therein. As mentioned above, the field equations (8.110) contain second-order derivatives on the trace of the energy-momentum tensor. This means that the field equations (8.110) contain third-order derivatives on the scalar field and it is not a convenient choice to solve the equations (8.110).

The work [103] gave a strategy which avoids solving the higher-derivative equations. Note that the original equations (8.107) and (8.108) are second order at most, hence are much easier to solve. For the assumption of the space-time metric (8.82), the auxiliary metric q_{MN} is given by

$$d\tilde{s}^2 = q_{MN}dx^M dx^N = u^2(y)\eta_{\mu\nu}dx^\mu dx^\nu + \frac{u^2(y)}{a^2(y)}dy^2, \tag{8.116}$$

where $u(y) = a(y)f_{\mathcal{R}}^{1/3}$. In terms of these variables, the field equations (8.107) are reduced to

$$\left(6\frac{u'^2}{u^2} - 3\frac{a'}{a}\frac{u'}{u} - 3\frac{u''}{u}\right)f_{\mathcal{R}} = \kappa_5^2\phi'^2, \tag{8.117}$$

$$5f_{\mathcal{R}}\left(\frac{a'}{a}\frac{u'}{u} + \frac{u''}{u}\right) - 2f_{\mathcal{R}}\frac{u'^2}{u^2} + f = 2\kappa_5^2 V. \tag{8.118}$$

The above two equations and the scalar field equation

$$\Box^{(5)}\phi = V_\phi \tag{8.119}$$

constitute the system to be solved. For the model $f(\mathcal{R}) = \mathcal{R} + \alpha\mathcal{R}^2$, one can assume the relation $u(y) = c_1 a^n(y)$ with $n \neq 0$, then the system can be solved [103]:

$$a(y) = \text{sech}^{\frac{2}{3(n-1)}}(ky), \tag{8.120}$$

$$V(y) = v_1\text{sech}^4(ky) + v_2\text{sech}^2(ky) + \frac{\Lambda_5}{2\kappa_5^2}, \tag{8.121}$$

$$\phi(y) = \phi_0\left[i\sqrt{3}\,\text{E}\left(iky, \frac{2}{3}\right) - i\sqrt{3}\,\text{F}\left(iky, \frac{2}{3}\right)\right.$$

$$\left. + \sqrt{2 + \cosh(2ky)}\,\tanh(ky)\right], \tag{8.122}$$

where $\text{F}(y, m)$ and $\text{E}(y, m)$ are the incomplete elliptic integrals of the first and second kinds, respectively. With some analyses on the energy density the parameters are constrained to be $\alpha > 0$ and $n < -2/3$ or $0 < n < 1/6$ or $n > 1$. Note that the warp factor is infinite at boundary for $0 < n < 1/6$ or $n < -2/3$. Usually, the scalar potential should be a function of the scalar field ϕ, but in this case it cannot be solved.

Different from the background solutions, it would be more convenient to consider the gravity fluctuations with the equation (8.110). The final equation is similar to the case of the metric $f(R)$ theory [103]:

$$\mathcal{K}\mathcal{K}^{\dagger}\Psi(z) = m^2\Psi(z), \tag{8.123}$$

where $\mathcal{K} = \partial_z + \frac{3}{2}\partial_z \ln a + \frac{1}{2}\partial_z \ln f_{\mathcal{R}}$ and $\Psi(z)$ is defined by $h_{\mu\nu}(x^\sigma, z) = \varepsilon_{\mu\nu}(x^\sigma)(a^3 f_{\mathcal{R}})^{1/2}\Psi(z)$. Again, $f_{\mathcal{R}}$ should be positive to keep the graviton zero mode real and to avoid the graviton ghost. It can be easily checked that the solutions (8.120) always support localized graviton zero mode $\Psi(z(y)) \propto \mathrm{sech}^{\frac{n}{n-1}}(ky)$. There is a new feature, which is different from the regular case. For $n < -2/3$, the effective potential is an infinitely deep potential well, for which all the states are bounded. However, the four-dimensional gravity can still recover, but with a tiny correction [103].

8.3.3 *Eddington inspired Born–Infeld theory*

There is another Palatini gravity theory which is widely considered in recent literatures, namely, the Eddington inspired Born–Infeld (EiBI) gravity theory [104]. It is an extension of Eddington's gravity theory, and contains the matter Lagrangian absent in Eddington's theory. The thick braneworld model was considered in Refs. [105, 106]. The action of the theory is [104]

$$S(g,\Gamma,\Phi) = \frac{1}{\kappa_5^2 b}\int d^5 x\left[\sqrt{-|g_{MN}+bR_{MN}(\Gamma)|} - \lambda\sqrt{-|g_{MN}|}\right] + S_M(g,\Phi), \tag{8.124}$$

where b and λ are some parameters, and $R_{MN}(\Gamma)$ is the Ricci tensor built from the independent connection Γ. As mentioned in the last subsection, the Palatini version of a gravitational theory abandons the priority of the metric, so the variation has to be made with respect to both of the metric and connection. The field equations are

$$\frac{\sqrt{-|g_{PQ}+bR_{PQ}|}}{\sqrt{-|g_{PQ}|}}[(g_{PQ}+bR_{PQ})^{-1}]^{MN} - \lambda g^{MN} = -\kappa_5 bT^{MN}, \tag{8.125}$$

$$\tilde{\nabla}_K(\sqrt{-q}q^{MN}) = 0, \tag{8.126}$$

where $q_{MN} \equiv g_{MN} + bR_{MN}$ is the new introduced auxiliary metric which satisfies $q_{MN}q^{MP} = \delta_N^P$, and the covariant derivative $\tilde{\nabla}$ is compatible with this metric. The work [105] constructed a flat brane generated by a canonical scalar field. By using the auxiliary metric and assuming the relation $\phi'(y) = Ka^2(y)$, the authors obtained an analytic domain wall solution:

$$a(y) = \mathrm{sech}^{\frac{3}{4}}(ky), \tag{8.127}$$

$$\phi(y) = \phi_0\left(i\mathrm{E}(\frac{ky}{2},2) + \mathrm{sech}^{\frac{1}{2}}(ky)\sinh(ky)\right), \tag{8.128}$$

$$V(\phi(y)) = \frac{7\sqrt{21}}{24 b\kappa_5}\mathrm{sech}^3(ky) - \frac{\lambda}{b\kappa_5}, \tag{8.129}$$

where $k = \frac{2}{\sqrt{21b}}$. Note that the scalar potential is expressed in terms of the space-time coordinates. However, according to the scalar field equation one can still find that the spatial boundaries $y = \pm\infty$ are mapped to the minimum of the scalar potential $V(\phi)$. Besides, it can be easily checked that the energy density localizes near the origin.

The above solution was generalised in Ref. [106] by imposing some more general assumptions $\phi'(y) = Ka^{2n}(y)$ and $\phi'(y) = K_1 a^2(y)(1 - K_2 a^2(y))$. Under these assumptions, some solutions with interesting features, like double kink solution, are allowed.

References [105, 106, 107] investigated linear perturbations of the EiBI brane system. It was found that the tensor perturbation is stable for the above models [105, 106] and the stability condition for the scalar perturbations of a known analytic domain wall solution with $e^{2A(y)} = \mathrm{sech}^{\frac{3}{2p}}$ is $0 < p < \sqrt{8}$ [107]. Quasi-localization of gravitational fluctuations was also studied [106].

8.3.4 Scalar-tensor theory

The scalar-tensor theory has long been considered as an alternative gravity theory that deviates from general relativity. It was first proposed from the inspiration of Mach's principle by Brans and Dicke [108] in 1961. Thin braneworld models in this theory were studied in Refs. [109, 110, 111], and thick braneworld models in Refs. [109, 112, 114, 115, 116, 117, 113]. Here we only review the thick brane models. We consider the following action in Jordon frame:

$$S = \int d^5x \sqrt{-g} \left(\frac{1}{2\kappa_5^2} F(\phi)R - \frac{1}{2}(\partial\phi)^2 - V(\phi) + \mathcal{L}_{\mathrm{M}}(g_{MN}, \psi) \right), \qquad (8.130)$$

where $F(\phi)$ is a positive function in order to avoid the problem of antigravity. In Jordon frame the scalar field ϕ non-minimally couples to the Ricci scalar. In Einstein frame ϕ does not couple to the Ricci scalar, but couples to the matter sector. For the flat brane assumption (8.82) and $\mathcal{L}_{\mathrm{M}}(g_{MN}, \psi) = 0$, the equations of motion are

$$3F(4A'^2 + A'') + 7A'F' + F'' = -2\kappa_5^2 V, \qquad (8.131)$$

$$3FA'' - A'F' + F'' = -\kappa_5^2 \phi'^2, \qquad (8.132)$$

$$\kappa_5^2 \phi'' + 4\kappa_5^2 A'\phi' - 2F_\phi(2A'' + 5A'^2) = \kappa_5^2 V_\phi. \qquad (8.133)$$

One of the analytic solutions with $F(\phi) = 1 - \alpha\kappa_5^2\phi^2$ was given in Ref. [109]:

$$\phi(y) = \frac{1}{\kappa_5}\sqrt{\frac{3(1-6\alpha)}{\alpha(1-2\alpha)}}\tanh(ky), \tag{8.134}$$

$$A(y) = -(\alpha^{-1} - 6)\ln\cosh(ky), \tag{8.135}$$

$$V(\phi) = \frac{k^2}{6\alpha}\left[\frac{9-54\alpha}{(1-2\alpha)\kappa_5^2} + 6(\alpha(7+24\alpha)-2)\phi^2 \right.$$
$$\left. + \frac{\alpha(1-2\alpha)(3-16\alpha)(4-12\alpha)}{1-6\alpha}\kappa_5^2\phi^4\right], \tag{8.136}$$

where the parameter α should satisfy $0 < \alpha < \frac{1}{6}$. Reference [113] found another interesting solution:

$$F(\phi) = \frac{v^2}{3} + \left(1 - \frac{v^2}{3}\right)\cosh\left(\frac{\sqrt{3}\kappa_5}{v}\phi\right), \tag{8.137}$$

$$\phi(y) = \frac{v}{\kappa_5}\arctan(\sinh ky), \tag{8.138}$$

$$A(y) = -\ln\cosh(ky), \tag{8.139}$$

$$V(\phi) = \frac{v^2 k^2}{2\kappa_5^2}\cos^2\left(\frac{\kappa_5\phi}{v}\right) + \frac{2k^2}{\sqrt{3}\kappa_5^2}\left(3-v^2\right)\sin\left(\frac{2\kappa_5\phi}{v}\right)\sinh\left(\frac{\sqrt{3}\kappa_5\phi}{v}\right)$$
$$-\frac{k^2}{\kappa_5^2}\left(2v^2 + (6-2v^2)\cosh\left(\frac{\sqrt{3}\kappa_5\phi}{v}\right)\right)\sin^2\left(\frac{\kappa_5\phi}{v}\right). \tag{8.140}$$

To avoid antigravity, it requires $F > 0$ and thus

$$0 < v^2 < \frac{3\cosh(\sqrt{3}\pi/2)}{3\cosh(\sqrt{3}\pi/2) - 1}. \tag{8.141}$$

All of these solutions constitute of domain walls, which can localize gravity and some matter fields. Such kind of works can be found in Refs. [114, 118].

The tensor perturbation and localization of gravity were investigated in Refs. [109, 114, 113], and we will not review this part. What is more interesting is the scalar perturbations, which have obvious differences with that of general relativity. However, the more general cases are the models with multiple scalars. Hence it would be more meaningful to consider multiple scalars.

The study of scalar perturbations of general relativity with multiple canonical scalars can be found in Refs. [119, 120]. For superpotential models, it has been shown that there is no tachyon instability, and only odd scalars can avoid the localized scalar zero mode. In particular, for the double field theory, there is always a normalizable zero mode.

The scalar perturbations of $\mathcal{N}$ nonminimally coupled scalars were systematically studied in Ref. [121] in Einstein frame, in which the theory is just general relativity with nonminimaly coupled scalar fields. The action is

$$S = \int d^D x\sqrt{-g}\left[\frac{1}{2\kappa_D^2}R + P\left(\mathcal{G}_{IJ}, X^{IJ}, \Phi^I\right)\right], \tag{8.142}$$

where $X^{IJ} = -\frac{1}{2}g^{MN}\partial_M\Phi^I\partial_N\Phi^J$ is the kinetic function, $\mathcal{G}_{IJ}$ is the field-space metric, and P is the Lagrangian of the scalars. Here, indices $I, J, K, L, \cdots (= 1, 2, \cdots, \mathcal{N})$ denote $\mathcal{N}$-dimensional field-space indices lowered or raised by the field-space metric $\mathcal{G}$ or its inverse, while $M, N, P, Q, \cdots$ run over D-dimensional ones of the space-time. Using the Arnowitt–Deser–Misner variables and some calculations in the flat gauge, the coupled equations of the independent $\mathcal{N}$ scalar modes ($\delta\Phi^I = Q^I$) can be obtained [121]:

$$\frac{1}{a^{D-1}}\mathcal{D}_y(a^{n-1}\mathcal{D}_yQ_I) - \frac{1}{a^2}m^2Q_I - \mathcal{M}_I^JQ_J = 0, \tag{8.143}$$

where

$$\mathcal{M}_{IJ} = V_{;IJ} - \mathcal{R}_{IKJL}u^Ku^L + \mathcal{U}_{IJ}, \tag{8.144}$$

$$\mathcal{U}_{IJ} = \frac{2}{(D-2)a^{D-1}}\mathcal{D}_y\left(\frac{a^{D-1}}{A'}u_Iu_J\right), \tag{8.145}$$

$$u^I \equiv \partial_y\Phi_0^I. \tag{8.146}$$

Here $\mathcal{D}_y = u^I\mathcal{D}_I$ with $\mathcal{D}_I$ the covariant derivative compatible with the field-space metric $\mathcal{G}_{IJ}$, and $\mathcal{R}_{IKJL}$ is the Riemann tensor constructed from the field space metric $\mathcal{G}_{IJ}$. If the field space is one-dimensional, Eq. (8.143) is an usual Schrödinger-like equation. However, for the field-space with multi-field, Eq. (8.143) becomes a series of coupled equations. Generally, Eq. (8.143) cannot be factorized, but there is an exception. If the background solution is obtained by using the superpotential method, then this equation can be factorized in a supersymmetric formalism:

$$\left(-\delta_J^I\mathcal{D}_y - Z_J^I + (D-1)\delta_J^IW\right)\left(\delta_K^J\mathcal{D}_y - Z_K^J\right)Q^K = \frac{m^2}{a^2}Q^I \tag{8.147}$$

with

$$Z_J^I = (D-2)\left(W_{;J}^I - \frac{W^IW_J}{W}\right). \tag{8.148}$$

Here $W = -A'(y)$ is the superpotential. The above equations mean that there is no tachyon instability since

$$\int dy e^{(D-3)A}m^2Q^IQ_I$$

$$= \int dy a^{D-1}Q_I\left(-\delta_J^I\mathcal{D}_y - Z_J^I + (D-1)W\delta_J^I\right)\left(\delta_K^J\mathcal{D}_y - Z_K^J\right)Q^K$$

$$= \int dy a^{D-1}|\mathcal{D}_yQ^I - Z_J^IQ^J|^2 \geq 0. \tag{8.149}$$

The scalar zero modes satisfy

$$\mathcal{D}_yQ^I - Z_J^IQ^J = 0. \tag{8.150}$$

Now it is clear that the zero mode solutions are totally determined by the background solutions. For singular scalar case, the zero mode solution is

$$Q_I = u_I/A'. \tag{8.151}$$

It cannot be localized on the brane for asymptotically AdS_5 domain wall solution. For multiple scalars case, it is more convenient to define the tetrad fields satisfying

$$e_I^i e_J^j \delta_{ij} = \mathcal{G}_{IJ}, \quad e_I^i e_j^I = \delta_j^i, \quad \mathcal{D}_y e_I^i = 0, \tag{8.152}$$

and make a decomposition $Q_I = \sum_i e_I^i Q_i(m^2, y)e^{ip_\mu x^\mu}$ with $\eta^{\mu\nu}p_\mu p_\nu = -m^2$. In the conformally flat coordinate z and in terms of the canonically normalized modes $\tilde{Q}_i \equiv a^{(D-2)/2}Q_i$, one gets the coupled Schrödinger-like equations [121]

$$-\partial_z^2 \tilde{Q}_i + \left[\left(\frac{(D-2)^2}{4}(\partial_z A)^2 - \frac{(D-2)}{2}\partial_z^2 A\right)\delta_i^j + a^2 \mathcal{M}_i^j\right]\tilde{Q}_j = m^2 \tilde{Q}_i. \tag{8.153}$$

Actually, if the potential matrix $\mathcal{M}_i^j$ is positive definite then there is no localized zero mode. The localized states should satisfy

$$\int_{-\infty}^{+\infty} dy\, a^{D-3} \mathcal{G}_{IJ} Q^I Q^J < \infty. \tag{8.154}$$

If one separates the field space into the background trajectory direction and its orthogonal space, then the perturbed modes are Q_σ and $\vec{Q}_s$, and the localization condition can be expressed as

$$\int_{-\infty}^{+\infty} dy\, a^{D-3}(\vec{Q}_s^2 + Q_\sigma^2) < \infty. \tag{8.155}$$

For the double-scalar superpotential case, for instance, $Q_\sigma = \frac{u^I}{|u^I|}Q_I$ and $Q_s = \frac{\mathcal{D}_y \sigma^I}{|\mathcal{D}_y \sigma^I|}Q_I$, and the zero modes are given by

$$Q_s = e^{(D-2)\int dy\, W_{ss}}, \tag{8.156}$$

$$Q_\sigma = \frac{\sqrt{W'}}{W}\int dy \frac{\omega W}{\sqrt{W'}}Q_s, \tag{8.157}$$

with $W_{ss} = W_{,IJ}s^I s^J$. The superpotential background solutions would lead to a normalized zero mode. It means that we will have a massless scalar field on the brane. This result is conflicted with observations for the fifth dimension and is not acceptable [120].

As mentioned in section 8.3.1, the metric $f(R)$ theory can be studied in the context of scalar-tensor theory, hence one of the application of the above analysis is the scalar perturbations of the metric $f(R)$ theory. Obviously, the metric $f(R)$ theory with a scalar field is equivalent to a scalar-tensor theory with two scalars in Jordon frame, or general relativity with two nonminimally coupled scalars in Einstein frame. Using the above results and considering the solution given in Ref. [92], we can show that the scalar perturbations are stable and no massless scalar mode can be localized on the brane [121].

8.3.5   *f(T) theory*

In this subsection we will give a brief review of the thick braneworld scenarios in the $f(T)$ gravity theory. Since the $f(T)$ theory is successful in explaining the acceleration of the universe [122], it has been investigated widely (for examples, see Refs. [123, 124, 125, 126, 127, 128] and therein). The braneworld models in the $f(T)$ theory have been studied in Refs. [129, 130, 131].

First, the $f(T)$ theory is the generalization of the teleparallel gravity, so we will review the basics of the teleparallel gravity briefly [132, 133]. The tangent space of any point in the spacetime with the coordinate x^M can be expanded based on the orthogonal basis which is formed by the vielbein fields $e_A(x^M)$. In this subsection, the Capital Latin letters $A, B, C, \cdots = 0, 1, 2, 3, 5$ label the tangent space, while M, N, O, P, $\cdots$ still represent the five-dimensional space-time indices. Obviously, the vielbein fields $e_A(x^M)$ are vectors in the tangent space, and their components in the coordinates of space-time are labeled as $e_A{}^M$. The relation between the metric and vielbein is

$$g_{MN} = e^A{}_M e^B{}_N \eta_{AB}, \tag{8.158}$$

where $\eta_{AB} = \mathrm{diag}(-1, 1, 1, 1, 1)$ is the Minkowski metric of the tangent space. From the relation (8.158) we can get

$$e_A{}^M e^A{}_N = \delta^M_N, \qquad e_A{}^M e^B{}_M = \delta^B_A. \tag{8.159}$$

The connection $\tilde{\Gamma}^P_{MN}$ (Weitzenböck connection) in the teleparallel gravity is defined as

$$\tilde{\Gamma}^P_{MN} \equiv e_A{}^P \partial_N e^A{}_M. \tag{8.160}$$

The torsion tensor is

$$T^P{}_{MN} = \tilde{\Gamma}^P{}_{MN} - \tilde{\Gamma}^P{}_{NM}. \tag{8.161}$$

The contortion tensor K^P_{MN} is defined as the difference between the Weitzenbök connection and Levi-Civita connection

$$K^P{}_{MN} \equiv \tilde{\Gamma}^P{}_{MN} - \Gamma^P{}_{MN} = \frac{1}{2} \left(T_M{}^P{}_N + T_N{}^P{}_M - T^P{}_{MN} \right). \tag{8.162}$$

The Lagrangian of the teleparallel gravity in five dimensions can be written as

$$L_T = -\frac{M_*^3}{4} e\, T \equiv -\frac{M_*^3}{4} e S_P{}^{MN} T^P{}_{MN}, \tag{8.163}$$

where e is the determinant of $e^A{}_M$, M_* is the fundamental Planck scale in five-dimensional space-time (we will set $M_* = 1$ in the following) and we have defined the tensor $S_P{}^{MN}$ as

$$S_P{}^{MN} \equiv \frac{1}{2} \left(K^{MN}{}_P - \delta^N_P T^{QM}{}_Q + \delta^M_P T^{QN}{}_Q \right). \tag{8.164}$$

The action of the $f(T)$ theory is

$$S = -\frac{M_*^3}{4} \int d^5x\, e\, f(T) + \int d^5x\, e\, \mathcal{L}_M, \tag{8.165}$$

where $f(T)$ is a function of the torsion scalar T and $\mathcal{L}_M$ denotes the Lagrangian of matters. The equations of motion can be obtained by varying the action with respect to the vielbein fields:

$$e^{-1} f_T g_{NP} \partial_Q \left(e\, S_M{}^{PQ} \right) + f_{TT} S_{MN}{}^Q \partial_Q T - f_T \tilde{\Gamma}^P{}_{QM} S_{PN}{}^Q$$
$$+ \frac{1}{4} g_{MN} f(T) = M_*^{-3} \mathcal{T}_{MN}. \tag{8.166}$$

The flat thick brane solutions for the $f(T)$ theory were obtained in Refs. [129, 130]. In the case of $f(T) = T + \alpha T^n$ and $\mathcal{L}_M = -\frac{1}{2}\partial^M \phi \partial_M \phi - V(\phi)$, the authors of Ref. [130] found the following solution:

$$A(y) = -\frac{2}{3} v^{1/n} \int dy \tanh^{\frac{1}{n-1}}(ky), \tag{8.167a}$$

$$\phi(y) = v \tanh^{\frac{n}{2(n-1)}}(ky), \tag{8.167b}$$

$$V(\phi) = \frac{2}{n^2}\phi^2 \left(\phi^{\frac{2-2n}{n}} - B_n \phi^{\frac{2n-2}{n}} \right) - \frac{4}{3}\left(\phi^{\frac{4}{n}} - \frac{B_n}{n}\phi^4 \right), \tag{8.167c}$$

where $B_n = (-1)^n 3^{1-n} 2^{4n-4} n(2n-1)\alpha$ and $k = 4(n-1)\sqrt{B_n}/n^2$, and the scalar potential has two global minima at $\phi = \pm v = \pm B_n^{\frac{n}{4(n-1)}}$ and a local minimum at $\phi = \phi_0 = 0$. In Ref. [129], two explicit analytical thick solutions were found. The first solution is for $n = 0$ or $n = \frac{1}{2}$:

$$e^{2A(y)} = \cosh^{-2b}(ky), \tag{8.168a}$$

$$\phi(y) = \sqrt{6b} M_*^{\frac{3}{2}} \arctan\left(\tanh\left(\frac{ky}{2} \right) \right), \tag{8.168b}$$

$$V(\phi) = \frac{3bk^2 M_*^3}{4} \left[(1 + 4b) \cos^2\left(\frac{2\phi}{\sqrt{6b} M_*^{\frac{3}{2}}} \right) - 4b \right]. \tag{8.168c}$$

The second one is for $n = 2$:

$$e^{2A(y)} = \cosh^{-2b}(ky), \tag{8.169a}$$

$$\phi(y) = \frac{\sqrt{3b}}{2} M_*^{3/2} \left[i\sqrt{2} E\left(iky; u \right) - i\sqrt{2} F\left(iky; u \right) \right.$$
$$\left. + \tanh(ky)\sqrt{72\alpha b^2 k^2 + u \cosh(2ky) + 1} \right], \tag{8.169b}$$

$$V(\phi(y)) = \frac{3}{4} bk^2 M_*^3 \left[144\alpha b^3 k^2 \tanh^4(ky) \right.$$
$$\left. - 4b \tanh^2(ky)\left(18\alpha bk^2 \text{sech}^2(ky) + 1 \right) + \text{sech}^2(ky) \right], \tag{8.169c}$$

where $u = 1 - 72\alpha b^2 k^2$ and b, k are positive parameters, and $F(y; q)$ and $E(y; q)$ are the first and second kind elliptic integrals, respectively. For the case $n = 2$, it requires that $72\alpha k^2 b^2 < 1$ in order to insure the reality of the scalar field ϕ.

The Lagrangian density of the matter was generalized to a generic form $\mathcal{L}_M = X + \lambda[(1 + \beta X)^p - 1] - V(\phi)$ in Ref. [130], where $X = -\frac{1}{2}\partial_M \phi \partial^M \phi$. The parameter β is positive and λ is real. The analytical solution was found for $p = 2, \lambda = -36\alpha k^7, \beta = \frac{4}{3}k^{-5}$:

$$A(y) = \frac{M_*^3}{k^3} \ln\left(\text{sech}\left(\frac{k^4}{M_*^3}y\right)\right), \tag{8.170a}$$

$$\phi(y) = \sqrt{\frac{3}{2}}\frac{M_*^3}{k^{3/2}}\arcsin\left(\tanh\left(\frac{k^4}{M_*^3}y\right)\right), \tag{8.170b}$$

$$V(\phi) = \frac{3}{8}k^2\left(C_1\cos\left(\frac{4\sqrt{\frac{2}{3}}k^{3/2}\phi}{M_*^3}\right) + C_2\cos\left(\frac{2\sqrt{\frac{2}{3}}k^{3/2}\phi}{M_*^3}\right) + C_3\right), \tag{8.170c}$$

where $C_1 = 9\alpha k^2\left(3k^3 + 4M_*^3\right)$, $C_2 = 36\alpha k^5 + k^3 - 144\alpha k^2 M_*^3 + 4M_*^3$ and $C_3 = 9\alpha k^5 + k^3 + 108\alpha k^2 M_*^3 - 4M_*^3$.

Besides, the localization of four-dimensional gravity was studied in Ref. [131] by analyzing linear tensor perturbation of the vielbein. It was found that the graviton KK modes of the tensor perturbation satisfy the following equation

$$\left(\partial_z + \mathcal{K}\right)\left(-\partial_z + \mathcal{H}\right)\psi = m^2\psi, \tag{8.171}$$

where $\mathcal{K} = \frac{3}{2}\partial_z A + 12e^{-2A}\left((\partial_z A)^3 - \partial_z^2 A \partial_z A\right)\frac{f_{TT}}{f_T}$, which means $m^2 \geq 0$, so any analytical thick brane solutions for the $f(T)$ theory are stable under the transverse-traceless tensor perturbation. The graviton zero mode has the following form

$$\psi_0 = N_0 e^{\frac{3}{2}A + 12\int e^{-2A}\left((\partial_z A)^3 - \partial_z^2 A \partial_z A\right)\frac{f_{TT}}{f_T}dz}, \tag{8.172}$$

where N_0 is the normalization coefficient. It is easy to show that the zero mode of the graviton can be localized on the brane for the above mentioned solutions.

Furthermore, there were some related work in other gravity theories, such as Weyl (pure geometrical) gravity [134, 135, 138, 139, 136, 137] and critical gravity [141, 140].

8.4   Localization of bulk matter fields

In the last section we know that the graviton zero mode of the tensor perturbation can be localized on the brane embedded in a five-dimensional AdS space-time and the Newtonian potential can be restored. As mentioned in the last section, all matter fields should be in the bulk in thick brane scenario. The zero modes of various bulk matter fields confined on the brane denote the particles or fields in the Standard Model, while the massive KK modes indicate new particles beyond the Standard Model. So a natural question is that whether various bulk matter fields can be localized on such brane. In order not to contradict the present observations, the zero modes of various matter fields should be confined on the brane, while the massive KK modes can be localized on the brane or propagate along extra dimensions. These massive KK modes give us the possibility of probing extra dimensions through their interactions with particles in the Standard Model [142, 143, 144, 145, 146, 147, 148]. Localization and resonances of various bulk matter fields on a brane have been investigated in five-dimensional brane models [46, 149, 150, 151, 55, 17, 174, 175, 176, 152, 177, 178, 179, 180, 181, 153, 182, 154, 155, 138, 184, 156, 157, 158, 159, 160, 183, 139, 185, 186, 161, 162, 163, 164, 165, 84, 166, 187, 167, 168, 169, 170, 171, 172, 173, 188, 189] and six-dimensional ones [151, 190, 191, 192, 193, 194, 195, 196]. In this section we will review some works about localization of bulk matters on thick branes with the following metric

$$ds^2 = e^{2A(z)}\left(\tilde{g}_{\mu\nu}(x)dx^\mu dx^\nu + dz^2\right). \tag{8.173}$$

8.4.1 Scalar fields

We first consider the case of scalar fields. We denote a bulk scalar field as $\Phi(x,y)$ to distinguish with the background scalar field $\phi(y)$. A main result is that a free massless scalar field can be localized on the brane if the gravity can.

The action of a free massless scalar field can be written as

$$S_0 = \int \sqrt{-g}\, d^5x \left(-\frac{1}{2}g^{MN}\partial_M\Phi\partial_N\Phi\right). \tag{8.174}$$

The equation of motion can be obtained by varying the above action with respect to the scalar field Φ as

$$\Box^{(5)}\Phi = \frac{1}{\sqrt{-g}}\nabla_M\left(\sqrt{-g}\,g^{MN}\nabla_N\Phi\right) = 0. \tag{8.175}$$

Combined with the conformal metric (8.173), Eq. (8.175) can be rewritten as

$$\left(\partial_z^2 + 3\left(\partial_z A\right)\partial_z + \Box^{(4)}\right)\Phi = 0, \tag{8.176}$$

where $\Box^{(4)} = \tilde{g}^{\mu\nu}\tilde{\nabla}_\mu\tilde{\nabla}_\nu$ is the four-dimensional d'Alembert operator. Then, we introduce the KK decomposition $\Phi(x^\mu, y) = \Sigma_n\varphi_n(x^\mu)\chi_n(y)$. By substituting this decomposition into Eq. (8.176) and making separation of variables, one can find that φ_n satisfies the four-dimensional Klein–Gordon equation $\left(\Box^{(4)} - m_n^2\right)\varphi_n = 0$, and the extra component χ_n satisfies the following equation of motion

$$\left(\partial_z^2 + 3(\partial_z A)\partial_z + m_n^2\right)\chi_n = 0. \tag{8.177}$$

At last, by integrating over the extra dimension and using Eq.(8.177) and the following normalization condition

$$\int_{-\infty}^{\infty} dz e^{3A}\chi_m(z)\chi_n(z) = \delta_{mn}, \tag{8.178}$$

one can reduce the fundamental five-dimensional action (8.174) of a free massless scalar field to the effective four-dimensional action of a massless ($m_0 = 0$) and a series of massive ($m_n > 0$) scalar fields:

$$S_0 = \sum_n \int d^4x \sqrt{-\tilde{g}} \left[-\frac{1}{2} \tilde{g}^{\mu\nu} \partial_\mu \varphi_n \partial_\nu \varphi_n - \frac{1}{2} m_n^2 \varphi_n^2 \right]. \tag{8.179}$$

The mass spectrum m_n of the KK modes is determined by the equation of motion (8.177). In order to investigate the mass spectrum, we introduce the following field redefinition

$$\tilde{\chi}_n(z) = e^{\frac{3}{2}A} \chi_n(z), \tag{8.180}$$

with which Eq. (8.177) turns to be the following Schrödinger-like equation

$$[-\partial_z^2 + V_0(z)]\tilde{\chi}_n(z) = m_n^2 \tilde{\chi}_n(z), \tag{8.181}$$

where m_n is the mass of the n-th KK excitation of the scalar field. The effective potential $V_0(z)$ takes the following form:

$$V_0(z) = \frac{3}{2}\partial_z^2 A + \frac{9}{4}(\partial_z A)^2. \tag{8.182}$$

Note that, the effective potential only depends on the warp factor A, and has the same form as the case of graviton KK modes in general relativity [197]. That is, the scalar zero mode will be localized on the brane on the condition that the gravity can be localized on the brane. The effective potential has different forms for different solutions of the warp factor A. We take an explicit form as example: $A(y) = \ln(\text{sech}(ky))$. This typical solution of the warp factor in the conformal coordinate has the following form:

$$A(z) = \ln\left(\frac{1}{1 + k^2 z^2}\right), \tag{8.183}$$

where k is a parameter. Then the effective potential reads as

$$V_0(z) = \frac{3k^2(4k^2z^2 - 1)}{(1 + k^2z^2)^2}. \tag{8.184}$$

From the above expression we can see that the potential V approaches to zero when z approaches to infinity, and the value of the potential at $z = 0$ is $-3k^2$. So the effective potential has the volcano shape. In this case, there is no mass gap between the zero mode and the massive KK modes, and the mass spectrum is continuous. Any massive KK mode cannot be localized on the brane since $V(|y| \to \infty) \to 0$. The solution of the zero mode with $m_0^2 = 0$ is given by

$$\tilde{\chi}_0(z) = N_0 e^{\frac{3}{2}A} = \frac{N_0}{(1 + k^2z^2)^{3/2}}, \tag{8.185}$$

where $N_0 = \left(\frac{8k}{3\pi}\right)^{\frac{1}{2}}$ is the normalization constant. This is the lowest energy eigenfunction for the Schrödinger-like equation (8.181), which indicates that there is no KK modes with negative m^2. In fact, this equation can be rewritten as

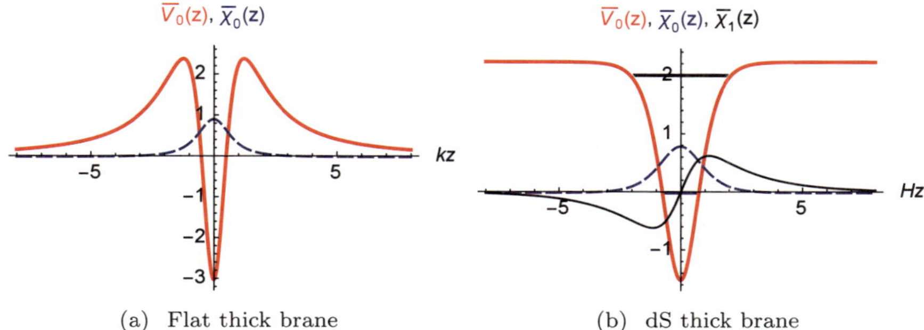

Fig. 8.5   The effective potential $\overline{V}_0$ and the bound KK modes $\overline{\chi}_n$ for a scalar field. (a) Flat thick brane: $\overline{V}_0 \equiv V_0/k^2$ (8.184) (red thick line) and $\overline{\chi}_0 \equiv \tilde{\chi}_0/\sqrt{k}$ (8.185) (blue dashed line).   (b) de Sitter thick brane [187]: $\overline{V}_0 \equiv V_0/H^2$ (8.186) (red thick line), $\overline{\chi}_0 \equiv \tilde{\chi}_0/\sqrt{H}$ (8.187) (blue dashed line), and $\overline{\chi}_1 \equiv \tilde{\chi}_1/\sqrt{H}$ (8.188) (solid black line). The masses for the two bound modes is given by $\overline{m}_0^2 \equiv m_0^2/H^2 = 0$ (blue segment) and $\overline{m}_1^2 \equiv m_1^2/H^2 = 2$ (black segment).

$\mathcal{K}\mathcal{K}^{\dagger}\tilde{\chi}_n = m_n^2\tilde{\chi}_n$, with $\mathcal{K} = \partial_z + \frac{3}{2}\partial_z A$, which ensures $m^2 \geq 0$, namely, there is no tachyonic scalar mode. Besides this scalar zero mode, there exist other continuous massive KK modes. The effective potential and the zero mode are shown in Fig. 8.5(a).

There are other shapes of the effective potential, such as the infinite deep well and the Pöschl–Teller potential [185,187]. For example, the effective potential of the KK modes of a massless scalar field on the de Sitter brane with $A(z) = \ln\left[\frac{H}{b}\text{sech}(Hz)\right]$ has the following Pöschl–Teller form [187]:

$$V_0(z) = \frac{3}{4}H^2\left[3 - 5\,\text{sech}^2(Hz)\right], \tag{8.186}$$

for which there are two bound KK states

$$\tilde{\chi}_0(z) = \sqrt{\frac{2H}{\pi}}\,\text{sech}^{3/2}(Hz), \tag{8.187}$$

$$\tilde{\chi}_1(z) = \sqrt{H}\,\text{sech}^{3/2}(Hz)\sinh(Hz), \tag{8.188}$$

and the mass spectrum of the bound states is given by

$$m_n^2 = n(3 - n)H^2, \quad n = 0, 1. \tag{8.189}$$

The effective potential, the zero mode, and the mass spectrum are shown in Fig. 8.5(b).

8.4.2   *Vector fields*

In this subsection, we review the localization of a bulk $U(1)$ gauge vector field on the brane in some five-dimensional thick brane models. We first consider the following five-dimensional action for a free bulk vector field:

$$S_A = -\frac{1}{4} \int d^5x \sqrt{-g}\, F^{MN} F_{MN}, \tag{8.190}$$

where $F_{MN} = \partial_M A_N - \partial_N A_M$ is the five-dimensional field strength. For the conformal metric (8.173), the equations of motion

$$\frac{1}{\sqrt{-g}} \partial_M \left(\sqrt{-g} g^{MN} g^{RS} F_{NS} \right) = 0 \tag{8.191}$$

can be written as the following component equations:

$$\frac{1}{\sqrt{-\tilde{g}}} \partial_\nu \left(\sqrt{-\tilde{g}}\, \tilde{g}^{\nu\rho} \tilde{g}^{\mu\lambda} F_{\rho\lambda} \right) + \tilde{g}^{\mu\lambda} e^{-A} \partial_z \left(e^A F_{5\lambda} \right) = 0, \tag{8.192}$$

$$\partial_\mu \left(\sqrt{-\tilde{g}}\, \tilde{g}^{\mu\nu} F_{\nu 5} \right) = 0. \tag{8.193}$$

The five-dimensional vector field $A_M(x^\lambda, z)$ can be decomposed as

$$A_M(x^\lambda, z) = \sum_n a_M^{(n)}(x^\lambda) \rho_n(z). \tag{8.194}$$

It can be seen that the action (8.190) is invariant under the following gauge transformation:

$$A_M(x^\lambda, z) \to \widetilde{A}_M(x^\lambda, z) = A_M(x^\lambda, z) + \partial_M F(x^\lambda, z), \tag{8.195}$$

or

$$A_\mu(x^\lambda, z) \to \widetilde{A}_\mu(x^\lambda, z) = A_\mu(x^\lambda, z) + \partial_\mu F(x^\lambda, z), \tag{8.196}$$

$$A_5(x^\lambda, z) \to \widetilde{A}_5(x^\lambda, z) = A_5(x^\lambda, z) + \partial_z F(x^\lambda, z), \tag{8.197}$$

where $F(x^\lambda, z)$ is an arbitrary regular scalar function. One can check that the gauge $A_5(x^\lambda, z) = 0$ is allowed with the above gauge transformation. For the KK theory with finite extra dimension, $A_5(x^\lambda, z)$ and $F(x^\lambda, z)$ should be periodic functions of the extra dimension. However, for the braneworld scenario with an infinite extra dimension, which is the case we are considering, there is no any constraint on $A_M(x^\lambda, z)$ and $F(x^\lambda, z)$. From the transformation (8.197) and the KK decomposition (8.194), one has [187]

$$A_5(x^\lambda, z) \to \widetilde{A}_5(x^\lambda, z) = \sum_n \widetilde{A}_5^{(n)}(x^\lambda, z) = \sum_n \widetilde{a}_5^{(n)}(x^\lambda) \rho_n(z)$$

$$= \sum_n a_5^{(n)}(x^\lambda) \rho_n(z) + \partial_z F(x^\lambda, z). \tag{8.198}$$

Therefore, if one chooses

$$F(x^\lambda, z) = \sum_n F_5^{(n)}(x^\lambda, z) = -\sum_n a_5^{(n)}(x^\lambda) \int \rho_n(z) dz, \tag{8.199}$$

then the fifth component $\widetilde{A}_5$ vanishes:

$$\widetilde{A}_5(x^\lambda, z) = 0, \tag{8.200}$$

which is just the gauge choice we will take. Note that, for the case of the KK theory one has $\rho_n(z) = \cos(nkz)$, which indicates that one can only take

$$F(x^\lambda, z) = \sum_n F_5^{(n)}(x^\lambda, z) = -a_5^{(0)}(x^\lambda) - \sum_{n \neq 0} a_5^{(n)}(x^\lambda) \int \rho_n(z) dz, \quad (8.201)$$

i.e., the zero mode $F_5^{(0)}$ is a function of x^λ only. Therefore, one can only choose the gauge condition $A_5^{(n)} = 0$ for massive KK modes ($n \neq 0$) instead of $A_5^{(0)} = 0$. Thus, one has no the gauge $\widetilde{A}_5(x^\lambda, z) = 0$ in the KK theory.

Now we choose the gauge $A_5 = 0$ and make the decomposition $A_\mu(x, z) = \sum_n a_\mu^{(n)}(x)\alpha_n(z)e^{-A/2}$. Then it is easy to find that the vector KK modes $\alpha_n(z)$ satisfies the following Schrödinger equation:

$$\left[-\partial_z^2 + V_1(z)\right]\alpha_n(z) = m_n^2 \alpha_n(z), \quad (8.202)$$

where the effective potential $V_1(z)$ is given by

$$V_1(z) = \frac{1}{2}\partial_z^2 A + \frac{1}{4}(\partial_z A)^2. \quad (8.203)$$

The vector zero mode with $m_0^2 = 0$ can be solved as

$$\alpha_0(z) = N_0 e^{A/2}. \quad (8.204)$$

By introducing the orthonormalization conditions

$$\int_{-\infty}^{+\infty} \alpha_m(z)\alpha_n(z)dz = \delta_{mn}, \quad (8.205)$$

the fundamental action (8.190) can be reduced to the effective one of a massless ($m_0 = 0$) and a series of massive ($m_n > 0$) four-dimensional vector fields:

$$S_1 = \sum_n \int d^4x \sqrt{-\tilde{g}} \left(-\frac{1}{4}\tilde{g}^{\mu\alpha}\tilde{g}^{\nu\beta} f_{\mu\nu}^{(n)} f_{\alpha\beta}^{(n)} - \frac{1}{2}m_n^2 \, \tilde{g}^{\mu\nu} a_\mu^{(n)} a_\nu^{(n)} \right), \quad (8.206)$$

where $f_{\mu\nu}^{(n)} = \partial_\mu a_\nu^{(n)} - \partial_\nu a_\mu^{(n)}$ is the four-dimensional field strength tensor.

The mass spectrum m_n and localization of the vector KK modes are also determined by the Schrödinger equation (8.202). For the RS-like solution with $A(z) = \ln\left(\frac{1}{1+k^2z^2}\right)$, the effective potential and vector zero mode are given by

$$V_1(z) = \frac{(2k^2z^2 - 1)}{(k^2z^2 + 1)^2}k^2, \quad (8.207)$$

$$\alpha_0(z) = \frac{N_0}{\sqrt{1 + k^2z^2}}, \quad (8.208)$$

which are shown in Fig. 8.6(a). Since

$$\int_{-\infty}^{+\infty} |\alpha_0(z)|^2 dz = \int_{-\infty}^{+\infty} N_0^2 e^A dz = N_0^2 \int_{-\infty}^{+\infty} dy = \infty, \quad (8.209)$$

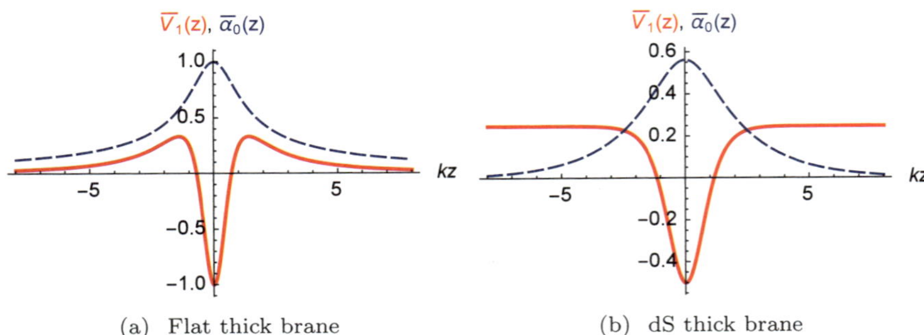

Fig. 8.6   The effective potential $\overline{V}_1(z)$ and the bound zero mode $\overline{\alpha}_0(z)$ for a vector field. (a) Flat thick brane: $\overline{V}_1(z) \equiv V_1(z)/k^2$ (8.207) (red thick line) and $\overline{\alpha}_0(z) \equiv \alpha_0(z)/\sqrt{k}$ (8.208) (blue dashed line).   (b) de Sitter thick brane [187]: $\overline{V}_1(z) \equiv V_1(z)/H^2$ (8.210) (red thick line) and $\overline{\alpha}_0(z) \equiv \alpha_0(z)/\sqrt{H}$ (8.211) (blue dashed line).

the vector zero mode $\alpha_0(z) = N_0 e^{A/2}$ with arbitrary $A(z)$, including (8.208), cannot satisfy the normalization condition $\int_{-\infty}^{+\infty} |\alpha_0(z)|^2 dz = 1$, and hence cannot be localized on the brane. So one needs some localization mechanisms for a bulk vector field for such brane models.

For the de Sitter brane model with $A(z) = \ln\left[\frac{H}{b}\mathrm{sech}(Hz)\right]$, the effective potential (8.210) for the vector KK modes has the following Pöschl–Teller form [187]:

$$V_1(z) = \frac{H^2}{4}\left[1 - 3\,\mathrm{sech}^2(Hz)\right]. \tag{8.210}$$

The above potential has a minimum $-H^2/2$ at $z = 0$ and a maximum $H^2/4$ at $z = \pm\infty$, which ensures the presence of a mass gap in the spectrum. There is only one bound state, i.e., the vector zero mode that can be localized on the de Sitter brane:

$$\alpha_0(z) = \sqrt{\frac{H}{\pi}}\,\mathrm{sech}(Hz). \tag{8.211}$$

The effective potential and the vector zero mode for the de Sitter brane model are shown in Fig. 8.6(b).

Besides, the zero mode of a free five-dimensional vector field can also be localized on the brane in some other special braneworld scenarios, such as AdS branes [187], Weyl Thick Branes [138], two-field thick branes with an finite extra dimension [198]. Note that if a RS-like brane has more than three space dimensions, then the vector zero mode can also be localized on the brane.

In order to localize the zero mode of a bulk vector field on a RS-like brane, some mechanisms were proposed. In the following, we give a brief review.

Kinetic energy term coupling. Inspired by the effective coupling of neutral scalar field to electromagnetic field and by the Friedberg–Lee model for hadrons [199], Chumbes, Hoff da Silva, and Hott [200] explored the coupling between the kinetic term of the vector field and the background scalar field ϕ to realize the localization of the vector field. The general action of the vector field nonminimally coupled with the background scalar field is given by [200]

$$S_A = -\frac{1}{4} \int d^5x \sqrt{-g}\, G(\phi) F^{MN} F_{MN}. \tag{8.212}$$

The localization condition for the vector zero mode is that the following integrate is finite:

$$\int_{-\infty}^{+\infty} G(\phi) dy < \infty. \tag{8.213}$$

For the solutions of the background scalar field $\phi = v\tanh(ky)$ and $\phi = v\arcsin(\tanh(ky))$, the corresponding coupling functions can be chosen as

$$G(\phi) = \left(1 - \frac{\phi^2}{v^2}\right)^{p/2} \qquad \text{for} \quad \phi = v\tanh(ky), \tag{8.214}$$

$$G(\phi) = \left(1 - \sin^2\left(\frac{\phi}{v}\right)\right)^{p/2} \qquad \text{for} \quad \phi = v\arcsin(\tanh(ky)), \tag{8.215}$$

where p is a positive constant. The above two couplings would result in the same function $G(\phi(y)) = \text{sech}^p(ky)$, which insures the normalization and hence the localization of the vector zero mode. References [201, 202] considered such localization mechanism for the Bloch branes [153] and found localized zero mode and quasi-localized massive KK modes of a bulk vector field.

For some two-field braneworld models [205, 203, 204, 198, 206], the above coupling can also be used to localize the vector field with $G = e^{\tau\pi(y)}$, where $\pi(y)$ is one of the two background scalars $\phi(y)$ and $\pi(y)$, and τ is the coupling constant (see Refs. [207, 153, 206, 208, 32]).

Yukawa-like coupling. An alternative approach to solve the localization problem of the gauge field is to introduce the Yukawa-like coupling [210, 171, 209], namely, consider the Stueckelberg-like gauge field action [171]:

$$S_A = \int d^5x \sqrt{-g} \left\{ -\frac{1}{4} F^{MN} F_{MN} - \frac{1}{2} G(\phi)(A_M - \partial_M B)(A^M - \partial^M B) \right\}, \tag{8.216}$$

where B is a dynamical scalar field just like in the Stueckelberg field [211], and $G(\phi)$ is the coupling function of the background scalar field ϕ. With the gauge transformation $A_M \to A_M + \partial_M \xi$, $B \to B + \xi$, the action (8.216) keeps gauge invariant. Through varying the action S_A, and parameterizing the five-dimensional field A_M as $A_M = (A_\mu, A_5) = (\widehat{A}_\mu + \partial_\mu \varphi, A_5)$ like the way in Ref. [212], one can obtain

$$\left[\Box^{(4)} + e^{2A}\left(\partial_y^2 + 2A'\partial_y - G\right)\right]\widehat{A}_\nu = 0, \tag{8.217}$$

$$\partial_y(e^{2A}\lambda) - e^{2A}G\rho = 0, \tag{8.218}$$

$$e^{2A}\Box^{(4)}\lambda + e^{4A}G\left(\rho' - \lambda\right) = 0, \tag{8.219}$$

$$e^{2A}G\Box^{(4)}\rho + \partial_y\left[e^{4A}G\left(\rho' - \lambda\right)\right] = 0, \tag{8.220}$$

with the two redefined gauge invariant scalar fields $\lambda = A_5 - \varphi$ and $\rho = B - \varphi$.

Then, by decomposing the gauge field as follows

$$\widehat{A}^\mu(x,y) = \sum_n a_n^\mu(x)\alpha_n(y), \tag{8.221}$$

one can reduce Eq. (8.217) as

$$\left[\partial_y^2 + 2A'\partial_y - G\right]\alpha_n(y) = -e^{-2A}m_n^2\alpha_n(y), \tag{8.222}$$

where $\Box^{(4)}a_n^\mu(x) = m_n^2 a_n^\mu(x)$.

In order to localize the gauge field on the brane, a proper function form of the coupling $G(\phi)$ should be chosen. The authors in Ref. [171] chose the following function:

$$G_{c_1,c_2}[\phi(y)] = c_1 A''(y) + c_2[A'(y)]^2. \tag{8.223}$$

For the solution of the warp factor $A(y) = A_0 - b\log[\cosh(ay)]$, the mass spectrum of the vector KK modes is continuous since the effective potential of the KK modes approaches to zero when $|z| \to \infty$. The vector zero mode $\alpha_0(y)$ turns out to be

$$\alpha_0(y) = k_0 e^{\xi A(y)}, \tag{8.224}$$

which can be normalizable.

Other mechanisms. Furthermore, the geometrical coupling with the gauge field was introduced by Alencar et al. in Ref. [210]. Zhao et al. [209] assumed that the five-dimensional gauge field has a dynamical mass term, which is proportional to the five-dimensional scalar curvature. Vaquera-Araujo et al. [171] added a brane-gauge coupling into the action. All these mechanisms are effective for the localization of a bulk vector field on the brane.

8.4.3 *Kalb–Ramond fields*

In this subsection, we review the localization of a bulk Kalb–Ramond field on a thick brane. It is known that a Kalb–Ramond field is an antisymmetric tensor field with higher spins proposed in string theory. The Kalb–Ramond field (NS-NS B-field) appears, together with the metric tensor and dilaton, as a set of massless excitations of a closed string. The action for a charged particle moving in an electromagnetic potential is given by $-q \int dx^M A_M$. While the action for a string coupled to a Kalb–Ramond field is $- \int dx^M dx^N B_{MN}$. This term in the action implies that the fundamental string of string theory is a source of the NS-NS B-field, much like charged particles are sources of the electromagnetic field. The Kalb–Ramond field is also used to describe the torsion of space-time in Einstein-Cartan theory.

The action of a free Kalb–Ramond field is

$$S_{\text{KR}} = - \int d^5 x \sqrt{-g} \, H_{MNL} H^{MNL}, \tag{8.225}$$

where $H_{MNL} = \partial_{[M} B_{NL]}$ is the field strength for the Kalb–Ramond field, and $H^{MNL} = g^{MO} g^{NP} g^{LQ} H_{OPQ}$. The field equations for the Kalb–Ramond field with the conformal metric (8.173) read as

$$\partial_\mu(\sqrt{-g} H^{\mu\alpha\beta}) + \partial_z(\sqrt{-g} H^{4\alpha\beta}) = 0, \tag{8.226}$$
$$\partial_\mu(\sqrt{-g} H^{\mu 4\beta}) = 0. \tag{8.227}$$

One can make a decomposition

$$B^{\alpha\beta}(x^\lambda, z) = \sum_n \hat{b}^{\alpha\beta}_{(n)}(x^\lambda) U_n(z) e^{-7A/2}, \tag{8.228}$$

and set the gauge $B_{\alpha 4} = 0$. Then it is not difficult to find that the KK mode $U_n(z)$ satisfies the following Schrödinger-like equation:

$$\left(- \partial_z^2 + V_{\text{KR}}(z) \right) U_n(z), = m_n^2 U_n(z) \tag{8.229}$$

where the effective potential $V_{\text{KR}}(z)$ is given by

$$V_{\text{KR}} = \frac{1}{4}(\partial_z A)^2 - \frac{1}{2}\partial_z^2 A. \tag{8.230}$$

By introducing the orthonormality conditions for the KK modes

$$\int dz \, U_m(z) U_n(z) = \delta_{mn}, \tag{8.231}$$

one can reduce the fundamental action (8.235) to the following four-dimensional one

$$S_{\text{KR}} = - \sum_n \int d^4 x \sqrt{-\tilde{g}} \left(\hat{h}^{(n)\mu\alpha\beta} \hat{h}^{(n)}_{\mu\alpha\beta} + \frac{1}{3} m_n^2 \hat{b}^{(n)\alpha\beta} \hat{b}^{(n)}_{\alpha\beta} \right) \tag{8.232}$$

with $\hat{h}^{(n)}_{\mu\alpha\beta} = \partial_{[\mu} \hat{b}^{(n)}_{\alpha\beta]}$ the four-dimensional field strength tensor. The solution of the Kalb–Ramond zero mode reads as

$$U_0(z) = e^{-A}, \tag{8.233}$$

and its normalization condition is

$$\int dz \, |U_0(z)|^2 = \int dz \, e^{-2A} = \int dy \, e^{-3A}. \tag{8.234}$$

It is clear that for the RS-like solution with $e^A = \text{sech}(ky)$, the zero mode of a free bulk Kalb–Ramond field cannot be localized on the brane.

Similar to the case of a vector field, one can also introduce a nonminimal coupling between the Kalb–Ramond field and the background scalar fields ϕ, π, $\cdots$:

$$S_{\text{KR}} = -\int d^5x\sqrt{-g}\, G(\phi, \pi, \cdots)H_{MNL}H^{MNL}. \tag{8.235}$$

The localization and resonances of such KR field have been investigated in Refs. [208, 198, 213, 161, 214]. For example, Ref. [198] considered $G = e^{\zeta\pi}$ in a two-field brane model, where the background scalar π is given by $\pi(z) = bA(z)$. The effective potential $V_{\text{KR}}(z)$ and the zero mode are given by

$$V_{\text{KR}}(z) = \frac{(1 - \sqrt{3b}\,\zeta)^2}{4}(\partial_z A)^2 + \frac{\sqrt{3b}\,\zeta - 1}{2}\partial_z^2 A, \tag{8.236}$$

$$U_0(z) = e^{(\sqrt{3b}\,\zeta - 1/2)A(z)}. \tag{8.237}$$

The localization condition is $\zeta > 1/\sqrt{3b}$ for $b \geq 1$ or $\zeta > (2-b)/\sqrt{3b}$ for $0 < b < 1$ [198]. The resonances of the KR field have also been investigated in Refs. [161, 214]. Other related work can be found in Refs. [215, 216, 203, 217, 218, 219].

We know that the scalar, vector, and Kalb–Ramond fields are the 0-form, 1-form, and 2-form fields, respectively. In fact, there are higher-form fields in a higher-dimensional space-time with dimension larger than four. In four-dimensional space-time, free q-form fields are equivalent to scalar or vector fields by a duality. In higher space-time, they correspond to new types of particles. Some early works for localization and Hodge duality of a q-form field were studied in Refs. [220, 221], where some gauges were chosen to make the localization mechanism simpler. However, these gauge choices only reflect parts of the whole localization information, including the Hodge duality of the KK modes. Recently, new localization mechanism, Hodge duality, and mass spectrum of a bulk massless q-form field on codimension-one branes (p-branes) were investigated in Refs. [222, 223] by using a new KK decomposition. There are two types of KK modes for the bulk q-form field: the q-form and $(q-1)$-form modes, which cannot be localized on the p-brane simultaneously. The Hodge duality in the bulk naturally becomes two dualities on the brane. Dualities in the bulk and on the brane are shown in Table 8.1. For the detail, see Ref. [222].

8.4.4 *Fermion fields*

In this subsection, we review the localization of bulk Dirac fermion fields on thick branes. In order to localize a fermione, one usually needs to introduce some interactions between the fermion and the background fields. Note that the general covariant equations for fields with arbitrary spin were derived by Y.-S. Duan [224],

		Duality
Bulk	Massless	$q-$form $\Leftrightarrow (p-q)-$form
Brane	KK modes $n=0$	$q-$form $\Leftrightarrow (p-q-1)-$form or $(q-1)-$form $\Leftrightarrow (p-q)-$form
	KK modes $n \geqslant 1$	$q-$form(with mass m_n)$+(q-1)-$form(massless) $\Updownarrow$ $(p-q)-$form(with mass m_n)$+(p-q-1)-$form(massless)

and the general covariant Dirac equation previously obtained by V. A. Fock and D. D. Ivanenko [225] is a special case of the general covariant equations.

Here, we consider a general action of the Dirac fermion [226]

$$S_{\frac{1}{2}} = \int d^5x \sqrt{-g} \left[F_1 \bar{\Psi} \Gamma^M D_M \Psi + \lambda F_2 \bar{\Psi}\Psi + \eta \bar{\Psi} \Gamma^M (\partial_M F_3) \gamma^5 \Psi \right]. \quad (8.238)$$

Here the functions F_1, F_2, and F_3 are functions of the background scalar fields ϕ^I and/or the Ricci scalar R, and λ and η are the coupling constants. In five-dimensional space-time, a Dirac fermion field is a four-component spinor and the corresponding gamma matrices Γ^M in curved space-time satisfy $\{\Gamma^M, \Gamma^N\} = 2g^{MN}$. The operator $D_M = \partial_M + \omega_M$ and the spin connection ω_M is defined as

$$\omega_M = \frac{1}{4} \omega_M{}^{\bar{M}\bar{N}} \Gamma_{\bar{M}} \Gamma_{\bar{N}} \quad (8.239)$$

with

$$\omega_M{}^{\bar{M}\bar{N}} = \frac{1}{2} E^{N\bar{M}} (\partial_M E_N{}^{\bar{N}} - \partial_N E_M{}^{\bar{N}}) - \frac{1}{2} E^{N\bar{N}} (\partial_M E_N{}^{\bar{M}} - \partial_N E_M{}^{\bar{M}})$$
$$- \frac{1}{2} E^{P\bar{M}} E^{Q\bar{N}} E_M{}^{\bar{R}} (\partial_P E_{Q\bar{R}} - \partial_Q E_{P\bar{R}}). \quad (8.240)$$

Here the letters with barrier $\bar{M}$, $\bar{N}$, $\cdots$ are the five-dimensional local Lorentz indices and the vielbein $E^M{}_{\bar{M}}$ satisfies $E^M{}_{\bar{M}} E^N{}_{\bar{N}} \eta^{\bar{M}\bar{N}} = g^{MN}$. The relation between the gamma matrices Γ^M and $\Gamma^{\bar{M}} = (\Gamma^{\bar{\mu}}, \Gamma^{\bar{5}}) = (\gamma^{\bar{\mu}}, \gamma^5)$ is given by $\Gamma^M = E^M{}_{\bar{M}} \Gamma^{\bar{M}}$.

For the metric (8.173), the non-vanishing components of the spin connection (8.239) are $\omega_\mu = \frac{1}{2}(\partial_z A)\gamma_\mu \gamma_5 + \hat{\omega}_\mu$, where $\hat{\omega}_\mu$ is derived from the four-dimensional metric $\tilde{g}_{\mu\nu}(x^\lambda)$. The five-dimensional Dirac equation reads as

$$\left[\gamma^\mu \partial_\mu + \hat{\omega}_\mu + \gamma^5(\partial_z + 2\partial_z A) + \mathcal{F}(z) \right] \Psi = 0, \quad (8.241)$$

where

$$\mathcal{F}(z) = \lambda e^{A(z)} \frac{F_2}{F_1} + \eta \frac{\partial_z F_3}{F_1}. \quad (8.242)$$

We make the following chiral decomposition for the five-dimensional Dirac field Ψ

$$\Psi(x,z) = e^{-2A(z)} \sum_n \left[\psi_{Ln}(x) f_{Ln}(z) + \psi_{Rn}(x) f_{Rn}(z) \right], \tag{8.243}$$

where $\psi_{Ln} = -\gamma^5 \psi_{Ln}$ and $\psi_{Rn} = \gamma^5 \psi_{Rn}$ are the left- and right-chiral components of the Dirac fermion field, respectively, and the four-dimensional Dirac fermion fields satisfy

$$\begin{aligned}
\gamma^\mu (\partial_\mu + \hat{\omega}_\mu) \psi_{Ln}(x) &= m_n \psi_{Rn}(x), \\
\gamma^\mu (\partial_\mu + \hat{\omega}_\mu) \psi_{Rn}(x) &= m_n \psi_{Ln}(x),
\end{aligned} \tag{8.244}$$

where m_n is the mass of the four-dimensional fermion fields $\psi_{Ln}(x)$ and $\psi_{Rn}(x)$. Substituting Eqs. (8.243) and (8.244) into Eq. (8.241) yields the coupling equations of the KK modes $f_{Ln,Rn}$:

$$(\partial_z - \mathcal{F}(z)) f_{Ln} = +m_n f_{Rn},$$

$$(\partial_z + \mathcal{F}(z)) f_{Rn} = -m_n f_{Ln}. \tag{8.245}$$

The above two equations can also be rewritten as the Schrödinger-like equations

$$[-\partial_z^2 + V_L(z)] f_{Ln} = m_n^2 f_{Ln}, \tag{8.246}$$

$$[-\partial_z^2 + V_R(z)] f_{Rn} = m_n^2 f_{Rn}, \tag{8.247}$$

where the effective potentials are given by

$$V_{L,R}(z) = \mathcal{F}^2(z) \pm \partial_z \mathcal{F}(z). \tag{8.248}$$

The Schrödinger-like equations (8.246) and (8.247) can be decomposed by using the supersymmetry quantum mechanics as

$$\begin{aligned}
\mathcal{K}^\dagger \mathcal{K} f_{Ln} &= m_n^2 f_{Ln} \\
\mathcal{K} \mathcal{K}^\dagger f_{Rn} &= m_n^2 f_{Rn}
\end{aligned} \tag{8.249}$$

with the operator $\mathcal{K} = \partial_z - \mathcal{F}(z)$, which insure that the mass square is non-negative, i.e., $m_n^2 \geq 0$. The corresponding chiral zero modes can be solved based on Eq. (8.245) with $m_0 = 0$:

$$f_{L0,R0} \propto e^{\pm \int dz \mathcal{F}(z)}. \tag{8.250}$$

By introducing the following orthonormality conditions for the KK modes $f_{Ln,Rn}$

$$\int_{-\infty}^{+\infty} F_1 f_{Lm} f_{Ln} dz = \int_{-\infty}^{+\infty} F_1 f_{Rm} f_{Rn} dz = \delta_{mn},$$

$$\int_{-\infty}^{+\infty} F_1 f_{Ln} f_{Rn} dz = 0, \tag{8.251}$$

one can derive the effective action of the four-dimensional massless and massive Dirac fermions from the five-dimensional Dirac action (8.238):

$$S_{\text{eff}} = \sum_n \int d^4 x \sqrt{-\hat{g}} \, \bar{\psi}_n \left[\gamma^\mu (\partial_\mu + \hat{\omega}_\mu) - m_n \right] \psi_n. \tag{8.252}$$

The conditions (8.251) can be used to check whether the fermion KK modes can be localized on the brane.

We know that there are two types of fermion localization mechanisms. The first one is the Yukawa coupling ($F_1 = 1, F_3 = 0$) between fermions and the background scalar fields [150, 227, 154, 228, 155, 184, 156, 157, 158, 159, 229, 230, 231, 186, 232, 233, 234, 164, 166, 235, 236, 237, 238, 239, 240, 241, 242], which does work when the background scalar fields are odd functions of the extra dimension. This form of coupling $\lambda F_2(\phi)\bar{\Psi}\Psi$ between the kink scalar ϕ and bulk fermions can be regarded as the coupling between a soliton and fermions in Ref. [243]. The corresponding effective potentials are (8.248)

$$V_{L,R}(z) = (\lambda e^A F_2)^2 \pm \partial_z(\lambda e^A F_2),\qquad(8.253)$$

and the chiral zero modes read

$$f_{L0,R0} \propto e^{\pm\lambda \int dz\ e^A F_2(\phi)} = e^{\pm\lambda \int dy\ F_2(\phi)}.\qquad(8.254)$$

We consider the brane models with the solution

$$e^{A(y\to\pm\infty)} \to e^{\mp ky} \quad\text{and}\quad \phi(y \to \pm\infty) \to \pm v.\qquad(8.255)$$

For the simplest Yukawa coupling with $F_2 = \phi(y)$ and strong enough but negative coupling ($\lambda < \lambda_0 \equiv -k/v$), the left-chiral fermion zero mode

$$f_{L0}(y \to \pm\infty) \to e^{\pm\lambda vy}\qquad(8.256)$$

satisfies the normalization condition $\int_{-\infty}^{+\infty} e^{-A(y)}|f_{L0}|^2 dy < \infty$, and hence can be localized on the brane. It is worth pointing out that the right-chiral fermion zero mode cannot be localized at the same time.

In Ref. [235], a "natural" ansatz for the Yukawa term $F_2\bar{\Psi}\Psi$ is proposed, where F_2 inherits its odd nature directly from the geometry shape of the warp factor $e^{A(z)}$. In order to guarantee the localization of the left-chiral fermion zero mode, the authors taken F_2 as $F_2(z) = M\partial_z e^{-A(z)}$, which is not arbitrariness and is independent of the braneworld model. With this choice, the localization of gravity on the brane implies the localization of spin-1/2 fermions as well.

If the background scalar field is an even function of the extra dimension, the Yukawa coupling mechanism will do not work, since the Z_2 reflection symmetry of the effective potentials for the fermion KK modes cannot be ensured [244]. In order to solve this problem, a new localization mechanism was presented in Ref. [244]. The coupling is given by $\eta\bar{\Psi}\Gamma^M\partial_M F_3(\phi)\gamma_5\Psi$ ($F_1 = 1,\ F_2 = 0$), which is used to describe the interaction between π-meson and nucleons in quantum field theory and is called as the derivative coupling.

For the above two mechanisms, the localization of bulk fermions depends on the coupling between bulk fermions and background scalar fields. For thick brane models without background scalar fields, the previous two mechanisms do not work any more. For such models, one can adopt the coupling between the bulk fermion fields and the scalar curvature R of the background space-time [245]. The form of coupling is the same as the derivative coupling $\mathcal{L}_{\text{int}} = \delta\bar{\Psi}\gamma_5\Gamma^M\partial_M F_3(R)\Psi$ [245] since the scalar curvature R is an even function of the extra dimension. With the derivative geometrical coupling, the corresponding effective potentials (8.248) and chiral zero modes become

$$V_{L,R}(z) = (\eta\partial_z F_3)^2 \pm \partial_z(\eta\partial_z F_3), \tag{8.257}$$

and

$$f_{L0,R0} \propto e^{\pm\eta \int dz\, \partial_z F_3} = e^{\pm\eta F_3}, \tag{8.258}$$

where F_3 is a function of the background scalar fields or the scalar curvature. The normalization conditions for the fermion zero modes are

$$\int_{-\infty}^{+\infty} e^{\pm 2\eta F_3} dz < \infty. \tag{8.259}$$

It can be seen that one of the left- and right-chiral fermion zero modes can be localized on the brane with some suitable choice of the function $F_3(\phi, R)$ (see Refs. [244, 245, 246, 247] for detail).

Here we should note that for a volcano-like effective potential, all the massive KK modes can escape to the extra dimension and the massive fermion KK resonances do not have contributions to the effective action (8.252) in four-dimensional space-time since the integral of the square of a massive KK mode is divergent along the extra dimension. Recently a new localization mechanism [226] was proposed by considering the non-minimal coupling between bulk fermions and background scalar fields (see (8.238)). Obviously, the localization of a bulk fermion on a brane is related to the function F_1 and this function has remarkable impacts on the normalization of the continuous massive KK modes (8.251). One can see that the continuous massive KK modes may have contributions to the four-dimensional effective fermion action (8.252) if one chooses a proper function F_1 (see Ref. [226]).

It is known that the shapes of the effective potential of the left- or right-chiral fermion KK mode can be classified as three types: volcano-like [154,227,229,230,184, 232, 234], finite- square-well-like [156, 248], and harmonic-potential-like [155, 186]. The corresponding spectra of the KK fermions are continuous, partially discrete and partially continuous, and discrete, respectively. For the volcano-like effective potential, all the massive KK modes can escape to the extra dimension, and one might obtain the fermion resonances by using the numerical methods presented in Refs. [229,230]. Inspired by the investigation of Ref. [155], Almeida et al. investigated the issue of localization of a bulk fermion on a brane, and firstly suggested that large peaks in the distribution of the normalized squared wave function $|f_{L,R}(0)|^2$ as a function of m would reveal the existence of fermion resonant states [229]. However, this method is suitable only for even fermion resonances because $f_{L,R}(0) = 0$ for any odd wavefunction. In order to find all fermion resonances, Liu et al. introduced the following relative probability [230]:

$$P = \frac{\int_{-|z_b|}^{|z_b|} |f_{Ln,Rn}(z)|^2 dz}{\int_{z_{max}}^{-z_{max}} |f_{Ln,Rn}(z)|^2 dz}, \tag{8.260}$$

where $z_{max} = 10|z_b|$ and the parameter z_b could be chosen as the coordinate that corresponds to the maximum of the effective potential V_L or V_R, which is also approximately the width of the brane. Here $|f_{Ln,Rn}(z)|^2$ can be explained as the probability density at z. If the relative probability (8.260) has a peak around $m = m_n$ and this peak has a full width at half maximum, then the KK mode with mass m_n is a fermion resonant mode. The total number of the peaks that have full width at half maximum is the number of the resonant modes. For the case of the symmetric potentials, the wave functions $f_{Ln,Rn}(z)$ are either even or odd. Hence, we can use the following boundary conditions to solve the differential equations (8.246) and (8.247) numerically [230]:

$$f_{Ln,Rn}(0) = 0, \ f'_{Ln,Rn}(0) = 1, \ \text{for odd KK modes}, \tag{8.261}$$

$$f_{Ln,Rn}(0) = 1, \ f'_{Ln,Rn}(0) = 0, \ \text{for even KK modes}. \tag{8.262}$$

One can also obtain the corresponding lifetime τ of a fermion resonance by the width (Γ) at half maximum of the peak with $\tau = \frac{1}{\Gamma}$ [17, 229]. Fermion resonances can also be obtained by using the transfer matrix method [161, 162, 163, 214, 249].

The localization and resonances of a bulk fermion have been investigated based on the Yukawa coupling mechanism [150, 155, 184, 229, 230, 231, 186, 232, 233, 234, 164, 166, 235, 236] and the derivative coupling mechanism [246, 247, 250]. Here we only list the results of the probabilities $P_{L,R}$ and the resonances of the left- and right-chiral KK fermions for the coupling with $F_1 = 1$, $F_2 = 0$, and $F_3 = \phi^2 \ln[\chi^2 + \rho^2]$ in a multi-scalar-field flat thick brane model in Figs. 8.7 and 8.8, respectively [250].

Before closing this subsection, we give some comments. Firstly, the mass spectra and lifetimes of the fermion resonances for both the left- and right-chiral fermions are the same [230, 246, 247, 250]. Secondly, the derivative coupling mechanism [244] can also be used for the branes generated by odd scalar fields [250]. Thirdly, the localized fermion zero mode is always chiral.

Besides the above mentioned fields, some other fields such as Gravitino Fields [251], Elko Spinors [252, 253, 254, 255], and new fermions [?] were also investigated in the content of extra dimensions and braneworlds.

8.5 Summary

In this review, we have given a brief introduction on several important extra dimension models and the five-dimensional thick brane models in extended theories of gravity. After introducing the KK theory, domain wall model, large extra dimension

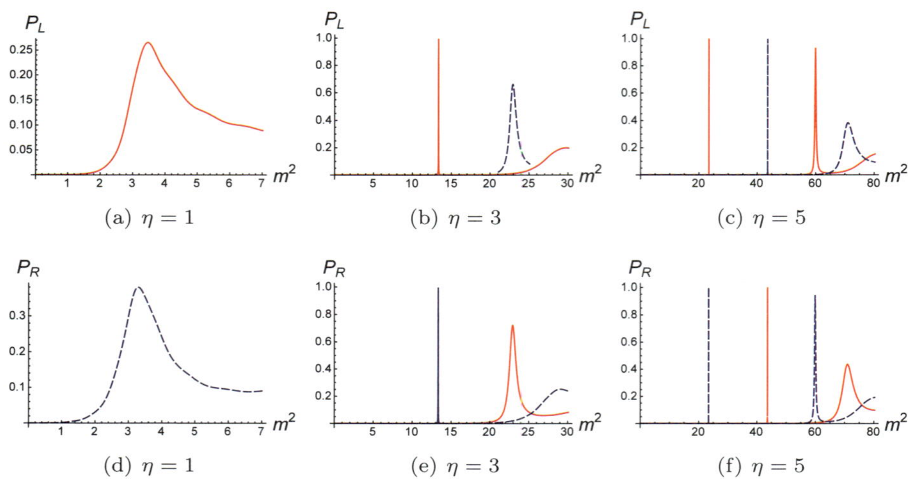

Fig. 8.7   Plots of the probabilities $P_{L,R}$ for the coupling with $F_1 = 1$, $F_2 = 0$, and $F_3 = \phi^2 \ln[\chi^2 + \rho^2]$ in a multi-scalar-field flat thick brane model [250]. Even parity and odd parity massive KK modes of the left-chiral (up channel) and right-chiral (down channel) fermions are denoted by blue dashed and red real lines, respectively. The pictures are taken from Ref. [250].

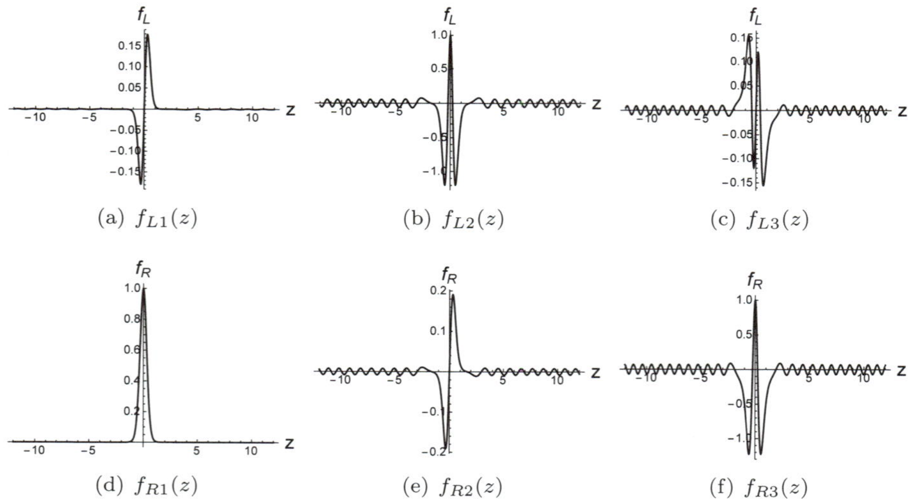

Fig. 8.8   Plots of the resonances of the left- and right-chiral KK fermions for the coupling with $F_1 = 1$, $F_2 = 0$, and $F_3 = \phi^2 \ln[\chi^2 + \rho^2]$ in a multi-scalar-field flat thick brane model [250]. The pictures are taken from Ref. [250].

model, and warped extra dimension models, we listed some thick brane solutions in extended theories of gravity, and reviewed localization of bulk matters on thick branes.

These extra dimension and/or braneworld models have been investigated, developed, or cited in thousands of literatures. But the study of extra dimensions and braneworld is far more than that. For other noteworthy extra dimension theories and related topics (including string theory, AdS/CFT correspondence, universal extra dimensions, multiple time dimensions, etc), interested readers can refer to the review papers or books mentioned earlier in this review.

In recent years, the study of extra dimensions has evolved from the early pure theory to the experimental stage [257, 258, 259, 260, 261]. Although there is no direct evidence that there are extra spatial dimensions, the idea of extra dimensions and braneworld could help us to understand the new physical phenomena and provide a candidate for explaining the past and new physical problems, which is one of the major motivations for people to study theories of extra dimensions. Of course, there are still some problems that have not been solved. Further researches (mainly for thick brane models) in the future may include but not limited to the following directions:

- Find analytic solutions of thick brane in new theories and study linear fluctuations of the solutions.
- Localization of matter fields and gravitational field in new theories.
- Intersecting brane models [262, 263, 264] and other new models.
- Physical effects of new particles in thick brane models in high-energy accelerators.
- Applications of braneworld models in cosmology (including neutrinos, black holes, inflation, dark energy and dark matter, and gravitational waves, etc) [265, 266, 267, 268, 269, 270, 271, 272, 273, 274].
- Evolution and formation of braneworlds [275, 276].
- Localized black-hole solutions in braneworld models [277, 278, 279, 281, 280].

Finally, we note that this short review cannot introduce the relevant researches comprehensively and we try our best to list the most relevant papers.

Acknowledgements

The author thanks M.L. Ge and R.G. Cai for inviting me to write this review paper. The author thanks C. Adam, C. Almeida, I. Antoniadis, N. Barbosa-Cendejas, D. Bazeia, M. Cvetic, V. Dzhunushaliev, A. Flachi, A. Herrera-Aguilar, L. Losano, I. Neupane, A. Salvio, S. SenGupta, A. Wereszczynski, as well as Z.-Q. Cui, Z.-C. Lin, T.-T. Sui, K. Yang, L. Zhao, and Y. Zhong for helpful corrections, comments and suggestions. The author also would like to express the special gratitude to B.-M. Gu, W.-D. Guo, Y.-Y. Li, H. Yu, and Y.-P. Zhang for preparing the draft of this manuscript. This work was supported by the National Natural Science Foundation of China (Grant No. 11522541 and No. 11375075) and the Fundamental Research Funds for the Central Universities (Grant No. lzujbky-2016-k04).

Bibliography

[1] G. Nordström, *On the possibility of unifying the electromagnetic and the gravitational fields*, Phys. Z. **15** (1914) 504–506, [physics/0702221].

[2] G. Nordström, *On a theory of electricity and gravitation*, Översigt af Finska Vetenskaps-Societetens Förhandlingar. Bd. **57** (1914-1915) 1–5, [physics/0702222].

[3] T. Kaluza, *On the Problem of Unity in Physics*, Sitzungsber. Preuss. Akad. Wiss. Berlin. (Math. Phys.) **96** (1921) 966–972.

[4] O. Klein, *Quantentheorie und fünfdimensionale Relativitätstheorie*, Zeitschrift für Physik A. **37** (1926) 895–906.

[5] O. Klein, *The Atomicity of Electricity as a Quantum Theory Law*, Nature **118** (1926) 516.

[6] K. Akama, *Pregeometry*, Lect. Notes Phys. **176** (1982) 267-271; K. Akama, *An Early Proposal of "Brane World"*, [hep-th/0001113].

[7] V.A. Rubakov and M.E. Shaposhnikov, *Do We Live Inside a Domain Wall?*, Phys. Lett. B **125** (1983) 136–138.

[8] V.A. Rubakov and M.E. Shaposhnikov, *Extra space-time dimensions: Towards a solution to the cosmological constant problem*, Phys. Lett. B **125** (1983) 139.

[9] I. Antoniadis, *A possible new dimension at a few TeV*, Phys. Lett. B **246** (1990) 377.

[10] N. Arkani-Hamed, S. Dimopoulos, and G. Dvali, *The hierarchy problem and new dimensions at a millimeter*, Phys. Lett. B **429** (1998) 263, [hep-ph/9803315].

[11] I. Antoniadis, N. Arkani-Hamed, S. Dimopoulos, and G. Dvali, *New dimensions at a millimeter to a Fermi and superstrings at a TeV*, Phys. Lett. B **426** (1998) 257-263, [hep-ph/9804398].

[12] L. Randall and R. Sundrum, *An Alternative to compactification*, Phys. Rev. Lett. **83** (1999) 4690, [hep-th/9906064].

[13] M. Gogberashvili, *Hierarchy problem in the shell universe model*, Int. J. Mod. Phys. D **11** (2002) 1635, [hep-ph/9812296].

[14] M. Gogberashvili, *Our world as an expanding shell*, Europhys. Lett. **49** (2000) 396–399, [hep-ph/9812365].

[15] M. Gogberashvili, *Four dimensionality in noncompact Kaluza–Klein model*, Mod. Phys. Lett. A **14** (1999) 2025–2032, [hep-ph/9904383].

[16] L. Randall and R. Sundrum, *A Large mass hierarchy from a small extra dimension*, Phys. Rev. Lett. **83** (1999) 3370, [hep-ph/9905221].

[17] R. Gregory, V.A. Rubakov, and S.M. Sibiryakov, *Opening up extra dimensions at ultralarge scales*, Phys. Rev. Lett. **84** (2000) 5928.

[18] G.R. Dvali, G. Gabadadze, and M. Porrati, *4-D gravity on a brane in 5-D Minkowski space*, Phys. Lett. B **485** (2000) 208–214, [hep-th/0005016].

[19] O. DeWolfe, D.Z. Freedman, S.S. Gubser, and A. Karch, *Modeling the fifth dimension with scalars and gravity*, Phys. Rev. D **62** (2000) 046008, [hep-th/9909134].

[20] M. Gremm, *Four-dimensional gravity on a thick domain wall*, Phys. Lett. B **478** (2000) 434–438, [hep-th/9912060].

[21] C. Csaki, J. Erlich, T.J. Hollowood, and Y. Shirman, *Universal aspects of gravity localized on thick branes*, Nucl. Phys. B **581** (2000) 309–338, [hep-th/0001033].

[22] T. Appelquist, H.-C. Cheng, and B.A. Dobrescu, *Bounds on universal extra dimensions*, Phys. Rev. D **64** (2001) 035002, [hep-ph/0012100].

[23] T.G. Rizzo, *Pedagogical introduction to extra dimensions*, hep-ph/0409309.

[24] C. Csaki, *TASI lectures on extra dimensions and branes*, hep-ph/0404096.

[25] G.D. Kribs, *TASI 2004 lectures on the phenomenology of extra dimensions*, hep-ph/0605325.

[26] R. Rattazzi, *Cargese lectures on extra-dimensions*, hep-ph/0607055.

[27] H.-C. Cheng, *2009 TASI Lecture —Introduction to Extra Dimensions*, arXiv:1003.1162.

[28] E. Ponton, *TASI 2011: Four Lectures on TeV Scale Extra Dimensions*, arXiv:1207.3827.

[29] R. Maartens, *Brane world gravity*, Living Rev. Rel. **7** (2004) 7, [gr-qc/0312059].

[30] P. Brax, C. van de Bruck, and A.-C. Davis, *Brane world cosmology*, Rept. Prog. Phys. **67** (2004) 2183–2232, [hep-th/0404011].

[31] R.A. Brown, *Brane world cosmology with Gauss-Bonnet and induced gravity terms*. PhD thesis, Portsmouth U., ICG, 2007, [gr-qc/0701083].

[32] V. Dzhunushaliev, V. Folomeev, and M. Minamitsuji, *Thick brane solutions*, Rept. Prog. Phys. **73** (2010) 066901, [arXiv:0904.1775].

[33] T.G. Rizzo, *Introduction to Extra Dimensions*, AIP Conf. Proc. **1256** (2010) 27–50, [arXiv:1003.1698].

[34] R. Maartens and K. Koyama, *Brane-World Gravity*, Living Rev. Rel. **13** (2010) 5, [arXiv:1004.3962].

[35] A. Ahmed and B. Grzadkowski, *Brane modeling in warped extra-dimension*, JHEP **1301** (2013) 177, [arXiv:1210.6708].

[36] S. Raychaudhuri and K. Sridhar, *Particle Physics of Brane worlds and Extra Dimensions*, Cambridge University Press, 2016.

[37] P.D. Mannheim, *Brane-localized gravity*, World Scientific Publishing, 2005.

[38] A. Herrera-Aguilar, D. Malagón-Morejón, R.R. Mora-Luna, and U. Nucamendi, *Aspects of thick brane worlds: 4D gravity localization, smoothness, and mass gap*, Mod. Phys. Lett. A **25** (2010) 2089-2097, [arXiv:0910.0363].

[39] Y.-X. Liu, Y. Zhong, and K. Yang, *An introduction to extra dimensions and thick brane models*, Progress in Physics **37** (2017) 41-74.

[40] L. O'Raifeartaigh and N. Straumann, *Early history of gauge theories and Kaluza–Klein theories*, hep-ph/9810524.

[41] M. Shifman, *Large Extra Dimensions: Becoming acquainted with an alternative*

paradigm, *Int. J. Mod. Phys.* A **25** (2010) 199–225, [arXiv:0907.3074].

[42] A. Salam and J.A. Strathdee, *On Kaluza–Klein Theory*, *Annals Phys.* **141** (1982) 316–352.

[43] D. Bailin and A. Love, *Kaluza–Klein Theories*, *Rept. Prog. Phys.* **50** (1987) 1087–1170.

[44] R. Coquereaux and G. Esposito-Farese, *The Theory of Kaluza–Klein-Jordan-Thiry revisited*, *Ann. Inst. H. Poincare Phys. Theor.* **52** (1990) 113–150.

[45] J.M. Overduin and P.S. Wesson, *Kaluza–Klein gravity*, *Phys. Rept.* **283** (1997) 303–380, [gr-qc/9805018].

[46] G.R. Dvali and M.A. Shifman, *Domain walls in strongly coupled theories*, *Phys. Lett. B* **396** (1997) 64.

[47] V.A. Rubakov, *Large and infinite extra dimensions*, *Phys. Usp.* **44** (2001) 871–893, [hep-ph/0104152].

[48] I.P. Volobuev and Y.A. Kubyshin, *Theor. Math. Phys.* **68** (1986) 788; *Theor. Math. Phys.* **68** (1986) 885; *JETP Lett.* **45** (1987) 581.

[49] J.D. Lykken, *Weak scale superstrings*, *Phys. Rev. D* **54** (1996) 3693, [hep-th/9603133].

[50] R. Franceschini, P.P. Giardino, G.F. Giudice, P. Lodone, and A. Strumia, *LHC bounds on large extra dimensions*, *JHEP* **1105** (2011) 092, [arXiv:1101.4919].

[51] R. Linares, H. A. Morales-Técotl, O. Pedraza, and L.O. Pimentel, *Gravitational potential of a point mass in a brane world*, *Phys. Rev. D* **89** (2014) 066002, [arXiv:1312.7060].

[52] W.-H. Tan, S.-Q. Yang, C.-G. Shao, J. Li, A.-B. Du, B.-F. Zhan et al., *New Test of the Gravitational Inverse-Square Law at the Submillimeter Range with Dual Modulation and Compensation*, *Phys. Rev. Lett.* **116** (2016) 131101.

[53] T. Han, J.D. Lykken, and R.-J. Zhang, *Kaluza–Klein states from large extra dimensions*, *Phys. Rev. D* **59** (1999) 105006, [hep-ph/9811350].

[54] Y.A. Kubyshin, *Models with extra dimensions and their phenomenology*, in *11th International School on Particles and Cosmology Karbardino-Balkaria, Russia, April 18-24, 2001*, 2001. hep-ph/0111027.

[55] M. Gremm, *Thick domain walls and singular spaces*, *Phys. Rev. D* **62** (2000) 044017, [hep-th/0002040].

[56] A.A. Andrianov, V.A. Andrianov, P. Giacconi, and R. Soldati, *Brane world generation by matter and gravity*, *JHEP* **0507** (2005) 003, [hep-th/0503115].

[57] V. Afonso, D. Bazeia, and L. Losano, *First-order formalism for bent brane*, *Phys. Lett. B* **634** (2006) 526, [hep-th/0601069].

[58] D. Bazeia, F.A. Brito, and L. Losano, *Scalar fields, bent branes, and RG flow*, *JHEP* **0611** (2006) 064, [hep-th/0610233].

[59] D. Bazeia, L. Losano, and C. Wotzasek, *Domain walls in three-field models*, *Phys. Rev. D* **66** (2002) 105025, [hep-th/0206031].

[60] V. Dzhunushaliev, *Thick brane solution in the presence of two interacting scalar fields*, *Grav. Cosmol.* **13** (2007) 302–307, [gr-qc/0603020].

[61] V. Dzhunushaliev, V. Folomeev, S. Myrzakul, and R. Myrzakulov, *Phantom thick brane in 5D bulk*, *Mod. Phys. Lett. A* **23** (2008) 2811, [arXiv:0804.0151].

[62] V. Dzhunushaliev, V. Folomeev, D. Singleton, and S. Aguilar-Rudametkin, *6D thick branes from interacting scalar fields*, *Phys. Rev. D* **77** (2008) 044006,

[hep-th/0703043].

[63] I.P. Neupane, *De Sitter brane-world, localization of gravity, and the cosmological constant*, Phys. Rev. D **83** (2011) 086004, [arXiv:1011.6357].

[64] A. de Souza Dutra, G.P. de Brito, and J.M. Hoff da Silva, *Method for obtaining thick brane models*, Phys. Rev. D **91** (2015) 086016, [arXiv:1412.5543].

[65] V. Dzhunushaliev and V. Folomeev, *Spinor brane*, Gen. Rel. Grav. **43** (2011) 1253, [arXiv:0909.2741]

[66] V. Dzhunushaliev and V. Folomeev, *Thick brane solutions supported by two spinor fields*, Gen. Rel. Grav. **44** (2012) 253, [arXiv:1104.2733]

[67] W.-J. Geng and H. Lu, *Einstein-Vector Gravity, Emerging Gauge Symmetry and de Sitter Bounce*, Phys. Rev. D **93** (2016) 044035, [arXiv:1511.03681].

[68] V. Dzhunushaliev, V. Folomeev, K. Myrzakulov, and R. Myrzakulov, *Thick brane in 7D and 8D spacetimes*, Gen. Rel. Grav. **41** (2009) 131, [arXiv:0705.4014].

[69] V. Dzhunushaliev, V. Folomeev, and M. Minamitsuji, *Thick de Sitter brane solutions in higher dimensions*, Phys. Rev. D **79** (2009) 024001, [arXiv:0809.4076].

[70] A. Herrera-Aguilar, D. Malagón-Morejón, and R.R. Mora-Luna, *Localization of gravity on a de Sitter thick braneworld without scalar fields*, JHEP **1011** (2010) 015, [arXiv:1009.1684].

[71] Y. Zhong and Y.-X. Liu, *Pure geometric thick $f(R)$-branes: stability and localization of gravity*, Eur. Phys. J. C **76** (2016) 321, [arXiv:1507.00630].

[72] H. Liu, H. Lu, and Z.-L. Wang, *$f(R)$ Gravities, Killing Spinor Equations, 'BPS' Domain Walls and Cosmology*, JHEP **1202** (2012) 083, [arXiv:1111.6602].

[73] O. DeWolfe, D.Z. Freedman, S.S. Gubser, and A. Karch, *Modeling the fifth-dimension with scalars and gravity*, Phys. Rev. D **62** (2000) 046008, [hep-th/9909134].

[74] A. Wang, *Thick de Sitter 3-branes, dynamic black holes and localization of gravity*, Phys. Rev. D **66** (2002) 024024, [hep-th/0201051].

[75] M. Giovannini, *Gauge-invariant fluctuations of scalar branes*, Phys. Rev. D **64** (2001) 064023, [hep-th/0106041].

[76] N. Barbosa-Cendejas, A. Herrera-Aguilar, U. Nucamendi, I. Quiros, and K. Kanakoglou, *Mass hierarchy, mass gap and corrections to Newton's law on thick branes with Poincare symmetry*, Gen. Rel. Grav. **46** (2014) 1631, [arXiv:0712.3098].

[77] J. Garriga and V.F. Mukhanov, *Perturbations in k-inflation*, Phys. Lett. B **458** (1999) 219–225, [hep-th/9904176].

[78] C. Armendariz-Picon, T. Damour, and V.F. Mukhanov, *k-Inflation*, Phys. Lett. B **458** (1999) 209–218, [hep-th/9904075].

[79] C. Adam, N. Grandi, J. Sanchez-Guillen, and A. Wereszczynski, *K fields, compactons, and thick branes*, J. Phys. A **41** (2008) 212004, [arXiv:0711.3550].

[80] D. Bazeia, A.R. Gomes, L. Losano, and R. Menezes, *Braneworld models of scalar fields with generalized dynamics*, Phys. Lett. B **671** (2008) 402, [arXiv:0808.1815].

[81] C. Adam, N. Grandi, P. Klimas, J. Sanchez-Guillen, and A. Wereszczynski, *Compact self-gravitating solutions of quartic (K) fields in brane cosmology*, J. Phys. A **41** (2008) 375401, [arXiv:0805.3278].

[82] Y. Zhong, Y.-X. Liu, and Z.-H. Zhao, *Non-perturbative procedure for stable*

K-*brane*, *Phys. Rev. D* **89** (2014) 104034, [arXiv:1401.0004].

[83] Y. Zhong and Y.-X. Liu, *Linearization of thick K-branes*, *Phys. Rev. D* **88** (2012) 024017, [arXiv:1212.1871].

[84] G. Germán, A. Herrera-Aguilar, D. Malagón-Morejón, R.R. Mora-Luna, and R. da Rocha, *A de Sitter tachyon thick braneworld*, *JCAP* **1302** (2013) 035, [arXiv:1210.0721].

[85] G. Germán, A. Herrera-Aguilar, A.M. Kuerten, D. Malagón-Morejón, and R. da Rocha, *Stability of a tachyon braneworld*, *JCAP* **1601** (2016) 047, [arXiv:1508.03867].

[86] R. Cartas-Fuentevilla, A. Escalante, G. Germán, A. Herrera-Aguilar, and R.R. Mora-Luna, *Coulomb's law corrections and fermion field localization in a tachyonic de Sitter thick braneworld*, *JCAP* **1605** (2016) 026, [arXiv:1412.8710].

[87] T.P. Sotiriou and V. Faraoni, $f(R)$ *theories of gravity*, *Rev. Mod. Phys.* **82** (2010) 451–497, [arXiv:0805.1726].

[88] R.P. Woodard, *Avoiding Dark Energy with 1/R Modifications of Gravity*, *Lect. Notes Phys.* **720** (2007) 403–433, [astro-ph/0601672].

[89] M. Parry, S. Pichler, and D. Deeg, *Higher-derivative gravity in brane world models*, *JCAP* **0504** (2005) 014, [hep-ph/0502048].

[90] V. Afonso, D. Bazeia, R. Menezes, and A.Y. Petrov, $f(R)$-*brane*, *Phys. Lett. B* **658** (2007) 71–76, [arXiv:0710.3790].

[91] V. Dzhunushaliev, V. Folomeev, B. Kleihaus, and J. Kunz, *Some thick brane solutions in* $f(R)$-*theory*, *JHEP* **1004** (2010) 130. [arXiv:0912.2812].

[92] Y.-X. Liu, Y. Zhong, Z.-H. Zhao, and H.-T. Li, *Domain wall brane in squared curvature gravity*, *JHEP* **1106** (2011) 135, [arXiv:1104.3188].

[93] T. Koivisto, *A note on covariant conservation of energy-momentum in modified gravities*, *Class. Quant. Grav.* **23** (2006) 4289–4296, [gr-qc/0505128].

[94] D. Bazeia, A.S. Lobao, Jr., R. Menezes, A. Yu. Petrov, and A.J. da Silva, *Braneworld solutions for F(R) models with non-constant curvature*, *Phys. Lett. B* **729** (2014) 127–135, [arXiv:1311.6294].

[95] D. Bazeia, R. Menezes, A.Yu. Petrov, and A.J. da Silva, *On the many-field* $f(R)$ *brane*, *Phys. Lett. B* **726** (2013) 523–526, [arXiv:1306.1847].

[96] Y. Zhong, Y.-X. Liu, and K. Yang, *Tensor perturbations of* $f(R)$-*branes*, *Phys. Lett. B* **699** (2011) 398–402, [arXiv:1010.3478].

[97] S. Chakraborty and S. SenGupta, *Spherically symmetric brane in a bulk of* $f(R)$ *and Gauss–Bonnet gravity*, *Class. Quant. Grav.* **33** (2016) 225001, [arXiv:1510.01953].

[98] Z.-G. Xu, Y. Zhong, Y. Yu, and Y.-X. Liu, *The structure of* $f(R)$-*brane model*, *Eur. Phys. J. C* **75** (2015) 368, [arXiv:1405.6277].

[99] W.D. Goldberger and M.B. Wise, *Modulus Stabilization with Bulk Fields*, *Phys. Rev. Lett.* **83** (1999) 4922–4925, [hep-ph/9907447].

[100] P. Pani and T.P. Sotiriou, *Surface Singularities in Eddington-Inspired Born–Infeld Gravity*, *Phys. Rev. Lett.* **109** (2012) 251102, [arXiv:1209.2972].

[101] T.P. Sotiriou, *Constraining* $f(R)$ *gravity in the Palatini formalism*, *Class. Quant. Grav.* **23** (2006) 1253–1267, [gr-qc/0512017].

[102] D. Bazeia, L. Losano, R. Menezes, G.J. Olmo, and D. Rubiera-Garcia, *Thick brane in* $f(R)$ *gravity with Palatini dynamics*, *Eur. Phys. J. C* **75** (2015) 569,

[arXiv:1411.0897].

[103] B.-M. Gu, B. Guo, H. Yu, and Y.-X. Liu, *Tensor perturbations of Palatini $f(\mathcal{R})$ branes*, Phys. Rev. D **92** (2015) 024011, [arXiv:1411.3241].

[104] M. Banados and P.G. Ferreira, *Eddington's Theory of Gravity and Its Progeny*, Phys. Rev. Lett. **105** (2010) 011101, [arXiv:1006.1769].

[105] Y.-X. Liu, K. Yang, H. Guo, and Y. Zhong, *Domain wall brane in Eddington-inspired Born–Infeld gravity*, Phys. Rev. D **85** (2012) 124053, [arXiv:1203.2349].

[106] Q.-M. Fu, L. Zhao, K. Yang, B.-M. Gu, and Y.-X. Liu, *Stability and (quasi)localization of gravitational fluctuations in an Eddington-inspired Born–Infeld brane system*, Phys. Rev. D **90** (2014) 104007, [arXiv:1407.6107].

[107] K. Yang, Y.-X. Liu, B. Guo, and X.-L. Du, *Scalar perturbations of Eddington-inspired Born–Infeld braneworld*, arXiv:1706.04818.

[108] C. Brans and R.H. Dicke, *Mach's Principle and a Relativistic Theory of Gravitation*, Phys. Rev. **124** (1961) 925–935.

[109] C. Bogdanos, A. Dimitriadis, and K. Tamvakis, *Brane models with a Ricci-coupled scalar field*, Phys. Rev. D **74** (2006) 045003, [hep-th/0604182].

[110] C. Bogdanos, *Exact Solutions in 5-D Brane Models With Scalar Fields*, J. Phys. Conf. Ser. **68** (2007) 012045, [hep-th/0609143].

[111] K. Yang, Y.-X. Liu, Y. Zhong, X.-L. Du, and S.-W. Wei, *Gravity localization and mass hierarchy in scalar-tensor branes*, Phys. Rev. D **86** (2012) 127502, [arXiv:1212.2735].

[112] K. Farakos, G. Koutsoumbas, and P. Pasipoularides, *Graviton localization and Newton's law for brane models with a non-minimally coupled bulk scalar field*, Phys. Rev. D **76** (2007) 064025, [arXiv:0705.2364].

[113] Y.-X. Liu, F.-W. Chen, H. Guo, and X.-N. Zhou, *Non-minimal coupling branes*, JHEP **1205** (2012) 108, [arXiv:1205.0210].

[114] H. Guo, Y.-X. Liu, Z.-H. Zhao, and F.-W. Chen, *Thick branes with a nonminimally coupled bulk-scalar field*, Phys. Rev. D **85** (2012) 124033, [arXiv:1106.5216].

[115] A. Herrera-Aguilar, D. Malagón-Morejón, R. Rigel Mora-Luna, and I. Quiros, *Thick braneworlds generated by a non-minimally coupled scalar field and a Gauss-Bonnet term: conditions for localization of gravity*, Class. Quant. Grav. **29** (2012) 035012, [arXiv:1105.5479].

[116] M. Gogberashvili, A. Herrera-Aguilar, D. Malagón-Morejón, R.R. Mora-Luna, and U. Nucamendi, *Thick brane isotropization in the 5D anisotropic standing wave braneworld model*, Phys. Rev. D **87** (2013) 084059, [arXiv:1201.4569].

[117] G. Germán, A. Herrera-Aguilar, D. Malagón-Morejón, I. Quiros, and R. da Rocha, *Study of field fluctuations and their localization in a thick braneworld generated by gravity nonminimally coupled to a scalar field with the Gauss-Bonnet term*, Phys. Rev. D **89** (2014) 026004, [arXiv:1301.6444].

[118] A. Tofighi, M. Moazzen, and A. Farokhtabar, *Gauge Field Localization on Deformed Branes*, Int. J. Theor. Phys. **55** (2016) 1105–1115.

[119] S.M. Aybat and D.P. George, *Stability of scalar fields in warped extra dimensions*, JHEP **1009** (2010) 010, [arXiv:1006.2827].

[120] D.P. George, *Survival of scalar zero modes in warped extra dimensions*, Phys. Rev. D **83** (2011) 104025, [arXiv:1102.0564].

[121] F.-W. Chen, B.-M. Gu, and Y.-X. Liu, *Stability of braneworlds with non-minimally coupled multi-scalar fields*, arXiv:1702.03497.

[122] G.R. Bengochea and R. Ferraro, *Dark torsion as the cosmic speed-up*, Phys. Rev. D **79** (2009) 124019, [arXiv:0812.1205].

[123] Y.-F. Cai, S. Capozziello, M. De Laurentis, and E.N. Saridakis, *$f(T)$ teleparallel gravity and cosmology*, Rept. Prog. Phys. **79** (2016) 106901, [arXiv:1511.07586].

[124] R. Ferraro and F. Fiorini, *Modified teleparallel gravity: Inflation without inflaton*, Phys. Rev. D **75** (2007) 084031, [gr-qc/0610067].

[125] R. Ferraro and F. Fiorini, *Born–Infeld gravity in Weitzenböck spacetime*, Phys. Rev. D **78** (2008) 124019, [arXiv:0812.1981].

[126] K. Bamba, R. Myrzakulov, S. Nojiri, and S.D. Odintsov, *Reconstruction of $f(T)$ gravity: Rip cosmology, finite-time future singularities and thermodynamics*, Phys. Rev. D **85** (2012) 104036, [arXiv:1202.4057].

[127] F. Fiorini, P.A. Gonzalez, and Y. Vasquez, *Compact extra dimensions in cosmologies with f(T) structure*, Phys. Rev. D **89** (2014) 024028, [arXiv:1304.1912].

[128] C.-Q. Geng, C. Lai, L.-W. Luo, and H.-H. Tseng, *Kaluza–Klein theory for teleparallel gravity*, Phys. Lett. B **737** (2014) 248–250, [arXiv:1409.1018].

[129] J. Yang, Y.-L. Li, Y. Zhong, and Y. Li, *Thick brane split caused by spacetime torsion*, Phys. Rev. D **85** (2012) 084033, [arXiv:1202.0129].

[130] R. Menezes, *First-order formalism for thick branes in modified teleparallel gravity*, Phys. Rev. D **89** (2014) 125007, [arXiv:1403.5587].

[131] W.-D. Guo, Q.-M. Fu, Y.-P. Zhang, and Y.-X. Liu, *Tensor perturbations of $f(T)$ branes*, Phys. Rev. D **93** (2016) 044002, [arXiv:1511.07143].

[132] K. Hayashi and T. Shirafuji, *New general relativity*, Phys. Rev. D **19** (1979) 3524–3553.

[133] R. Aldrovandi and J.G. Pereira, *Teleparallel Gravity*, vol. 173. Springer, Dordrecht, 2013, 10.1007/978-94-007-5143-9.

[134] N. Barbosa-Cendejas and A. Herrera-Aguilar, *Localization of 4-D gravity on pure geometrical thick branes*, Phys. Rev. D **73** (2006) 084022; Erratum:*Phys. Rev. D* 77 (2008) 049901, [hep-th/0603184].

[135] N. Barbosa-Cendejas, A. Herrera-Aguilar, M.A.R. Santos, and C. Schubert, *Mass gap for gravity localized on Weyl thick branes*, Phys. Rev. D **77** (2008) 126013, [arXiv:0709.3552].

[136] R.A.C. Correa, D.M. Dantas, C.A.S. Almeida, and P.H.R.S. Moraes, *Refinements of the Weyl pure geometrical thick branes from information-entropic measure*, arXiv:1607.01710.

[137] T.-T. Sui, L. Zhao, Y.-P. Zhang, and Q.-Y. Xie, *Localization and mass spectra of various matter fields on Weyl thin brane*, Eur. Phys. J. C **77** (2017) 411, [arXiv:1701.04957].

[138] Y.-X. Liu, L.-D. Zhang, S.-W. Wei, and Y.-S. Duan, *Localization and mass spectrum of matters on Weyl thick branes*, JHEP **0808** (2008) 041, [arXiv:0803.0098].

[139] Y.-X. Liu, K. Yang, and Y. Zhong, *de Sitter Thick Brane Solution in Weyl Geometry*, JHEP **1010** (2010) 069, [arXiv:0911.0269].

[140] Y. Zhong, Y.-X. Liu, F.-W. Chen, and Q.-Y. Xie, *Warped brane worlds in critical*

gravity, *Eur. Phys. J. C* **74** (2014) 3185, [arXiv:1403.5109].

[141] F.-W. Chen, Y. Zhong, Y.-Q. Wang, S.-F. Wu, and Y.-X. Liu, *Brane worlds in critical gravity*, *Phys. Rev. D* **88** (2013) 104033, [arXiv:1201.5922].

[142] CDF collaboration, T. Aaltonen et al., *Search for New Dielectron Resonances and Randall-Sundrum Gravitons at the Collider Detector at Fermilab*, *Phys. Rev. Lett.* **107** (2011) 051801, [arXiv:1103.4650].

[143] ATLAS collaboration, G. Aad et al., *Search for extra dimensions using diphoton events in 7 TeV proton-proton collisions with the ATLAS detector*, *Phys. Lett. B* **710** (2012) 538–556, [arXiv:1112.2194].

[144] I. Sahin, M. Koksal, S. Inan, A. Billur, B. Sahin, and R.Y.P. Tektas, E. Alici, *Graviton production through photon-quark scattering at the LHC*, *Phys. Rev. D* **91** (2015) 035017, [arXiv:1409.1796].

[145] M. Williams, C. P. Burgess, L. van Nierop, and A. Salvio, *Running with rugby balls: bulk renormalization of codimension-2 branes*, *JHEP* **1301** (2013) 102 [arXiv:1210.3753].

[146] A. Salvio, *Brane gravitational interactions from 6D supergravity*, *Phys. Lett. B* **681** (2009) 166 [arXiv:0909.0023].

[147] S. L. Parameswaran, S. Randjbar-Daemi, and A. Salvio, *General Perturbations for Braneworld Compactifications and the Six Dimensional Case*, *JHEP* **0903** (2009) 136 [arXiv:0902.0375].

[148] A. Salvio and M. Shaposhnikov, *Chiral asymmetry from a 5D Higgs mechanism*, *JHEP* **0711** (2007) 037 [arXiv:0707.2455].

[149] A. Pomarol, *Gauge bosons in a five-dimensional theory with localized gravity*, *Phys. Lett. B* **486** (2000) 153, [hep-ph/9911294].

[150] B. Bajc and G. Gabadadze, *Localization of matter and cosmological constant on a brane in Anti-de Sitter space*, *Phys. Lett. B* **474** (2000) 282–291, [hep-th/9912232].

[151] I. Oda, *Localization of matters on a string-like defect*, *Phys. Lett. B* **496** (2000) 113. [hep-th/0006203]

[152] K. Ghoroku and A. Nakamura, *Massive vector trapping as a gauge boson on a brane*, *Phys. Rev. D* **65** (2002) 084017, [hep-th/0106145].

[153] D. Bazeia and A.R. Gomes, *Bloch brane*, *JHEP* **0405** (2004) 012, [hep-th/0403141].

[154] A. Melfo, N. Pantoja, and J.D. Tempo, *Fermion localization on thick branes*, *Phys. Rev. D* **73** (2006) 044033, [hep-th/0601161].

[155] Y.-X. Liu, L.-D. Zhang, L.-J. Zhang, and Y.-S. Duan, *Fermions on thick branes in background of sine-Gordon kinks*, *Phys. Rev. D* **78** (2008) 065025, [arXiv:0804.4553].

[156] J. Liang and Y.-S. Duan, *Localization of matters on the bent brane in AdS$_5$ bulk*, *Phys. Lett. B* **680** (2009) 489–495.

[157] J. Liang and Y.-S. Duan, *Localization of matters on thick branes*, *Phys. Lett. B* **678** (2009) 491–496.

[158] J. Liang and Y.-S. Duan, *Localization and mass spectrum of spin-1/2 fermionic field on a thick brane with Poincare symmetry*, *Europhys. Lett.* **87** (2009) 40005.

[159] J. Liang and Y.-S. Duan, *Localization of matter and fermion resonances on double walls*, *Phys. Lett. B* **681** (2009) 172–178.

[160] R. Guerrero, A. Melfo, N. Pantoja, and R.O. Rodriguez, *Gauge field localization on*

brane worlds, Phys. Rev. D **81** (2010) 086004, [arXiv:0912.0463].

[161] R.R. Landim, G. Alencar, M.O. Tahim, and R.N. Costa Filho, *A transfer matrix method for resonances in Randall-Sundrum models*, JHEP **1108** (2011) 071, [arXiv:1105.5573].

[162] R.R. Landim, G. Alencar, M.O. Tahim, and R.N. Costa Filho, *A transfer matrix method for resonances in Randall-Sundrum models II: the deformed case*, JHEP **1202** (2012) 073, [arXiv:1110.5855].

[163] G. Alencar, R.R. Landim, M.O. Tahim, and R.N. Costa Filho, *A transfer matrix method for resonances in Randall-Sundrum models III: an analytical comparison*, JHEP **1301** (2013) 050, [arXiv:1207.3054].

[164] O. Castillo-Felisola and I. Schmidt, *Localization of fermions in different domain wall models*, Phys. Rev. D **86** (2012) 024014, [arXiv:1202.4734].

[165] P. Jones, G. Munoz, D. Singleton, and Triyanta, *Field localization and Nambu Jona-Lasinio mass generation mechanism in an alternative 5-dimensional brane model*, Phys. Rev. D **88** (2013) 025048, [arXiv:1307.3599].

[166] A.A. Andrianov, V.A. Andrianov, and O.O. Novikov, *CP Violation in the models of fermion localization on a domain wall (brane)*, Theor. Math. Phys. **175** (2013) 735–743, [arXiv:1304.0182].

[167] L.J.S. Sousa, C.A.S. Silva, and C.A.S. Almeida, *Brane bounce-type configurations in a string-like scenario*, Phys. Lett. B **718** (2013) 579, [arXiv:1209.6016].

[168] F.W.V. Costa, J.E.G. Silva, and C.A.S. Almeida, *Gauge vector field localization on a 3-brane placed in a warped transverse resolved conifold*, Phys. Rev. D **87** (2013) 125010, [arXiv:1304.7825].

[169] L.J.S. Sousa, C.A.S. Silva, D.M. Dantas, and C.A.S. Almeida, *Vector and fermion fields on a bouncing brane with a decreasing warp factor in a string-like defect*, Phys. Lett. B **731** (2014) 64, [arXiv:1402.1855].

[170] S.G. Rubin, *Scalar field localization on deformed extra space*, Eur. Phys. J. C **75** (2015) 333, [arXiv:1503.05011].

[171] C.A. Vaquera-Araujo and O. Corradini, *Localization of abelian gauge fields on thick branes*, Eur. Phys. J. C **75** (2015) 48, [arXiv:1406.2892].

[172] S. Choudhury, J. Mitra, and S. SenGupta, *Fermion localization and flavour hierarchy in higher curvature spacetime*, arXiv:1503.07287.

[173] I.C. Jardim, G. Alencar, R.R. Landim, and R. Costa Filho, *Solutions to the problem of Elko spinor localization in brane models*, Phys. Rev. D **91** (2015) 085008, [arXiv:1411.6962].

[174] T. Gherghetta and A. Pomarol, *Bulk fields and supersymmetry in a slice of AdS*, Nucl. Phys. B **586** (2000) 141–162, [hep-ph/0003129].

[175] D. Youm, *Bulk fields in dilatonic and selftuning flat domain walls*, Nucl. Phys. B **589** (2000) 315–336, [hep-th/0002147].

[176] S.L. Dubovsky and V.A. Rubakov, *On models of gauge field localization on a brane*, Int. J. Mod. Phys. A **16** (2001) 4331–4350, [hep-th/0105243].

[177] H. Abe, T. Kobayashi, N. Maru, and K. Yoshioka, *Field localization in warped gauge theories*, Phys. Rev. D **67** (2003) 045019, [hep-ph/0205344].

[178] K. Ghoroku and M. Yahiro, *Scalar field localization on a brane with cosmological constant*, Class. Quant. Grav. **20** (2003) 3717–3728, [hep-th/0211112].

[179] Y.S. Myung, *Localization of graviphoton and graviscalar on the brane*, J. Korean

Phys. Soc. **43** (2003) 663–669, [hep-th/0009117].

[180] I. Oda, *Gravitational localization of all local fields on the brane, Phys. Lett. B* **571** (2003) 235–244, [hep-th/0307119].

[181] M. Laine, H. B. Meyer, K. Rummukainen, and M. Shaposhnikov, *Effective gauge theories on domain walls via bulk confinement?, JHEP* **0404** (2004) 027, [hep-ph/0404058].

[182] R. Koley and S. Kar, *Scalar kinks and fermion localisation in warped spacetimes, Class. Quant. Grav.* **22** (2005) 753–768, [hep-th/0407158].

[183] Y.-X. Liu, X.-H. Zhang, L.-D. Zhang, and Y.-S. Duan, *Localization of matters on pure geometrical thick branes, JHEP* **0802** (2008) 067, [arXiv:0708.0065].

[184] Y.-X. Liu, H.-T. Li, Z.-H. Zhao, J.-X. Li, and J.-R. Ren, *Fermion resonances on multi-field thick branes, JHEP* **0910** (2009) 091, [arXiv:0909.2312].

[185] Y.-X. Liu, H. Guo, C.-E. Fu, and J.-R. Ren, *Localization of Matters on Anti-de Sitter Thick Branes, JHEP* **1002** (2010) 080, [arXiv:0907.4424].

[186] Y.-X. Liu, H. Guo, C.-E. Fu, and H.-T. Li, *Localization of gravity and bulk matters on a thick anti–de Sitter brane, Phys. Rev. D* **84** (2011) 044033 [arXiv:1101.4145].

[187] H. Guo, A. Herrera-Aguilar, Y.-X. Liu, D. Malagón-Morejón, and R.R. Mora-Luna, *Localization of bulk matter fields, the hierarchy problem and corrections to Coulomb's law on a pure de Sitter thick braneworld, Phys. Rev. D* **87** (2013) 095011, [arXiv:1103.2430].

[188] A. Herrera-Aguilar, A.D. Rojas, and E. Santos-Rodriguez, *Localization of gauge fields in a tachyonic de Sitter thick braneworld, Eur. Phys. J. C* **74** (2014) 2770, [arXiv:1401.0999].

[189] A. Díaz-Furlong, A. Herrera-Aguilar, R. Linares, R.R. Mora-Luna, and H.A. Morales-Técotl, *On localization of universal scalar fields in a tachyonic de Sitter thick braneworld, Gen. Rel. Grav.* **46** (2014) 1815, [arXiv:1407.0131].

[190] S.L. Parameswaran, S. Randjbar-Daemi, and A. Salvio, *Gauge fields, fermions and mass gaps in 6D brane worlds, Nucl. Phys. B* **767** (2007) 54, [hep-th/0608074].

[191] S.L. Parameswaran, S. Randjbar-Daemi, and A. Salvio, *Stability and negative tensions in 6D brane worlds, JHEP* **0801** (2008) 051, [arXiv:0706.1893].

[192] M. Gogberashvili, P. Midodashvili, and D. Singleton, *Fermion generations from 'apple-shaped' extra dimensions, JHEP* **0708** (2007) 033, [arXiv:0706.0676].

[193] A. Flachi and M. Minamitsuji, *Field localization on a brane intersection in anti-de Sitter spacetime, Phys. Rev. D* **79** (2009) 104021, [arXiv:0903.0133].

[194] F.W.V. Costa, J.E.G. Silva, D.F.S. Veras, and C.A.S. Almeida, *Gauge fields in a string-cigar braneworld, Phys. Lett. B* **747** (2015) 517, [arXiv:1501.00632].

[195] M.T. Arun and D. Choudhury, *Bulk gauge and matter fields in nested warping: I. The formalism, JHEP* **1509** (2015) 202, [arXiv:1501.06118].

[196] D.M. Dantas, D.F.S. Veras, J.E.G. Silva, and C.A.S. Almeida, *Fermionic Kaluza–Klein modes in the string-cigar braneworld, Phys. Rev. D* **92** (2015) 104007, [arXiv:1506.07228].

[197] D. Bazeia, A.R. Gomes, and L. Losano, *Gravity localization on thick branes: A Numerical approach, Int. J. Mod. Phys. A* **24** (2009) 1135–1160, [arXiv:0708.3530].

[198] C.-E. Fu, Y.-X. Liu, and H. Guo, *Bulk matter fields on two-field thick branes, Phys. Rev. D* **84** (2011) 044036, [arXiv:1101.0336].

[199] R. Friedberg and T.D. Lee, *Fermion-field nontopological solitons*, Phys. Rev. D **15** (Mar, 1977) 1694–1711.

[200] A.E.R. Chumbes, J.M. Hoff da Silva, and M.B. Hott, *Model to localize gauge and tensor fields on thick branes*, Phys. Rev. D **85** (2012) 085003, [arXiv:1108.3821].

[201] W.T. Cruz, A.R.P. Lima, and C.A.S. Almeida, *Gauge field localization on the Bloch Brane*, Phys. Rev. D **87** (2013) 045018, [arXiv:1211.7355].

[202] Z.-H. Zhao, Y.-X. Liu, and Y. Zhong, *$U(1)$ gauge field localization on a Bloch brane with Chumbes-Holf da Silva-Hott mechanism*, Phys. Rev. D **90** (2014) 045031, [arXiv:1402.6480].

[203] H.R. Christiansen, M.S. Cunha, and M.O. Tahim, *Exact solutions for a Maxwell-Kalb–Ramond action with dilaton: Localization of massless and massive modes in a sine-Gordon brane-world*, Phys. Rev. D **82** (2010) 085023, [arXiv:1006.1366].

[204] W.T. Cruz, M.O. Tahim, and C.A.S. Almeida, *Gauge field localization on a dilatonic deformed brane*, Phys. Lett. B **686** (2010) 259–263.

[205] G. Alencar, R.R. Landim, M.O. Tahim, C.R. Muniz, and R.N. Costa Filho, *Antisymmetric tensor fields in Randall-Sundrum thick branes*, Phys. Lett. B **693** (2010) 503–508, [arXiv:1008.0678].

[206] A. Kehagias and K. Tamvakis, *Localized gravitons, gauge bosons and chiral fermions in smooth spaces generated by a bounce*, Phys. Lett. B **504** (2001) 38–46, [hep-th/0010112].

[207] V. Dzhunushaliev, V. Folomeev, D. Singleton, and S. Aguilar-Rudametkin, *6d thick branes from interacting scalar fields*, Phys. Rev. D **77** (Feb, 2008) 044006.

[208] M.O. Tahim, W.T. Cruz, and C.A.S. Almeida, *Tensor gauge field localization in branes*, Phys. Rev. D **79** (2009) 085022, [arXiv:0808.2199].

[209] Z.-H. Zhao, Q.-Y. Xie, and Y. Zhong, *New localization method of $U(1)$ gauge vector field on flat branes in (asymptotic) AdS_5 spacetime*, Class. Quant. Grav. **32** (2015) 035020, [arXiv:1406.3098].

[210] G. Alencar, R.R. Landim, M.O. Tahim, and R.N. Costa Filho, *Gauge field localization on the brane through geometrical coupling*, Phys. Lett. B **739** (2014) 125–127, [arXiv:1409.4396].

[211] H. Ruegg and M. Ruiz-Altaba, *The Stueckelberg field*, Int. J. Mod. Phys. A **19** (2004) 3265–3348, [hep-th/0304245].

[212] B. Batell and T. Gherghetta, *Localized U(1) gauge fields, millicharged particles, and holography*, Phys. Rev. D **73** (2006) 045016, [hep-ph/0512356].

[213] Y.-X. Liu, C.-E. Fu, H. Guo, and H.-T. Li, *Deformed brane with finite extra dimension*, Phys. Rev. D **85** (2012) 084023, [arXiv:1102.4500].

[214] Y.-Z. Du, L. Zhao, Y. Zhong, C.-E. Fu, and H. Guo, *Resonances of Kalb–Ramond field on symmetric and asymmetric thick branes*, Phys. Rev. D **88** (2013) 024009, [arXiv:1301.3204].

[215] B. Mukhopadhyaya, S. Sen, and S. SenGupta, *Does a Randall-Sundrum Scenario Create the Illusion of a Torsion Free Universe?*, Phys. Rev. Lett. **89** (2002) 121101; Erratum: Phys. Rev. Lett. **89** (2002) 259902, [hep-th/0204242].

[216] B. Mukhopadhyaya, S. Sen, S. Sen, and S. SenGupta, *Bulk Kalb–Ramond field in Randall-Sundrum scenario*, Phys. Rev. D **70** (2004) 066009, [hep-th/0403098].

[217] H.R. Christiansen and M.S. Cunha, *Kalb–Ramond excitations in a thick-brane*

scenario with dilaton, Eur. Phys. J. C **72** (2012) 1942, [arXiv:1203.2172].

[218] W.T. Cruz, R.V. Maluf, and C.A.S. Almeida, *Kalb–Ramond field localization on the Bloch brane*, Eur. Phys. J. C **73** (2013) 2523, [arXiv:1303.1096].

[219] W.T. Cruz, M.O. Tahim, and C.A.S. Almeida, *Results in Kalb–Ramond field localization and resonances on deformed brane*, Europhys. Lett. **88** (2009) 41001, [arXiv:0912.1029].

[220] N. Kaloper, E. Silverstein, and L. Susskind, *Gauge symmetry and localized gravity in M-theory*, JHEP **0105** (2001) 031, [hep-th/0006192].

[221] M. Duff and J.T. Liu, *Hodge duality on the brane*, Phys. Lett. B **508** (2001) 381, [hep-th/0010171].

[222] C.-E. Fu, Y.-X. Liu, H. Guo, and S.-L. Zhang, *New localization mechanism and Hodge duality for q-form field*, Phys. Rev. D **93** (2016) 064007, [arXiv:1502.05456].

[223] C.-E. Fu, Y. Zhong, Q.-Y. Xie, and Y.-X. Liu, *Localization and mass spectrum of q-form fields on branes*, Phys. Lett. B **757** (2016) 180-186, [arXiv:1601.07118].

[224] Y.-S. Duan, *General Covariant Equations for Fields of Arbitrary Spin*, J. Exptl. Theoret. Phys. **34** (1958) 632, [arXiv:1707.02217].

[225] V.A. Fock and D.D. Ivanenko, Z. Physik **30** (1929) 678.

[226] Y.-P. Zhang and Y.-X. Liu, *Non-minimumal coupling localization mechanism of fermions on thick branes*, arXiv:1708.xxxxx.

[227] C. Ringeval, P. Peter, and J.-P. Uzan, *Localization of massive fermions on the brane*, Phys. Rev. D **65** (2002) 044016, [hep-th/0109194].

[228] T.R. Slatyer and R.R. Volkas, *Cosmology and fermion confinement in a scalar-field-generated domain wall brane in five dimensions*, JHEP **0704** (2007) 062, [hep-ph/0609003].

[229] C. Almeida, J. Ferreira, M.M., A. Gomes, and R. Casana, *Fermion localization and resonances on two-field thick branes*, Phys. Rev. D **79** (2009) 125022, [arXiv:0901.3543].

[230] Y.-X. Liu, J. Yang, Z.-H. Zhao, C.-E. Fu, and Y.-S. Duan, *Fermion localization and resonances on a de Sitter thick brane*, Phys. Rev. D **80** (2009) 065019, [arXiv:0904.1785].

[231] A.E.R. Chumbes, A.E.O. Vasquez, and M.B. Hott, *Fermion localization on a split brane*, Phys. Rev. D **83** (2011) 105010, [arXiv:1012.1480].

[232] W.T. Cruz, A.R. Gomes, and C.A.S. Almeida, *Fermions on deformed thick branes*, Eur. Phys. J. C **71** (2011) 1790, [arXiv:1110.4651].

[233] R.A.C. Correa, A. de Souza Dutra, and M.B. Hott, *Fermion localization on degenerate and critical branes*, Class. Quant. Grav. **28** (2011) 155012, [arXiv:1011.1849].

[234] L.B. Castro, *Fermion localization on two-field thick branes*, Phys. Rev. D **83** (2011) 045002, [arXiv:1008.3665].

[235] N. Barbosa-Cendejas, D. Malagón-Morejón, and R. R. Mora-Luna, *Universal spin-1/2 fermion field localization on a 5D braneworld*, Gen. Rel. Grav. **47** (2015) 77, [arXiv:1503.07900].

[236] K. Agashe, A. Azatov, Y. Cui, L. Randall, and M. Son, *Warped dipole completed, with a tower of higgs bosons*, JHEP **1506** (2015) 196, [arXiv:1412.6468].

[237] D. Bazeia and A. Mohammadi, *Fermionic bound states in distinct kinklike*

backgrounds, Eur. Phys. J. C **77** (2017) 203, [arXiv:1702.00891].

[238] T. Paul and S. SenGupta, *Fermion localization in a backreacted warped spacetime*, *Phys. Rev. D* **95** (2017) 115011, [arXiv:1704.06115].

[239] J. Mitra, T. Paul and S. SenGupta, *Fermion localization in higher curvature gravity*, arXiv:1707.06532.

[240] R. S. Hundi and S. SenGupta, *Fermion mass hierarchy in a multiple warped braneworld model, J. Phys. G* **40** (2013) 075002, [arXiv:1111.1106].

[241] R. Koley, J. Mitra and S. SenGupta, *Fermion localization in generalised Randall Sundrum model, Phys. Rev. D* **79** (2009) 041902, [arXiv:0806.0455].

[242] R. Koley, J. Mitra and S. SenGupta, *Chiral fermions in a spacetime with multiple warping, Phys. Rev. D* **78** (2008) 045005, [arXiv:0804.1019].

[243] R. Jackiw and C. Rebbi, *Solitons with fermion number 1/2, Phys. Rev. D* **13** (1976) 3398–3409.

[244] Y.-X. Liu, Z.-G. Xu, F.-W. Chen, and S.-W. Wei, *New localization mechanism of fermions on braneworlds, Phys. Rev. D* **89** (2014) 086001, [arXiv:1312.4145].

[245] Y.-Y. Li, Y.-P. Zhang, W.-D. Guo, and Y.-X. Liu, *Fermion localization mechanism with derivative geometrical coupling on branes, Phys. Rev. D* **95** (2017) 115003, [arXiv:1701.02429].

[246] H. Guo, Q.-Y. Xie, and C.-E. Fu, *Localization and quasilocalization of a spin-1/2 fermion field on a two-field thick braneworld, Phys. Rev. D* **92** (2015) 106007, [arXiv:1408.6155].

[247] Q.-Y. Xie, H. Guo, Z.-H. Zhao, Y.-Z. Du, and Y.-P. Zhang, *Spectrum structure of a fermion on Bloch branes with two scalar-fermion couplings, Class. Quant. Grav.* **34** (2017) 055007, [arXiv:1510.03345].

[248] Z.-H. Zhao, Y.-X. Liu, H.-T. Li, and Y.-Q. Wang, *Effects of the variation of mass on fermion localization and resonances on thick branes, Phys. Rev. D* **82** (2010) 084030, [arXiv:1004.2181].

[249] Q.-Y. Xie, J. Yang, and L. Zhao, *Resonance mass spectra of gravity and fermion on Bloch branes, Phys. Rev. D* **88** (2013) 105014, [arXiv:1310.4585].

[250] Y.-P. Zhang, Y.-Z. Du, W.-D. Guo, and Y.-X. Liu, *Resonance spectrum of a bulk fermion on branes, Phys. Rev. D* **93** (2016) 065042, [arXiv:1601.05852].

[251] X.-N. Zhou, Y.-Z. Du, H. Yu, and Y.-X. Liu, *Localization of Gravitino Field on $f(R)$ Thick Branes*, arXiv:1703.10805.

[252] Yu-Xiao Liu, Xiang-Nan Zhou, Ke Yang, Feng-Wei Chen, *Localization of 5D Elko spinors on Minkowski branes, Phys. Rev. D* **86** (2012) 064012, [arXiv:1107.2506].

[253] I.C. Jardim, G. Alencar, R.R. Landim,, and R.N. Costa Filho, *Comment on "Localization of 5D Elko Spinors on Minkowski Branes", Phys. Rev. D* **91** (2015) 048501, [arXiv:1411.5980].

[254] I.C. Jardim, G. Alencar, R.R. Landim, R. N. Costa Filho, *Solutions to the problem of ELKO spinor localization in brane models, Phys. Rev. D* **91** (2015) 085008, [arXiv:1411.6962].

[255] D.M. Dantas, R. da Rocha, C.A.S. Almeida, *Exotic Elko on string-like defects in six dimensions, Europhys. Lett.* **117** (2017) 51001, [arXiv:1512.07888].

[256] K.P.S. de Brito and R. da Rocha, *New fermions in the bulk, J. Phys. A* **49** (2016) 415403, [arXiv:1609.06495].

[257] D.E. Morrissey, T. Plehn, and T.M.P. Tait, *Physics searches at the LHC, Phys.*

Rept. **515** (2012) 1–113, [arXiv:0912.3259].

[258] ATLAS collaboration, G. Aad et al., *Search for dark matter candidates and large extra dimensions in events with a jet and missing transverse momentum with the ATLAS detector*, *JHEP* **1304** (2013) 075, [arXiv:1210.4491].

[259] CMS collaboration, V. Khachatryan et al., *Search for dark matter, extra dimensions, and unparticles in monojet events in proton-proton collisions at $\sqrt{s} = 8$ TeV*, *Eur. Phys. J. C* **75** (2015) 235, [arXiv:1408.3583].

[260] S.-Q. Yang, B.-F. Zhan, Q.-L. Wang, C.-G. Shao, L.-C. Tu, W.-H. Tan, and J. Luo, *Test of the Gravitational Inverse Square Law at Millimeter Ranges*, *Phys. Rev. Lett.* **108** (2012) 081101.

[261] W.-H. Tan, S.-Q. Yang, C.-G. Shao, J. Li, A.-B. Du, B.-F. Zhan, Q.-L. Wang, P.-S. Luo, L.-C. Tu, and J. Luo, *New Test of the Gravitational Inverse-Square Law at the Submillimeter Range with Dual Modulation and Compensation*, *Phys. Rev. Lett.* **116** (2016) 131101.

[262] M. Cvetic, G. Shiu, and A.M. Uranga, *Three-Family Supersymmetric Standardlike Models from Intersecting Brane Worlds*, *Phys. Rev. Lett.* **87** (2001) 201801, [hep-th/0107143].

[263] H. Steinacker and J. Zahn, *An Index for Intersecting Branes in Matrix Models*, *SIGMA* **9** (2013) 067, [arXiv:1309.0650].

[264] T. Li, D.V. Nanopoulos, S. Raza, and X.C. Wang, *A realistic intersecting D6-brane model after the first LHC run*, *JHEP* **1408** (2014) 128, [arXiv:1406.5574].

[265] I.P. Neupane, *Dark Energy and Dark Matter in Models with Warped Extra Dimensions*, Proceedings of the 7th International Heidelberg Conference on Dark 2009 Christchurch, New Zealand, 18-24 January 2009, *Dark Matter in Astrophysics and Particle Physics* (2009) 116-133.

[266] C. Doolin and I.P. Neupane, *Cosmology of a Friedmann-Lamaitre-Robertson-Walker 3-Brane, Late-Time Cosmic Acceleration, and the Cosmic Coincidence*, *Phys. Rev. Lett.* **110** (2013) 141301, [arXiv:1211.3410].

[267] I.P. Neupane, *Natural braneworld inflation in light of recent results from Planck and BICEP2*, *Phys. Rev. D* **90** (2014) 123502, [arXiv:1409.8647].

[268] V.A. Gani, A.E. Dmitriev, and S.G. Rubin, *Deformed compact extra space as dark matter candidate*, *Int. J. Mod. Phys. D* **24** (2015) 1545001, [arXiv:1411.4828].

[269] V.A. Gani, A.E. Dmitriev, and S.G. Rubin, *Two-dimensional Manifold with Point-like Defects*, *Phys. ProcediaD* **74** (2015) 32, [arXiv:1511.01869].

[270] D. Andriot, G. Cacciapaglia, A. Deandrea, N. Deutschmann, and D. Tsimpis, *Towards Kaluza–Klein Dark Matter on Nilmanifolds*, *JHEP* **1606** (2016) 169, [arXiv:1603.02289].

[271] H. Yu, B.-M. Gu, F.-P. Huang, Y.-Q. Wang, X.-H. Meng, and Y.-X. Liu, *Probing extra dimension through gravitational wave observations of compact binaries and their electromagnetic counterparts*, *JCAP* **1702** (2017) 039, [arXiv:1607.03388].

[272] D. Andriot and G. Lucena Gómez, *Signatures of extra dimensions in gravitational waves*, *JCAP* **1706** (2017) 048, [arXiv:1704.07392].

[273] N. Banerjee and T. Paul, *Inflationary scenario from higher curvature warped spacetime*, arXiv:1706.05964.

[274] A. Das, D. Maity, T. Paul, and S. SenGupta, *Bouncing cosmology from warped*

extra dimensional scenario, arXiv:1706.00950.

[275] T.N. Tomaras, *Brane-world evolution with brane-bulk energy exchange*, PoS *jhw* **2003** (2003) 013, [hep-th/0404142].

[276] D. Wang, *Gravitational Collapse and Black Hole Formation in a Braneworld*, PhD thesis, arXiv:1505.00093.

[277] P. Kanti and K. Tamvakis, *Quest for localized 4D black holes in brane worlds*, *Phys. Rev. D* **65** (2002) 084010, [hep-th/0110298].

[278] P. Kanti, N. Pappas, and K. Zuleta, *On the localization of four-dimensional brane-world black holes*, *Class. Quant. Grav.* **30** (2013) 235017, [arXiv:1309.7642].

[279] P. Kanti, N. Pappas, and T. Pappas, *On the localisation of four-dimensional brane-world black holes: II. The general case*, *Class. Quant. Grav.* **33** (2016) 015003, [arXiv:1507.02625].

[280] D. Wang and M.W. Choptuik, *Black Hole Formation in Randall-Sundrum II Braneworlds*, *Phys. Rev. Lett.* **117** (2016) 011102, [arXiv:1604.04832].

[281] T. Nakas, *Searching for Localized Black-Hole solutions in Brane-World models*, arXiv:1707.08351.

Appendix A

Chronicle of Prof. Yi-Shi Duan's life

- 1927–1934 (Birth–Age 7)

  Prof. Yi-Shi Duan was a Chinese theoretical physicist. He was born on 17 July, 1927 in Beijing, native of Wusheng County, Sichuan province. His father, Yu-Ling Duan, was an architect, who graduated from the Department of Civil Engineering of Peking University in September 1921. During the liberation of Nanjing in 1949, his father, invited by Marshal Bo-Cheng Liu, participated in the People's Liberation Army (in the Second Southwest Army Corps) contributing to the bridge and road construction. Yi-Shi Duan's Mother, Yue-Qing Liu, was a primary school teacher. The highly educated family background and patriotism had a tremendous influence on him.

- September 1934–August 1937 (Age 7–10)

  Yi-Shi Duan studied in Beijing Yuying Primary School and he was a straight A student. In his class, the Japanese students often bullied others but the teacher warned Chinese students not to fight back for fear of the traitors' confiscation. Duan and his classmates refused this humiliation and fought back with slingshot after school.

- September 1937–August 1939 (Age 10–12)

  In 1937, the Lu Gou Bridge Incident marked the outbreak of the Anti-Japanese War. Duan left Beijing with his parents and moved to Chengdu, capital of Sichuan province. When they migrated, the Japanese aircraft had been bombing all the way. There, he continued his primary study in Nongcheng School.

- September 1939–August 1940 (Age 12–13)

  Duan studied in the Affiliated Primary School of Ziyang Normal School of Sichuan province. After graduation, he was determined to be an air force pilot to defend China and fight against the Japanese invaders.

- September 1940–July 1945 (Age13–18)

  Duan was soon admitted to Youth School of the Air Force located in the Guan County of Sichuan province. By that time, many university professors had been transferred to the Air Force Youth School to avoid the Japanese air raid, which offered him an opportunity to learn calculus even in the middle school. It sparked him to learn physics and mathematics, and also helped him build a solid foundation for the study of theoretical mechanics and atomic physics. In his early age, Duan resolved one of the three classical Greek geometric problems with refined solutions, *how to trisect an angle*, which had caused serious technical difficulty in mechanical drawing. At that time Western countries used a *T-shaped* instrument to solve this problem, which was inaccurate and complicated in use. Duan invented another instrument which was accurate and easy to manufacture. After 1949, a factory in Northeast China adopted his design for mass production. (from *Xinhua Daily*, 13 October, 1951, 2nd edition.)

- August 1945–September 1946 (Age 18–19)

  Duan entered the Air Force Academy after graduation from the Air Force School.

- September 1946–February 1947 (Age 19–20)

  In 1946, Duan studied in the Air Force Officer School in Jianqiao, Hangzhou China, where he became an independent pilot of the air force. By that time the Anti-Japanese War had ended, so he left the Air Force Academy and prepared for a university study with his passion for physics.

- 1947–1951 (Age 20–24)

  In 1947, with excellent grades, Duan was admitted to Department of Physics, Nanjing University (once called Jinling University).

  In 1948, he took piano as his elective course, and playing the piano became his lifetime hobby.

  In 1949, Nanjing was liberated and his father was working in the Southwest Service Corps of the Second Field Army of the PLA. Marshal Yi Chen, as authorized by Marshal Bo-Cheng Liu, found Yi-Shi Duan and siad, "Your father is assisting us to liberate the Southwest of China. We will be away from the city and resources are limited so you have to take care of your mother, and earn your own tuition and living expenses. If you run into any difficulty, please let me know." Since then, Duan began a self-sufficient life *study as well as newspaper-selling*. Life was hard during the war. But the war didn't prevent him from learning. He taught himself general relativity and quantum mechanics.

In 1951, one of his friends posted Duan's design of *trisecting an angle* instrument to a Northeast factory. The factory replied that the instrument improved productivity dramatically, and soon Duan's design was adopted. Moreover, Duan simplified a complicated integral operation in complex analysis by proposing a new concise method to solve the V function from a U function. The method has been widely applied in electromagnetics and hydrodynamics. His achievements aroused attention of the Chinese Academy of Sciences, who invited him to attend the 15th National Association of Student Delegates. (from *Xinhua Daily*, 13 October 1951, 2nd edition.)

In 1951, as the disciple of Prof. Ru-Lin Wu, Duan completed his bachelor's thesis, *A Study on the Motion of Electronic Wanderers under Non-Uniform Electromagnetic Fields*. Using the Lorentz force formula and the Fermat's theorem, he derived a differential equation of a charged particle (electron wanderer) in a non-uniform electromagnetic field, which became a theoretical basis for further studies.

In June, 1951, Duan accomplished his university study with honors.

- 1951–1953 (Age 24–26)

In 1951, as a rewarding for his outstanding academic performance, Duan was offered a teaching assistant job and began to work in Department of Physics, Nanjing University. Then he took the examination to study in the Soviet Union, supported by the government of new China. He was admitted to Moscow State University with outstanding merit. Before that, he also earned a postgraduate opportunity with Prof. San-Qiang Qian, a famous Chinese nuclear physicist who encouraged him to become a nuclear physicist.

- 1953–1956 (Year 26–29)

In 1953, before his departure, Chairman Mao had a meeting with the students and encouraged them to gain more cutting-edge knowledges to be able to serve the country. Premier En-Lai Zhou also invited them for a dinner. The meat soup noodles was kept in his wonderful memories.

In 1953, Duan studied in Moscow State University under the supervision of Prof. M.F. Shirokov, an authority of general relativity and aerodynamics. He also studied under the guidance of Professors N.N. Bogolyubov, L.D. Landau (Nobel Prize winner) and D. Ivanenko. He was once a research assistant of Landau in academic survey of English literature. Landau invited Duan to integrate his academic achievement into his famous theoretical physics series. In Moscow State University Duan conducted frontier study in the areas of particle physics, general relativity and quantum field theory. Later, his supervisor offered him an opportunity to learn the *Theory of Explosive Detonation, Inner Ballistics of Rockets* and *Automatic Control of Missiles.*

In 1954, having mastered the general relativity, he conducted a study on the mass of an electron as a point-particle in the gravitational field. He came up with a general non-singular solution to a model with gravitational-electromagnetic coupling. Obtaining a general formula for the mass of an elementary particle, he examined the possible relationship between the radius and mass of an elementary particle. This investigation was a remarkable generalization of the work of Shirokov. Duan's paper was published in *Journal of Theoretical and Experimental Physics* (JETP, a top journal of the Soviet Union) and it was an attempt to solve point-particle divergence by studying the coupling of elementary particles and gravitational fields.

In 1955, Duan was awarded an *Outstanding Graduate Studentship* in Moscow State University.

In 1956, Duan focused on the mass and self-energy problem of scalar mesons within the framework of general relativity, by proposing a finite self-energy scheme and achieving a general Yukawa interaction between mesons. This work was published in JETP (English version).

In Moscow, Duan also took part in some social activities, serving as an interpreter for some visiting professors, such as Xue-Sen Qian, Pei-Yuan

Zhou and Long-Ji Jiang (Vice President of Peking University). He was also assigned to play host to Peng Li, who studied in Moscow Power Institute and later became Premier of China. A short meeting between Duan and Prof. Long-Ji Jiang led to their lifelong friendship. (*Later, Jiang and Duan both came to Lanzhou University. Due to Duan's academic achievements and frankness, Jiang appointed him as Head of Department of Physics, Lanzhou University.*)

- 1956–1957 (Age 29–30)

In 1956, Duan conducted the research on generalized covariant equations of scalar, pseudo-scalar and vector fields, with generally covariant interactions. With the achievement in his doctoral dissertation, he received his vice-doctorate of Moscow State University in July 1956.

In September 1956, after graduation Duan went to the Dubna Joint Nuclear Research Institute of the Soviet Union and worked as an intermediate researcher, in particle physics and quantum field theory. There, he became acquainted with Prof. Gan-Chang Wang, a famous scientist of China, who was the vice president of the Dubna Institute. The favorable atmosphere of scientific research also inspired people to discuss the latest breakthroughs of nuclear science.

In 1957, Duan studied the conservation law of elementary particles. He obtained the generally covariant conservation law by generalizing the Noether theorem to the curved spacetime. According to the new conservation theorem, the usual conservation law is just a special case under the specific transformations for energy, momentum, the angular momentum, charge etc. This work appeared in the preprint of JINR, with number JINR-Preprint-1957-P-65.

In July 1957 Duan submitted his paper "General Covariant Equations for Fields of Arbitrary Spin" to JETP, which was published in 1958 and translated into English. In this paper, Duan obtained the Gel'fand–Iaglom field equations for particles with arbitrary spin in the curved spacetime with the help of semimetric (or vielbein). This pioneering work on the gauge theory of general relativity impressed Laudan deeply. N.D. Birrel also took this work as one of the important references in "Quantum Field in Curved Space" in 1982.

In summer 1957, Duan married Ms. You-Mei Huang who was a graduate student of the Institute of Physical Chemistry. Ms. Huang graduated from Department of Chemistry, Nanjing University in 1953 and received her PhD from the Institute of Physical Chemistry (USSR) in 1960. Later she worked in Lanzhou Institute of Chemical Physics (CAS) as a professor, and also as a group leader for many years. Her father, Prof. Rui-Cai Huang, was a famous soil scientist and tenure of Nanjing Agricultural University, and her mother was a housewife.

- 1957–1958 (Age 30–31)

In 1957, Duan returned China and taught in Lanzhou University. He founded the theoretical physics division in Lanzhou University, and continued his studies in the frontiers of theoretical physics including particle physics, general relativity and topological field theory etc.

In 1958, Duan studied the nucleon-nucleon interaction. He obtained the integral equation of Green function and Bethe–Salpeter equation for two nucleons with bounded states, which was published in "Journal of Lanzhou University".

In "The Great Leap Forward" movement in 1958, Duan proposed to study the analog computer and he guided the students to learn electronics, electromagnetism, quantum mechanics etc. On 1 August, 1958, the first electronic computer in China labeled as "103" was born at CAS. Only 4 months later, the electronic computer with the capacity to be used for theoretical physics was successfully developed by Duan and Yu-Ping Kuang et al. at Lanzhou University.

Note: The frame of this machine is 2 meter high and 5 meter long, with 200 vacuum tubes, 6000 electric resistances, and other precision instruments which totally cost only 10,000 CNY. This computer simulated the approximate difference equations of the differential equations by the complex electronic circuits, and finally was able to solve various differential equations and integral differential equations within 1% errors. The method to solve Schrodinger equation with this computer was introduced in "Journal of Lanzhou University" with title "Several Schemes to Solve Schrodinger Equation Based on the Analog Computer". The first Party Secretary of Gansu Province inspected the computer on 4 January 1959.

- 1959 (Age 32)

  During the "Three Years of Natural Disasters" (1959–1961), Duan continued working diligently on the general covariant theory of element particles, ferromagnetic properties based on the double-time Green functions and Heisenberg nonlinear field theory, under the extremely difficult living conditions. In addition, he taught several courses, such as particle physics (1959–1960), quantum field theory (1960–1961), group theory (1960–1961), renormalization theory (1960) and he also lectured on CPT theorem in the particle theory.

  In 1959, Long-Ji Jiang (President of Lanzhou University), Gong-Ou Xu (Dean of the Department of Physics), Nai-Fu Cui (Provost of Lanzhou University) recognized Prof. Duan to be an expert on general relativity, gauge field theory and particle physics, thus appointed him as Head of Theoretical Physics Division as well as the graduate supervisor. At that time, he was the youngest division Head and supervisor in Lanzhou University.

- 1960 (Age 33)

  In 1960, Duan focused on the chiral symmetry of interaction between the element particles, and discussed the relation between chiral symmetry and mass inverse transformation. Furthermore, he obtained all the possible coupling between different fields, including four fermions interaction for weak interaction, coupling between electromagnetic field and spinor field, and interaction between meson and nucleons. The paper "Mass Inverse Transformation and Element Particles' Interaction" was published in "Journal of Lanzhou University".

  In 1960, with Duan's research work of the Duffin–Kemmer theory of general relativity, the generally covariant equations for particles with spin 0 and 1 with the help of vielbein was derived. He also compared different couplings between boson or fermion with gravitational field. The paper "Duffin–Kermmer Theory in the General Relativity" was published in "Journal of Lanzhou University" (with Peng-Cheng Zou).

  In 1960, having studied the covariant equation for pseudoscalar field Duan thus obtained the generally covariant Lagangian for spinor and pseudoscalar field. The effect of gravitational field on nucleon structure was discussed. The paper "Generally covariant equation of pseudoscalar field" was published in "Journal of Lanzhou University" (with Jing-Ye Zhang).

- 1961 (Age 34)

In 1961, Duan researched the phenomena of ferromagnetic resonance based on the double-time Green functions, and deduced the frequency of ferromagnetic resonance at the temperature T. With enough magnetic field and special parameters, the Oguchi equation for the frequency of ferromagnetic resonance was obtained in accordance with the experimental data well. The paper "On Magnetic Resonance Theory (Effect of Magnetocrystalline Anisotropy on the Frequency of Ferromagnetic Resonance)" was published in "Journal of Lanzhou University" (with Jing-Ye Zhang). It is one of the earliest works which studied condensed matter physics based on double-time Green functions in China.

In 1961, Duan studied the Heisenberg's nonlinear field theory, and discussed the relation between generally covariant theory and Heisenberg's theory, providing the theoretical basis for Heisenberg's unify field theory from the viewpoint of general relativity. The paper "The Generally Covariant Theory of Element Particles and Heisenberg Theory" was published in "Journal of Lanzhou University" (with Jing-Ye Zhang).

In 1961, in order to fulfill the requirement of Ministry of Education and strengthen the fundamental courses, President Long-Ji Jiang asked the core teachers to teach the basic courses. This year Duan gave lectures on mechanics for undergraduate students, which is part of the fundamental courses among Mechanics, Molecule, Electromagnetism, Atom Physics and Optics.

In September, Xiao-Ning Duan, the eldest daughter of Duan was born at Gulou Hospital in Nanjing.

- 1962 (Age 35)

In 1962, Duan carried out the study on the Fock's coordinate conditions in the general relativity. With semimetric he proved that the Fock's coordinate conditions are sufficient for the inertial reference frames, however, they are not the unique conditions. The paper "Notes of Fock's Coordinate Conditions in the General Relativity" (in cooperation with his student Jing-Ye Zhang) was published in Acta Physica Sinica.

In 1962, Duan studied the decay of Pion. Investigating the analytical properties of Lehmann Green functions, he found a new dispersion relation which infers the decay amplitude of pion. The paper "Decay of Pion Meson" was published in "Journal of Lanzhou University" (with Zhong-Qi Ma).

Duan was awarded the "Excellent Teacher Award" in Lanzhou University.

- 1963 (Age 36)

In 1963, Duan illuminated the necessity of adopting vielbein representation and made a detailed expatiation of conservation law in general relativity by utilizing vielbein. Starting with general coordinates transformation, Duan worked out the general covariant energy-momentum conservation law, which strictly possesses general covariance of Riemannian coordinates. This is an important development in general relativity. The new covariant conservation law overcomes the disadvantages of the energy-momentum law proposed by Einstein and Landau, which only adapts in the Galilean coordinates and has many shortcomings. Because Møller's energy formulation cannot explain the nonzero energy density of Bondi plane wave, the general covariant energy-momentum conservation law theory successfully overcomes the problem of Møller's energy formulation and leads to the reasonable gravitational radiation formula. The paper "The Energy-Momentum Conservation Law in General Relativity" was published in "Acta Physica Sinica" (with Jing-Ye Zhang). This research won the first prize of Gansu Provincial Prize for Scientific and Technological Progress in Higher Education. Later, this research was specifically introduced in the book "General Relativity" by Prof. Liao Liu and was named as "Duan-Expression of energy-momentum conservation law".

(Note: This work is different from the general covariant theory of elementary particles obtained in 1957. The work in 1957 mainly discussed the conserved quantities of elementary particles in gravitational field. Whereas, here the vielbein representation was adopted in order to calculate the local conserved energy-momentum of gravitational field.)

- 1964 (Age 37)

In 1964, Duan was awarded the "Excellent Teacher Award" in Lanzhou University again.

- 1965–1967 (Age 38–40)

  In 1965, Duan studied the Kaon meson decay in the case of strong interacting K-π intermediate state by using the dispersion relation, calculated the branching ratio of different decay channels and obtained the results which were consistent with experiments. The paper "K_3 Meson Decay" was published in "Acta Physica Sinica" (with his student Zhong-Yuan Zhu).

  Duan studied the problem of π-π resonance state. Starting with analyticity and unitarity, the form of sub-wave amplitude with resonance behavior of π-π scattering was generally discussed, and the resonance parameters of p wave and d wave was calculated. The papers "On π-π Resonance State" and "π-π Resonance States" were published in "Acta Physica Sinica" (with his student Mo-Lin Ge).

- 1968 (Age 41)

  In December 1968, Duan's youngest daughter Xiao-Jing Duan was born in the General Hospital of Nanjing Military Region.

- 1969–1973 (Age 42–46)

  In March 1969, after the Zhenbao Island Incident, Duan commenced to carry out national defense research. In Department of Physics, Lanzhou University, the Technical Improvement Group was formed and he guided to accomplish the military project on non-contact fuze of anti-tank rocket. The group members were Chong-Li Nie, Lin-Sheng Cheng, Yao-Gang Qi, Mo-Lin Ge, Ning Kang, Cheng-Li Zhang, Chun-Guang Zhang, Chang-Rong Luo etc.. During Counterattack in Self-Defence in Zhenbao Island, all rockets of mainland China were contact fuze. However, because the armoured vehicles of Soviet Union had slope, the rocket projectiles immediately ricochet after contact with the tank. Using small transistor to make non-contacting fuze, simulating tank's amour with tinfoil, detonating the bulbs lighting up or not, Duan verified the principle of capacitive sensing non-contact electronic fuze. Base on the principle, Duan undertook the research appointed by the Military Commission. After many experiments, the ricochet was finally overcome. By calculating the effect on trajectories caused by wind speed, altitude, air pressure, temperature, the hit rate was greatly improved. When the project was finished, Duan went to Beijing

Nankou to test shoot and he verified that the non-contact electronic fuze antitank rocket projectiles had extremely strong armor piercing lethality. He recalled, "They (the staffs) draw a limy circle, just one shooting, the rocket projectile hit the circle, pierced the armor immediately, and killed the dog inside the tank." Marshal Jian-Ying Ye watched the shooting experiment and highly praised him.

In 1972, Academician Kai-Jia Cheng passed Lanzhou and visited Duan in Lanzhou University. They had an academic discussion. (Note: In 1951, When Academician Kai-Jia Cheng was a associate professor in Department of Physics, Nanjing University, Duan was a teaching assistant.)

- 1974 (Age 47)

In 1974, Duan published the paper "About Gauge Field Theory" in Journal of Lanzhou University (in cooperation with his student Mo-Lin Ge). Starting from the general form of gauge covariant derivative, the integral expression of "parallel displacement" operator of arbitrary field function ψ was obtained. Taking this and the structure of Lie group and Lie algebra into consideration, Duan made a further and deeper discussion on gauge field theory and provided a new strategy to further develop gauge invariant theory.

Duan published the paper "The Scaling of Deep Nonelastic Electron-Nucleon Scattering (Single Internal Line Lightcone Approximation)" (in cooperation with his student Mo-Lin Ge) in Journal of Lanzhou University, discussing the scaling properties in the deep electron-nucleon scattering by using single internal line lightcone approximation theory. This theory possesses strict gauge invariance, and addresses the bound state problems.

- 1975 (Age 48)

In 1975, Duan systematically studied the magnetic monopole problem, and published the paper "Magnetic Monopole and Higgs Field" (co-authored with his student Mo-Lin Ge) in "Journal of Lanzhou University". They pointed out that the essence of the magnetic monopole theory which was proposed by G.'tHooft is that gauge potential could partly expressed by Higgs field. Moreover, the existence of magnetic monopole was tightly related to the topological properties of Higgs field in isospin space, and it is unnecessary to solve its gauge field equation. This work has profound significance for building the magnetic monopole theory in nonabelian gauge field theory. This theory was proposed almost at the same time when the similar theory was proposed by Arafune, but it was more clear and definite. This was the first time that the decomposition idea of gauge potential was introduced worldwide.

Duan published the paper "The Dual Charge of the Gauge Invariant Components in Nonabelian Gauge Field" (in cooperation with Mo-Lin Ge and Bo-Yu Hou).

- 1976 (Age 49)

  In 1976, Duan published "Magnetic Monopole and Higgs Field" (in cooperation with Mo-Lin Ge and Bo-Yu Hou) in Chinese Science Bulletin, making a further discussion on the relationship between the magnetic monopole and gauge field.

  Duan published "The Dual Charge of Nonabelian Gauge Field Theory" (in cooperation with Mo-Lin Ge and Bo-Yu Hou) in Acta Physica Sinica. In the paper, Duan studied the formulations and relations of gauge invariant quantities in nonabelian $SU(2)$ gauge field, such as electric charges, its dual charges (magnetic charges), electromagnetic field and the massive vector particles, and he specifically researched the relationship between the dual charges (magnetic charges) and the large scale topological properties of the electric charge isospin space.

- 1977 (Age 50)

  In 1977, Duan published the papers "$SU(2)$ Gauge Theory and Magnetic Monopole" and "The Geometric Representation of Magnetic Monopole Theory in Isospin Space" (in cooperation with his student Mo-Lin Ge).

  In summer, the famous theoretical physicist Chen-Ning Yang visited Lanzhou University. After Duan's report on gauge field and magnetic monopole, Prof. Yang complimented "Wonderful! Excellent! Truly excellent!!"

  The research "Anti-Tank Rocket Projectiles Non-Contact Electronic Fuze" won the first prize for the National Defense Artillery Research granted by Chinese Military Committee.

- 1978 (Age 51)

  In 1978, Duan published the paper "The Geometric Representation of Nonabelian Gauge Potential" (in cooperation with his student Mo-Lin Ge). From non-perturbing Cartan geometry, the decomposition of gauge potential was discussed. In virtue of gamma matrices and the commutation relations of the generators of Lorentz group, the expression of gauge potential in the isospin space was obtained.

  In 1978, Duan published the paper "The Decomposition and Reduction of Gauge Field and Dual Charge Solution of Abelianizable Field" (co-authored with Bo-Yu Hou and Mo-Lin Ge) in "Sinica", systematically introducing the decomposition theory of gauge potential. And the paper generalized the viewpoint of gauge potential decomposition and topological current conservation law to the semi-simple group gauge theory, and proposed a new theory that gives gauge particles masses which differs from Higgs mechanism. This work gained widespread attention from international colleagues. When Duan lectured in the United States, a series of relevant academic reports were delivered in five universities including Illinois University and University of Missouri.

  The research on magnetic monopole won the Significant Achievement Prize in Natural Sciences on the National Science Conference in 1978; "Non-contact electronic fuze" won the Significant Achievement Prize in Military Defense on the National Science Conference in 1978.

- 1979 (Age 52)

  In 1979, "The SU(2) Gauge Theory and the Electric Dynamics of N Monopoles" (in cooperation with his student Mo-Lin Ge) was published in the journal "Chinese Science". In this paper, the authors expounded that the gauge potential can be decomposed and has inner structure. This paper became the classic literature in the areas of gauge field theories, and this theory is called the "Duan–Ge–Cho Decomposition". The theory was proposed twenty years earlier than the similar work done by L. Faddeev (L. Faddeev and A. Niemi, Phys. Rev. Lett. 82, 1624 (1999)) and has wide applications in cosmology, condensate matter physics, particle physics, differential geometry and topology.

  Duan was invited to publish the paper "On Gauge Field Theory" (in cooperation with his student Mo-Lin Ge) in Journal of Sun Yat-sen University. The review started with the survey of the gauge field theory, then promoted the necessity and importance of the gauge field study in China.

The review "General Relativity and It's Experimental Basement" was published in Chinese Journal of Nature.

Duan published "The Decomposition and Reduction of Gauge Field and the Dual Charge Solution of Abelizable Field" (in cooperation with Bo-Yu Hou and Mo-Lin Ge) in Scientia Sinica (in Chinese). In the paper, according to the isotropic direction of the charge operator, the gauge potential and the corresponding field strength are decomposed covariantly into the fields originated from the charge and the dual charge and the vector particle field with the corresponding charge.

After carrying out the degrees system in China, Duan was appointed as one of the first PhD advisors by The Academic Degrees Committee of the State Council.

From 1979 to 2002, based on the latest development of theoretical physics and the graduates' academic level, Prof. Duan taught the following courses: General Relativity, Quantum Field Theory, Particle Physics, Lie Group and Lie Algebras, Gauge Field Theory, Gravitational Gauge Theory, The Standard Unified Electroweak Theory, QCD, Modern Differential Geometry and Topology, Topological Field Theory, The SU(N) Gauge Potential Decomposition Theory, Non-Commutative Geometry, Abstract Algebra, Modern Cosmology, Super Symmetric and Super String Theory, Space-Time Defects Theory, The Topological Current and Applications, and The Complexity Manifold and The Calabi–Yau Manifold.

From 1979 to 1994, Prof. Duan was Dean of Department of Physics, Lanzhou University.

The first Standing Director of High Energy Physics Society; The first Director of Chinese Gravity and Relativistic Astrophysics Society; Chairman of Physics Society of Gansu Province.

In July 1979, being invited to Geneva, Switzerland to participate in the International Conference on High Energy Physics, Duan gave a talk on gauge field theory. Subsequently, Duan visited the European Center for Nuclear Research (CERN).

In December 1979, commissioned by the Ministry of Education, Duan held the Workshop on General Relativity in Lanzhou University. More than 60 university teachers from China attended the workshop.

- 1980 (Age 53)

In 1980, Duan attended the National Symposium on Particle Physics held in Conghua, Guangzhou and the paper "Exceptional Algebra and Topological Properties of Isospace of Constant Curvature" (in cooperation with his student Mo-Lin Ge) were published in the conference proceedings.

Duan published the paper "Constant Curvature Space with SU(2) Dual Charge" (in cooperation with Mo-Lin Ge) in the High Energy Physics and Nuclear Physics. In this paper, the relation between the topological properties of the 4-dimensional constant curvature space and the SU(2) dual charge was discussed, and the conclusion was the natural extension of the 3-dimensional isospin space topological charge.

The paper "The Gauge Vacuum Effect in the SU(2) Gauge Theory" (in cooperation with Mo-Lin Ge) was published in the High Energy Physics and Nuclear Physics

The research on "anti-tank rocket electronic fuze" won the second prize for the Scientific and Technological Achievement granted by the People's Liberation Army.

Invited to give talks on gauge field theory in Missouri University.

1980–1990, Duan was appointed as member of the Physics Teaching Instruction Committee under the State Education Commission.

In the early 1980s, China held the entrance examination for Sino-USA joint cultivating physics students (CUSPEA), Duan, Prof. Bo-Chu Qian, Prof. Zhi-Cheng Wang and other professors made great efforts to assist the students in Department of Physics, Lanzhou University to achieve the highest rankings, which made Lanzhou University famous both in China and abroad.

- 1981 (Age 54)

In 1981, the research "Conservation Theory in General Relativity" won the first prize for Scientific and Technological Achievements of Higher Education in Gansu province.

The certification of PhD for theoretical physics in Lanzhou University was approved, and Duan was appointed as director.

Director of Gansu Dialectics Research Institute

- 1982 (Age 55)

In 1982, Duan published the paper "Infrared Behavior of the Running Coupling Constant in QCD" (in cooperation with his student Shan-Gao Cai) in The Communication of Theoretical Physics. Based on the Slavnov–Taylor identity, the infrared behavior of QCD running coupling constants has been studied by making use of Schwinger–Dyson equation of gluon propagator.

Partial achievement in the previous project from Defensive Department was published in Journal of Ordnance, with the title of "The Theoretical Research of Capacitive Induced an-touched Trigger Fuze" (in cooperation with Cheng-Li Zhang, and Chong-Guang Zhang).

From September to December, invited by Mr. Jin-Tao Hu, Secretary of the Communist Youth League of Gansu Province, later, General Secretary of the Chinese Communist Party, Duan held popular science lectures to secondary school students.

In 1982, cooperating with scholars from Fudan University, Sun Yat-sen University, Northwestern University, Institute of High Energy Physics CAS and Institute of Theoretical Physics CAS etc., Duan completed the research project "Classic Gauge Field Theory". This project won the Third Prize of China Natural Science Award (note: Duan and Mo-Lin Ge contributed to put forward the theory of decomposition and internal structure of gauge potential, and the general theory of topological flow in semi-simple group norm field theory. These theories have been applied in issues of the Chern class and Euler class. Duan and Mo-Lin Ge further developed the topological current and topological particle theories).

As a member of the committee that organized the "Third International Marra Grossmann Meetings" in Shanghai, Duan's lecture was entitled "The Conservation Law of Energy and Momentum in General Relativity".

- 1983 (Age 56)

In 1983, during his visit to Stanford Linear Accelerator Center (S-LAC), Duan finished the paper "The Running Coupling Constant in QCD"

(PRAC-PUB-3261). The properties of the coupling constant of QCD were discussed by employing the renormalization group theory, as well, the relationship between the coupling constant and the transfer momentum was exposed. He also composed the paper "Harmonic Maps and Their Application to General Relativity" (PRAC-PUB-3265). A new method for studying Euler equations based on harmonic mapping was proposed and it was proved that the solution of Eulerian equations corresponding to different harmonic maps satisfies the conformal transformation in 2-dimensional case. This method was also used to study the Ernst equation in general relativity.

In 1983, Duan published the paper "A Quantum Effect in General Relativity" (co-authored with Zhong-Yuan Yu. SLAC-PUB-3158). By making use of the generalized covariant Dirac equation, the gravitational correction of the electron anomalous magnetic moment was calculated.

"Conservation Law of Energy and Momentum in General Relativity" (with You-Tang Wang) was published in Science in China. The formula of the quadrupole moment of gravity radiation is given by the conservation of energy and momentum in General Relativity. This formula was confirmed by the data of the gravitational radiation attenuation effect of the PSR1913 + 16 in the pulsar binary. The paper was published both in the Chinese and English versions of Chinese Science.

"The Decomposition of the Gauge Potential and Its Quantization Theory" (co-authored with Shu-Cheng Zhao) in Communication in Theoretical Physics. The problem of quantization and renormalization of Yang–Mills field was studied from the perspective of gauge potential decomposition.

Prof. Duan was appointed as the third Director of Chinese Physical Society, the first executive council member, the second Director of Chinese Society for Gravitation and Relativistic Astrophysics, and Vice Director of the National Theoretical Physics Teaching Materials Editorial Board.

1983–1984, Duan was invited to visit SLAC as a visiting professor. During his visit, he also gave a lecture on gravitational gauge theory at Stanford University and attended an international academic conference in Baltimore.

- 1984 (Age 57)

  In 1984, Prof. Duan first proposed and established the ϕ-mapping topological current theory during his visit to Stanford University. This theory is an important development of topological quantum mechanics and topological field theory. The paper "The Structure of the Topological Current" was preprinted at Stanford linear accelerator center with SLAC-PUB-3301. The theory was later known as "Duan's topological current theory", which is widely studied in the Bose–Einstein condensation, solid spin and dislocation, cosmic strings, superconductor, spiral waves and other aspects.

  The paper "Harmonic Mapping Theory Euler Equation Solution and its Application in General Relativity" was published in Acta Physica Sinica in which the solution of the Euler equation in the harmonic mapping theory was discussed.

  Prof. Duan first studied the spin and dislocation in solids from the perspective of gauge field theory, and he published the papers "Gauge Field Theory of Dislocation and Disclination Continuum" (pre-printed number SLAC-PUB-3286) and "Gauge Field Theory and Conservation Laws in Elastic Dislocation and Disclination Continuum Mechanics" (pre-printed number SLAC-PUB-3302).

  The paper "A New Theory of Massive Gauge Field" (in cooperation with his student Shu-Cheng Zhao) was published in Journal of Lanzhou University. In this work a non-spontaneous symmetry breaking of the massive gauge field theory was proposed.

  The paper "QCD in the Effective Coupling Constant Dispersion and Infrared Behavior" (co-authored with his student Kong-Qing Yang) was published in Journal of Lanzhou University. Based on the formal theory of the effective coupling constant of QCD, the dispersion relation of the function was studied by using the non-perturbation method of the dispersion relation, and the infrared coupling of the effective coupling constant in QCD was obtained.

  Duan was awarded "National Distinguished Teacher Award".

  Standing Committee Member of the International Symposium on Hopkins' Theory of Particle Theory.

- 1985 (Age 58)

In 1985, Duan published the paper "Gauge Theories of Gravitation" (Proceedings of Symposium on Yang–Mills Gauge theories) in Communications in Theoretical Physics.

In Journal of Lanzhou University, Duan published papers "QCD in the Effective Coupling Constant Padé Approximation" (co-authored with his students Kong-Qing Yang, Wen-Fu Wang and Cheng-Ming Chen) and "The Effective Charges in Quantum Electrodynamics" (in cooperation with his student Chun-Rong Luo).

The research "The General Conservation Law of Energy and Momentum in General Relativity" won the First Prize for Scientific and Technological Progress granted by the State Education Commission.

The second Director of High Energy Physics Society.

Invited by the Graduate School of Chinese Academy of Sciences, Duan held a workshop on general relativity and cosmology in Chinese Academy of Sciences, Yuquan Road, Beijing.

Duan attended in the international conference in Florence, Italy.

- 1986 (Age 59)

In 1986, Duan published the paper "Abel Gauge Field Theory of Massive Vector Meson Field" (in cooperation with his student Shu-Cheng Zhao) in Journal of Lanzhou University.

Duan was awarded with "Excellent Teacher Award" in Lanzhou University for the third time.

Duan attended the international conference in Bonn, Germany.

- 1987 (Age 60)

January 1987–December 1989, Duan presided over the National Science Foundation project "The Development and Application of Gauge Field Theory" (No. 18670133).

September 1987–September 1989, Duan presided over the Doctoral Program Foundation of Institutions of Higher Education of China "Grand Unified Theory of Conservation Law and Topological Current" (No. 32860393).

In 1987, Duan published the paper "Higgs Mechanism and Background

Field Method" (in cooperation with his student Shu-Cheng Zhao) in Journal of Lanzhou University.

In Acta Physica Sinica, Duan published the paper "The Generalized Covariant Conservation Law in Einstein–Cartan Gravitational Theory" (in cooperation with his students Ji-Cheng Liu and Xue-Geng Dong). From the generalized Noether theorem, the conservation law corresponding to the general Lagrangian in the Einstein–Cartan gravitational theory was discussed in general, and the general Lagrangian density with torsion was obtained by the generalized displacement transformation. The conservation law of the generalized covariant Energy-Momentum was discussed as well as the necessity of the existence of the superpotential.

On 17 June, 1987, the 11th Johns Hopkins Workshop on Current Problem in Particle Theory was successfully held at Lanzhou University. Duan offered a report entitled "ϕ-mapping Method, Topological Current, and String Theory", in which ϕ-mapping topological current theory was explained thoroughly. The participants include P. West at the European Center for Nuclear Research, A. Strominger from Harvard University, and G. Domokos at Johns Hopkins University. The provincial Party Secretary Zi-Qi Li, Governor Zhi-Jie Jia congratulated him on the site. As the Host, Duan composed the poem, "Everything began in oneness, and Tao follows after nature. Observing, one can realize how marvelous the cosmos is; researching, one can find how mysterious particles are."

- 1988 (Age 61)

In 1988, the conference proceeding "Frontier in Particle Physics Proceedings, 11th Johns Hopkins Workshop on Current Problems in Particle Theory" was edited by Duan, G. Domokos and S. Kovesi-Domokos and was published by World Scientific Publishing Co Pte Ltd.

The fourth Director of the Chinese Physical Society.

From 1988 to 1998, Member of the Sixth and Seventh Committee of the Chinese People's Political Consultative Conference of Gansu Province.

Duan attended the international conference in Baltimore.

- 1989 (Age 62)

From 1989 to 2007, Duan served as Honorary Chairman of Gansu Institute of Physics.

Duan was honored with the National Distinguished Teacher Award.

Duan attended the international conference in Florence, Italy.

- 1990 (Age 63)

January 1990–December 1992, Duan presided over the National Science Foundation project "Topological Field Theory and Topological Current Theory" (No. 18975017).

September 1990–September 1992, Duan presided over the Doctoral Program Foundation of Institutions of Higher Education of China "Topological Gauge Field Theory and Topological Gravitational Theory" (No. 9073004).

In 1990, Duan carried out the study of dislocation topology and established the gauge field theory of dislocation and disclination, which attracted the international counterparts. "Topological Structure of Dislocation in the Gauge Field Theory of Dislocations and Disclinations Continuum" (in cooperation with his student Sheng-Li Zhang) was published in Int. J. Engng. Sci..

Duan was awarded with "National Advanced Worker in Science and Technology".

- 1991 (Age 64)

In 1991, Duan won the Outstanding Contribution Award for Developing China's Higher Education granted by the State Council and he also obtained the special allowance of the State Council.

Duan finished his research on topological current of dislocation, and developed the topological current theory. His paper "Topological Current Structure of Dislocations in the 4-Dimensional Gauge Field Theory of Dislocation and Disclination Continuum" (in cooperation with his student Sheng-Li Zhang) was published in Int. J. Engng. Sci..

In Journal of Lanzhou University, Duan published the papers "Weyl Theory and Gravity Charge, Inertia Factor and Cosmological Constant" (in cooperation with his student Shu-Cheng Zhao) and "Generalized Harmonic Mapping Equation and its Soliton Solution".

In May, Duan attended the Fourth Representative Meeting of China Association for Science and Technology.

In August, Duan attended the international conference in Baltimore.

- 1992 (Age 65)

January 1992–December 1993, Duan presided over the National Science Foundation project "The Application of the Gauge Field Theory and Topology Theory in Solid Physics" (No.19172030).

January 1992–September 1994, Duan presided over the Doctoral Program Foundation of Institutions of Higher Education of China "The Application of Topological Current Theory and Topological Field Theory on Materials Science".

In 1992, Duan carried out the research on quantum group. His paper "The Quantum Group $SU_q(2)$ and the q-Analogue of Angular Momenta in Classical Mechanics"(in cooperation with his student Sheng-Li Zhang) was published in J. Phys. A.

Duan also studied a new coordinate condition in general relativity, and his paper "A New Coordinate Condition in General Relativity" (in cooperation with his student Sheng-Li Zhang) was published in Gen. Rel. Grav..

He carried out the study of conformal field theory and later published the paper "Conformal (Weyl) Invariance and Higgs Mechanism" (in cooperation with his student Shu-Cheng Zhao) in IL Nuovo Cimento.

Duan also focused on the topological current theory of disclination. His paper "Topological Current Structure of Disclinations in the 4-Dimensional Gauge Field Theory of Dislocation and Disclination Continuum" (in cooperation with his student Sheng-Li Zhang) was published in Int. J. Engng. Sci..

(Notes: From 1984 to 1992, Duan and his collaborators published several papers in Int. J. Engng. Sci and other magazines. By applying the gauge field theory, moving-vielbein theory, gauge potential decomposition theory and the theory with internal structures into solid defects, they proposed the strict non-linear gauge theory of dislocations and disclinations continuum as well as established the topological theory of dislocations and disclinations with parallel field theory and topological current theory. They directly calculated the strict description of the local and global topological properties of dislocations and disclinations from the expression of the curvature and torsion. "This research is an important development in this field, and it strictly unified the geometrical and topological theory of solid defects directly from the gauge theory itself," said Prof. D.G.B. Edelen, authority of American Mechanics and Applied Mathematics, "The articles published by Duan and his collaborators in Int. J. Engng. Sci. will be the most important and sustainable papers in the field", and he also used Duan's research as an important reference in his monographs Gauge Theory and Defects in Solids.

- 1993 (Age 66)

In 1993, Duan studied the topological structure and topological current of the Gauss–Bonnet–Chern theorem. The paper "Topological Structure of Gauss–Bonnet–Chern Density and Its Topological Current" (in cooperation with his student Xing-He Meng) was published in J. Math. Phys..

In November, Duan initiated lectures on new effects of physics in order to further improve the academic level of Lanzhou University physics research and teaching talent, to introduce latest developments in physics to the senior students and graduate students of physics and related disciplines, and to build an academic atmosphere for physics teachers and students. At the same time, the lectures could serve for physics teaching and academic exchanges of scientific research personnel in the Lanzhou area. The main content of the Nobel Prize in Physics was the new effects for the development of physics and the new physics effects on other natural science and technology applications. Since its inception, Duan has presided over more than 80 lectures. The presentations he has given include: gravity radiation of a binary pulsar, quark model, the discovery of TOP quark, atomic bomb, modern cosmology, the discovery of Tau lepton and its significance, superstring and quark of unified theory, nano-science and nanotechnology, depleted uranium weapons, modern cosmology dark energy, gradual free quantum chromodynamics, Einstein and general relativity, cosmic microwave background radiation and cosmology, the top ten problems of physics in 21 century, Higgs particle and the mechanism to produce the basic particle quality. Support units for the lectures include: the new effect of physics lectures Foundation, Institute of Theoretical Physics of Lanzhou University, School of Physics and Technology of Lanzhou University, Institute of Modern Physics, Chinese Academy of Sciences and Gansu

Institute of Physics.

1993–2004, Duan assumed the office of Director of Institute of Theoretical Physics, Lanzhou University.

He went to Budapest in Hungary to attend the international conference.

- 1994 (Age 67)

In 1994, Duan carried out the research on the demarcation of normative and space-time defects, and his paper "Gauge Potential Decomposition, Space-Time Defects and Planck's" (in cooperation with his students Sheng-Li Zhang and Shi-Xiang Feng) was published in J. Math.Phys.. He was appointed as Honorary Dean of Department of Physics, Lanzhou University.

He went to Florence, Italy to participate in the international conference.

- 1995 (Age 68)

January 1995–December 1997, Duan presided over the National Science Foundation project "The Divergence Theory of Topological Flow and Topological Charge" (No.19475018).

September 1995–September 1997, Duan presided over the Doctoral Program Foundation of Institutions of Higher Education of China "Material Structure and the Theory of Field Topology and Topology Flow" (No.9573007).

In 1995, Duan published "The Theory of General Relativity in Generalized Covariant Law of Conservation of Angular Momentum" (in cooperation with his student Shi-Xiang Feng) in Journal of Physics. The law of conservation of angular momentum for generalized covariance was obtained by employing the local Lorentz invariance of the Einstein gravitational one material system. The rationality of this law was illustrated by two typical examples.

In the same year, Duan published the paper "General Decomposition Theory of Spin Connection Structure of Gauss–Bounnet–Chern Density and Morse Theory" (in cooperation with his student Xi-Guo Li) in Helv. Phys. Acta.

He was invited to Baltimore, Maryland, USA to attend the international conference.

- 1996 (Age 69)

In 1996, his paper "Conservative Angular Momentum as SU(2) Charges in Complex Gravity" was published in Commun. Theor. Phys., and "About the Energy of the Universe"(in cooperation with his student Shi-Xiang Feng) was published in Chin. Phys. Lett..

He went to Heidelberg, Germany to participate in the international conference.

- 1997 (Age 70)

In 1997, Duan studied the topology of the space-time defect theory and established a complete theory of ϕ-mapping topological flow bifurcation. His paper "The Origin and Bifurcation of the Space-Time Defects in the Early Universe" (in cooperation with his students Guo-Hong Yang, Ying Jiang) was published in Gen. Rel. Grav..

In July, domestic and foreign researchers gathered in Lanzhou to attend the 21st Johns Hopkins Workshop on Particle Physics hosted by Duan to discuss the topology of particles and field theory. Duan composed, "Experts on topology and gauge gather together in such an age following your heart instead of following the social trend and it is time for us to innovate our careers".

- 1998 (Age 71)

January 1998–December 2000, Duan presided over the National Science Foundation project "Internal Structure and the New Topological Invariant of the Gauge Field" (No. 19775021).

In 1998, Duan studied the topology problem in condensed matter system, and opened up a new field of topological state. "Topological structure of the London equation"(in cooperation with his students Hong Zhang, Sheng Li) was published in Phys. Rev. B.

He started researches on Gauss–Bonnet–Chern topological bifurcation and the connection with Morse theory. "The Bifurcation Theory of the Gauss–Bonnet–Chern Topological Current and Morse Function" (in cooperation with his students Sheng Li, Guo-Hong Yang) was published in Nucl. Plys. B.

Duan was invited to publish a paper entitled "Decomposition Theory of Spin Connection, Topological Structure of Gauss–Bounet–Chern Topological Current and Morse Theory" (in cooperation with his student Sheng Li) in the Jingshin theoretical physics symposium in honor of professor Ta-You Wu's 90th birthday.

In July, Duan met Prof. Shing-Tung Yau, the internationally renowned mathematician, at Lanzhou University. They discussed mathematical problems about which they both were concerned.

Duan went to Goteborg, Sweden to attend the international conference.

- 1999 (Age 72)

January 1999–December 2001, Duan presided over the Doctoral Program Foundation of Institutions of Higher Education of China "Gauge Field Decomposition and Internal Structure and New Topological Invariants" (No. 98073009), which has been awarded as Excellent Achievement.

In 1999, Duan studied the point defects of a three-dimensional vector order parameter field, and his paper "Point Defects of a Three-Dimensional Vector Order Parameter" (in cooperation with his student Hong Zhang) was published in Phys. Rev. E.

Duan studied the branch conditions for wave dislocations in light beams, and the paper "Branch Conditions for Wave Dislocations in Light Beams"(in cooperation with his students Guang Jia and Hong Zhang) was published in J. Phys. A: Math. Gen..

- 2000 (Age73)

In 2000, Duan studied the evolution of the Chern–Simons vortices, and the paper "The Evolution of the Chern–Simons Vortices"(in cooperation with his student Li-Bin Fu) was published in Phys. Rev. D.

Duan researched the topological aspect of the arbitrary dimensional topological defects, and the paper "A New Topological Aspect of the Arbitrary Dimensional Topological Defects" (in cooperation with his student Yin Jiang) was published in J. Math. Phys..

In 2000, Duan attended the conference held by The Quantum Mechanics Society in Nanjing, and presented a report entitled "The New Development of Topological Quantum Mechanics and Its Applications in the

Frontiers of Physics".

- 2001 (Age 74)

  January 2001–December 2004, Duan presided over the National Science Foundation project "The New Development of Topological Gauge Field Theory and Its Application in the Frontier Science of Physics" (No. 10175028).

  In 2001, Duan studied the circulation condition of the two-component BEC, and the paper "Circulation Condition of the Two-Component Bose–Einstein Condensate"(in cooperation with his student Duo-Jie Jia) was published in Phys. Lett. A.

  Duan studied the topological excitation in the fractional quantum Hall system, and the paper "Topological Excitation in the Fractional Quantum Hall System" (in cooperation with his students Peng-Ming Zhang and Hong Zhang) was published in Phys. Lett. A.

  Duan also carried out the study on the decomposition of $SO(n)$ gauge potential and the paper was published in *High Energy Physics and Nuclear Physics* (in cooperation with his students Xi-Guo Li and Jian-Jun Song).

  In 2001, Duan was invited to hold a training course for teachers at Qinghai Normal University, and his lectures focused on particle physics, topological physics and cosmology.

  Duan attended the international conference in Florence, Italy.

- 2002 (Age 75)

  January 2002–December 2004, Duan presided over the Doctoral Program Foundation of Institutions of Higher Education of China "New Developments and Applications of Topological Gauge Field Theory and Topological Quantum Mechanics" (No. 20010730007).

  In 2002, Duan studied the decomposition of the SU(N) connection and the Wu–Yang potential. His paper "Decomposition of the SU(N) Connection and the Wu–Yang Potential" (in cooperation with his student Peng-Ming Zhang) was published in Mod. Phys. Lett. A.

  In 2002, Duan studied the topological aspects of liquid crystals, and the paper "Topological Aspects of Liquid Crystals" (in cooperation with his student Guo-Hong Yang) was published in Int. J. Theor. Phys..

In 2002, Duan concentrated on the torsion structure in Riemann–Cartan manifold and the topological property of the dislocation, and the paper "Torsion Structure in Riemann–Cartan Manifold and Dislocation" (in cooperation with his student Xi-Guo Li) was published in Gen. Rel. Grav..

In 2002, Duan attended the International String Physics Conference in Beijing.

On 24 April, Duan was invited to Nankai University. He met the famous mathematician Shiing-Shen Chern, and they discussed the GBC theorem and his new insights, as well as the application of the GBC theorem in topological field theory. Duan proved that the invariants formed by the tensor product of the gauge field in the GBC theorem can be strictly expressed as the topological charge of the ϕ-mapping of the vector field on the corresponding Riemann manifold.

On 24 October, 2002, Duan was invited to Zhejiang University. He attended the conference "Facing the Challenge of Physics in the Twenty-First Century", which was held to celebrate the 10th anniversary of the Zhejiang Modern Physics Center. He presented a report entitled "Cosmic Strings and Dark Energy".

- 2003 (Age 76)

In 2003, Duan studied the topological structure of the knots in Chern–Simons field theory, and the paper "Many Knots in Chern–Simons Field Theory" (in cooperation with his students Xing Liu and Li-Bin Fu) was published in Phys. Rev. D.

Duan studied the topological properties of the monopoles and vortex lines in three-component spinor BEC system, and the paper "Mermin–Ho Vortices and Monopoles in Three-Component Spinor BEC" (in cooperation with his students Xing Liu and Peng-Ming Zhang) was published in J. Phys. A.

Duan proposed that the complex scalar order parameter field is one of the dark energy candidates, and the topological defect of the order parameter field is the cosmic strings, and the paper "Cosmic Strings and Quintessence" (in cooperation with his students Ji-Rong Ren and Jie Yang) was published in Chin. Phys. Lett..

In 2004 Duan was awarded because his publications were among the top three in SCI and EI and in authoritative journals in 2003 in Lanzhou University.

Although 76 years old, he held several courses and lectures for graduate students, such as Electro-Weak Theory (January), Homotopy Theory (March–May), Riemann–Cartan Manifold and Torsion (June), Noether Conservation Flow and Natural Conservation Flow, ϕ-Mapping Current Theory (June–December), Abstract Algebra, Lie Group and Lie Algebra, the Classical Gauge Field Theory and Topological Theory (September 2003–March 2004).

Duan was invited to attend the symposium on the development of theoretical physics in Hunan Normal University, and he presented a report entitled "Scientific Research, Learning and Teaching". In the practice of teaching and scientific research for many years, Duan summarized his teaching and studying philosophy as follows:

- be brief in language and to the point
- be excellent combination
- induce new doctrines from a collection of ideas
- recover the original simplicity
- discard the dregs
- extract the essence
- eliminate the false and retain the true
- arrive at supreme goodness

- **2004 (Age 77)**

    In 2004, Duan published a paper entitled "Knotlike Cosmic Strings in the Early Universe" (in cooperation with his student Xin Liu) in JHEP. In this paper, through the action of Chern–Simons, Duan studied the topological properties of cosmic strings with a knot structure in the early universe.

    Based on the ϕ-mapping topological current theory, Duan studied the transverse force on a moving vortex with the acoustic geometry, and the paper "Transverse Force on a Moving Vortex with the Acoustic Geometry" (in cooperation with his student Peng-Ming Zhang, Li-Ming Cao, and Chen-Kui Zhong) was published in Phys. Lett. A.

    Duan was awarded as the No. 1 whose papers were published in SCI-one region in Lanzhou University in 2004.

Duan was honored with the Senior Professor of Lanzhou University.

Duan held several lectures and courses for graduate students: SO(N) Gauge Field Theory (March–June), Matrices and Determinants (October), General Relativity and Gauge Theory of Gravitation (September 2004–June 2005).

Duan was invited to Qinghai Normal University, and his lectures were on modern cosmology, dark energy, big bang.

2004–2016, Honorary Director of Institute of Theoretical Physics, Lanzhou University.

"Topological Gauge Field Theory in Condensed Matter and Cosmological Space-Time Defect" won the First Prize for Gansu Science and Technology Progress.

"The New Development of Topological Gauge Field Theory and Its Applications in the Frontier science of Physics" was awarded the Second Prize for National Science granted by the Ministry of Education.

Duan et. al. first proposed the point that the gauge potential can be decomposed and has internal structure, and proposed the ϕ-mapping topological current theory. On the basis, Duan established the new topological field theory and new topological quantum mechanics, as well as the topological gauge field theory for the study of solid defects. Duan also revealed the internal relations between the geometrical properties and topological structures of the fields. Moreover, after 1990s, the following important problems have been solved:

(1) Based on the torsion of the Riemann–Cartan manifold, the new Abel gauge field theory and the corresponding topological invariants were established, and the topological defects theory of the solid dislocations and disclinations were founded.

(2) By employing the new Abel gauge field formed by the torsion tensor of the Riemann–Cartan manifold, the defects of the cosmological space-time were studied. For the first time, the physical mechanism of the generation of the cosmic strings in the universe was proposed and the topological structure of cosmic string was investigated. The order parameter fields for generating the cosmic strings was identify as the dark energy that accelerates the expansion of the universe.

(3) With the decomposition of the gauge potential and the ϕ-mapping topological current theory, the London theory in the superconducting theory was proved for the first time, and the quantitative relationship between the topological quantum number of superconductor defect and the zero mapping degree of order parameter wave function was obtained.

(4) In order to establish the topological quantum mechanics, the important assumptions in the quantum mechanics that the curl of the velocity vector field is non-zero and has strict singularity was strictly proved for the first time. It pointed out that the topological quantum number of the vortices is determined by the zero point index of the ϕ-mapping of the wave function. The topological structure of vortices in Bose Einstein condensates was studied.

(5) With the ϕ-mapping topological current theory, a new theory of topological bifurcation in the defect systems was established. The theory was also applied into the superconducting, Bose Einstein condensation and cosmic string systems, thus critical conditions of the corresponding defect bifurcations were obtained.

(6) With U(1) gauge potential decomposition theory and ϕ-mapping topological current theory, that the Chern–Simons current is the topological flow of the corresponding string, and the Chern–Simons action is the topological invariant of knot strings have been proved. And the proof is more concise and rigorous than that of E. Witten, a famous mathematician and physicist.

(7) With the assistance of the topological field theory established through gauge potential decomposition and ϕ-mapping topological current theory, the topological structures of spiral wave, wave dislocation and sunspot vortices were also studied.

- 2005 (Age 78)

  January 2005–December 2007, Duan held the National Natural Science Foundation Program "Topological gauge field theories of cosmological string, dark energy, extra dimensional manifolds and p-branes" (No. 10475034).

  In 2005, on the basis of the Abelian-Higgs model, Duan obtained the generalized Bogomol'nyi equation for the self-dual vortex. The paper entitled "Fermionic zero modes in self-dual vortex background" (in cooperation with his students Yong-Qiang Wang, Tie-Yan Si and Yu-Xiao Liu) was published in Mod. Phys. Lett. A.

Duan taught Particle Physics and Modern Cosmology to the graduate students (September 2005–January 2006).

Invited by Sichuan University, he delivered a series of lectures including "Einstein and Relativity", "Research Study and Teaching", "From Particles to the Universe", "Key Problems in Cosmology and Particle Physics", "The Principle of Atomic Bombs" etc.

- 2006 (Age 79)

In 2006, employing the ϕ-mapping topological flux theory, Duan studied the topological properties of the vortex line and the instantons with knotted structure in a charged two-condensate Bose system. The paper "Knotted soliton in a charged two-condensate Bose system" (in cooperation with his students Xin-Hui Zhang, Yu-Xiao Liu and Li Zhao) was published in Phys. Rev. B.

Having studied the higher-dimensional topological defects in local non-abelian topological tensor flux, Duan published the paper "Higher-dimensional knotlike topological defects in local non-Abelian topological tensor currents" (in cooperation with his students Shao-Feng Wu and Peng-Ming Zhang) in J. Math. Phys.

Having investigated the properties of topological vortex in a two-gap superconductor, Duan's paper "Topological vortices in a two-gap superconductor" (with his students Li Zhao and Xin-Hui Zhang) was published in Phys. Lett. A.

Taught postgraduate courses: Theory of Extra Dimensions (March–May), Complex Manifolds and Calabi–Yau Manifolds (September–December) and so on.

Invited as Supervisor of the Long-Ji classes in Department of Physics, Lanzhou University, Duan lectured "Enjoy physics, study and creation" and "Cosmological microwave background radiation and cosmology".

Invited to attend the "Academician and Experts Forum" held by the Youth League Committee and Department of Physics, Lanzhou University,

Duan offered the fifth lecture entitled "Ten Forefront Issues of the 21st Century".

Invited to attend the academic activities for the celebration of the 85th Anniversary of Xiamen University, Duan gave a talk entitled "A Centenary of Relativity".

Duan was invited by Prof. Shing-Tung Yau to hold the Lanzhou satellite conference, branch of the International String Theory Conference. Participants including Prof. Liao Liu from Beijing Normal University, Prof. Andrew Strominger from Harvard University, Prof. Yong-Shi Wu from Utah University and Prof. Mu-Lin Yan from University of Science and Technology of China.

In September, Duan communicated with the Nobel Laureate Samuel Ting while Ting was visiting Lanzhou University.

- 2007 (age 80)

January 2007–December 2009, Duan presided over the National Natural Science Foundation project "Extra dimensions and the topological field theory of p-branes" (No. 10610301019).

In 2007, Duan studied the localization of fermions on a string-like defect. The paper "Localization of fermions on a string-like defect" (in cooperation with his students Yu-Xiao Liu and Li Zhao) was published in JHEP.

Having studied the inner topological structure of Hopf invariant, Duan published "Inner topological structure of Hopf invariant" (in cooperation with his students Ji-Rong Ren and Ran Li) in J. Math. Phys..

After his research on the topological properties and the fermion absorption of spherically symmetric black holes, Duan published "Fermion absorption cross section and topology of spherically symmetric black holes" (in cooperation with his students Yu-Xiao Liu, Li Zhao and Zhen-Bin Cao) in Phys. Lett. B.

From the research of the fermions in self-dual vortex background on a string-like defect, Duan published "Fermions in self-dual vortex background on a string-like defect" (in cooperation with his students Yu-Xiao Liu, Li Zhao and Xin-Hui Zhang) in Nucl. Phys. B.

Duan taught courses or lectures for postgraduates, including "The

Threshold Energy of Particle Reactions" (March), "Superstring Theory/Conformal Field and Supersymmetry" (March–May).

Duan lectured to the students of Long-Ji Classes with titles "Long-Ji Jiang—an outstanding educator in our country", "Einstein and relativity", "The intimate relation between mathematics and physics".

In May, the Nobel laureate David Gross visited Lanzhou University and discussed academic problems with Duan.

In July, with his students and colleagues home and abroad, Duan held a forefront seminar on theoretical physics. He was invited to give a lecture on "My insights at age eighty". His students spoke highly of him, "Duan's scientific works have a specific feature: he never blindly goes with trends. Importantly, some of his viewpoints are on the lead throughout the world". The Institute of Theoretical Physics at the Chinese Academy of Sciences complimented him "Duan is a talent in researching, a model in educating".

- 2008 (Age 81)

In 2008, Duan published the paper "Localization of matters on pure geometrical thick branes" (in cooperation with his students Yu-Xiao Liu, Xin-Hui Zhang and Li-Da Zhang) in JHEP, which concentrated on the localization of matter fields on pure geometric thick brane.

Duan published the paper "Vortices-induced quantum Rontgen effect in BEC: a consistent approach" (in cooperation with his student Ru-Nan Huang) in J. Phys. A., in which quantum Rontgen effect induced by vortices in Bose–Einstein condensation was investigated.

In the paper "Magnetic monopoles in ferromagnetic spin-triplet superconductors"(in cooperation with his students Li-Da Zhang and Yu-Xiao Liu) in J. Phys. Condens. Mat., Duan studied Magnetic monopoles in ferromagnetic spin-triplet superconductors.

The paper "Localization and mass spectrum of matters on Weyl thick branes" (in cooperation with his students Yu-Xiao Liu, Li-Da Zhang and Shao-Wen Wei) in JHEP, which focused on the localization and mass spectrum of various kinds of matter fields on Weyl thick branes.

His paper "Fermions on thick branes in the background of sine-Gordon kinks" (in cooperation with his students Yu-Xiao Liu, Li-Da Zhang and Li-Jie Zhang) in Phys. Rev. D, investigated the fermions on thick branes in the background of sine-Gordon kinks.

In July, Duan was invited to hold the Annual meeting of Chinese Society of Gravitation and Relativity Astrophysics, and he delivered an address entitled "Gravitational Gauge Theory".

In September and October, he was invited to offer two lectures to the students of the Long-Ji Jiang Classes of Lanzhou University. One was "Scale the heights of science—learn the wisdom and philosophy of creation", and the other was "The breaking of symmetry".

- 2009 (Age 82)

In 2009, Duan investigated the localization and mass spectra of various bulk matter fields on symmetric and asymmetric de Sitter thick branes. The paper "Bulk matters on symmetric and asymmetric de Sitter thick branes" (in cooperation with his students Yu-Xiao Liu, Zhen-Hua Zhao, Shao-Wen Wei) was published in J. Cosmol. Astropart. Phys..

Focusing on the localization and resonance spectrum of fermions on a one-scalar-generated de Sitter thick brane, Duan published "Fermion localization and resonances on a de Sitter thick brane" (in cooperation with his students Yu-Xiao Liu, Jie Yang, Zhen-Hua Zhao, Chun-E Fu) in Phys. Rev. D.

Duan studied the possibility of localizing various matter fields on a bent AdS_4 (dS_4) thick brane in AdS_5 and his paper "Localization of matters on the bent brane in AdS_5 bulk" (in cooperation with his student Jun Liang) was published in Phys. Lett. B.

When CCTV's famous host Junyi Shui presided over the Centennial Anniversary Evening Party of Lanzhou University, he interviewed Duan. Duan said, Lanzhou University has cultivated many doctoral tutors, deans, headmasters and academicians, and he hoped the students of Lanzhou University would win the Nobel Prize in the future.

At the end of the year, the reporter of China Education Television interviewed Duan.

- 2010 (Age 83)

In 2010, Duan carried out the study on the evolution of multi-vortices in weak coupled ABJM theory by employing the ϕ-mapping topological current theory. His paper "Multi-vortices evolution in weak coupled ABJM theory" (in cooperation with his students Jie Yang, Yang Li and Yu-Xiao Liu) was published in Int. J. Mod. Phys. A.

In March, Duan was interviewed by the reporter for "South Reviews" at home.

In May, Han-Song Wang, Secretary of the Party Committee of Lanzhou University contacted Tianjin Ophthalmic Center to arrange cataract surgery for Duan. After the surgery, academician Mo-Lin Ge invited Duan to recuperate in Nankai University. They had academic exchanges at the Institute of Mathematics in Nankai University after his rehabilitation.

On 1 July, 2011, the famous Korean physicist Prof. Yong-Min Cho paid Duan a special visit.

- 2012 (Age 85)

In 2012, After investigating the inner topological structure of the vortex lines in a ferromagnetic spin-triplet superconductor, Duan published his paper "Vortex lines in a ferromagnetic spin-triplet superconductor" (in cooperation with his students Li Zhao and Jie Yang) in Chin. Phys. B.

- 2013 (Age 86)

In May 2013, Lanzhou University interviewed Duan, and compiled the interview "My life in Science" into "Cuiying Memory Project" series: "My Lanzhou University: Personal Interviews (Part 1)".

In November, Duan delivered a lecture on the new physical effects entitled "Higgs particles and the mechanism for generating mass of element particles" to the teachers and students of Lanzhou University.

- 2015 (Age 88)

In December 2015, Duan's Quantum Field Theory was published by Beijing Higher Education Press. The book consists of lectures from Duan's long-term teaching of quantum field theory in Lanzhou University, and it is one of the earliest lecture notes on quantum field theory in China.

After returning from Moscow State University in 1957, Duan prepared the lectures on quantum field theory, and began to teach this course at Lanzhou University. He also compiled textbooks and offered courses for group theory and general relativity for undergraduates. The characteristic tradition has continued ever since. By establishing the conservation laws and conserved quantities of various fields under the corresponding transformations based on the generalized conservation theorems I and II discovered by himself, Duan simplified the content of the quantum field theory and made it more concise and comprehensible.

- 2016 (Age 89)

  Mathematics and Humanities (Vol. 21): Mathematics Herb Garden contained the article "Researching without drifting, innovating with natural Tao: Duan's academic life" (Written by Ji-Rong Ren).

  In order to learn Duan's rigorous scientific spirit and precious teaching experience, inherit the tradition of theoretical physics in Lanzhou University, and promote the development of frontier disciplines of theoretical physics further, the School of Physical Science and Technology from 17 July to 19 July at Lanzhou University held a symposium on frontiers in theoretical physics as well as in honor of Prof. Duan's 65th anniversary of service in teaching. At the opening ceremony, the host, academician Chang-Pu Sun mentioned, "Landau said that Prof. Yi-Shi Duan is the most intelligent Chinese youth", "Prof. Chen-Ning Yang said that the gauge field theory can also be made in the ravine." Academician Mo-Lin Ge said in the report that his pioneering and understanding of scientific research grew up from the root of Prof. Duan. Prof. Fan Wang of Nanjing University introduced Prof. Duan's work on magnetic monopole, and he remarked that a professor from England first mentioned the phrase of "Duan School" and cited four papers of Prof. Duan about gauge potential decomposition between 1979–2002. Prof. Rong-Gen Cai from Institute of Theoretical Physics, Chinese Academy of Sciences (ITP-CAS), commented that Prof. Duan's law of conservation of energy and momentum was included in Prof. Liao Liu's textbook "general theory of relativity". Prof. Duan himself spoke at the opening ceremony, he encouraged everyone: "indigo blue is extract from the indigo plant, but is bluer than the plant it comes from", and he hoped that all teachers and students work harder in the future and make more contributions to our country.

Appendix B

段一士先生年谱

- 1927—1934（出生—7岁）

    段一士先生1927年7月17日出生于北京，祖籍四川武胜县。父亲段毓灵是建筑工程师，1921年9月毕业于北京大学土木学系。1949 年南京解放，父亲受刘伯承元帅邀请，随军参加中国人民解放军第二野战军西南服务团，在前线为解放军解放四川铺路架桥。母亲刘月卿是小学教师。段一士先生小时候受到父母的良好教育和爱国主义熏陶。

- 1934.9—1937.8（7岁—10岁）

    段一士先生在北京育英小学学习，成绩优秀。小学时，班里日本小孩欺负人，老师不允许中国小孩还手，以免被汉奸抄家。放学后，他和同学躲在墙后用弹弓打接小孩回家的日本人。

- 1937.9—1939.8（10岁—12岁）

    1937年七七卢沟桥事变，抗日战争爆发，段一士先生随父母离开北京，一路遭到日本飞机的轰炸。来到四川后，在四川成都农城小学学习。

- 1939.9—1940.8（12岁—13岁）

    段一士先生在四川资阳师范附属小学学习。小学毕业以后，他立志要当空军飞行员，保卫祖国，消灭日寇。

- 1940.9—1945.7（13岁—18岁）

    段一士先生小学毕业后，考入位于四川灌县空军幼年学校，梦想学成后驾机轰炸东京。他从小热爱物理和数学。这时为了躲避日本飞机轰炸，成都的一些大学教授转到四川灌县空军幼年学校任教，这使得段一士先生在初中阶段就有机会学习微积分，以此为基础，他进一步学习了理论力学和原子物理。在数学课上，他知道了古希腊三大几何难题之一："一角三等分"，该难题在机械制图上形成了严重的困难。那时西方用T式仪解决此问题，但这种仪器使用时既麻烦又不完全精确。段一士先生发明了一种易于制造、使用便捷准确的仪器。解放后，东北某厂采用这种仪器。（注：摘自新华日报1951年10月13日第二版。）

- 1945.8—1946.8（18岁—19岁）

段一士先生在空军幼年学校毕业后升入空军军官学校学习飞行。

- 1946.9—1947.2（19岁—20岁）

1946年，段一士先生升入杭州笕桥空军军官学校学习，并学会驾机单飞。由于抗战已经胜利，抱着学习物理的强烈愿望，他主动离开航校，准备考大学。

- 1947—1951（20岁—24岁）

1947年，段一士先生以优秀成绩考入南京大学（前金陵大学）物理系学习。

1948年，段一士先生选修了钢琴课，从此弹钢琴成为他学习、工作后的休息放松方式之一。1949年南京解放，由于段一士先生的父亲参加中国人民解放军第二野战军西南服务团，陈毅元帅受刘伯承元帅的委托找到段一士先生，并说"你父亲协助解放西南，部队实行供给制，今后你要照顾好母亲，要自己挣学费和生活费，有什么事情可以找我。"于是，段一士先生一边卖报谋生，一边勤奋学习，并自学了广义相对论和量子力学。

1951年，段一士先生的朋友将他的一角三等分仪器的设计图表寄给东北某仪器制造厂。不久，工厂回信，他的一角三等分仪器经过工厂试验，结果大大改进和提高了生产效率，工厂立刻采取了。段一士先生在电学与流体力学上创造了复变函数由U求V的新方法，用一个简明的公式简化了复杂的积分运算。段一士先生在学习上的成就，受到中国科学院的重视，并受邀出席第十五届全国学代会。（注：摘自新华日报1951年10月13日第二版。）

1951年，在吴汝麟教授的指导下，段一士先生完成了大学毕业论文《电子游子在非均匀电场与磁场中运动之研究》。利用洛伦兹力公式和费马定理，推导了带电粒子（电子游子）在非均匀电磁场中运动的微分方程，该方程是研究带电粒子在非均匀电磁场中运动的理论基础。

1951年6月，段一士先生以优异成绩完成大学学业。

- 1951—1953（24岁—26岁）

1951年，由于学习成绩优秀，段一士先生留南京大学物理系任助教。不久，他参加了新中国第一批派遣留苏学生考试，以优秀成绩被录取。之前他还考取了著名科学家钱三强先生的研究生，但钱先生鼓励他赴前苏联留学。

- 1953—1956（26岁—29岁）

1953年，出国留学前，毛主席接见段一士先生等留学生，并鼓励留学生们为祖国多学知识，周总理自己掏钱请留学生们吃臊子面。

同年，段一士先生赴莫斯科大学留学，导师为广义相对论和空气动力学专家施乐可夫（M.F. Shirokov）教授。他还师从波戈留波夫（N.N. Bogolyubov）教授、诺贝尔奖获得者朗道（L.D. Landau）院士和伊凡宁柯（D. Ivanenko）教授，曾作为朗道院士的助教帮助查阅英文资料。朗道院士让段一士先生将学术论文编入他著名的理论物理丛书中。他在莫斯科大学进行了粒子物理、广义相对论和量子场论等领域前沿问题的研究。研究生后期，导师安排段一士先生兼学炸药爆轰理论、火箭内弹道学和导弹自动控制。

1954年，段一士先生刚到莫斯科大学攻读研究生不久，利用熟练掌握的广义相对论理论，开展了引力场中作为点粒子的电子质量问题研究，给出了引力场和电磁场耦合系统一般性的非奇异解。得到了基本粒子质量的一般性公式，并考虑了基本粒子半径与质量的可能关系。该工作是将Shirokov的工作做了进一步的推广。文章发表于前苏联JETP杂志。这是段一士先生试图通过研究基本粒子与引力场的耦合来寻找解决点粒子发散问题的新途径。

1955年，段一士先生被评为莫斯科大学优秀研究生。

1956年，段一士先生在广义相对论框架下研究了标量介子场的质量和自能问题，给出了有限自能方案，并得到了介子间推广的Yukawa相互作用。这篇工作被翻译成英文，并发表于英文版的JETP。

在莫斯科大学攻读研究生期间，段一士先生作为翻译接待和陪同钱学森教授、周培源教授、北京大学副校长江隆基等对莫斯科大学的访问。此外，还受组织委托接待了在苏联莫斯科动力学院学习并到莫斯科大学参观的李鹏（后任国务院总理）。这次和江隆基副校长的"一面之缘，却结下一世之好"。后来，江校长和段一士先生都来到兰州大学工作。江校长欣赏这位年轻人的专业水平，喜欢他的率直和幽默。

- 1956—1957（29岁—30岁）

1956年，段一士先生研究了标量场、赝标量场和矢量场的广义协变方程，并考虑了这些场之间的广义协变的相互作用。这部分工作被写入博士论文。1956年7月获得副博士学位。

　　同年9月，段一士先生研究生毕业以后，赴前苏联Dubna联合核子研究所工作并任中级研究员，研究粒子物理和量子场论。在此期间，他结识了我国著名的科学家王淦昌先生（时任杜布纳研究所的副院长）。杜布纳研究所的科学研究气氛很浓厚，他们经常开展自由讨论，学习国际上核子科学研究的最新成果。

　　1957年，段一士先生研究了基本粒子的广义协变理论，将诺特（E. Noether）定理推广到包含引力的广义协变形式，得到了广义协变守恒定理。在特殊变换下该守恒定理给出通常的能量、动量、角动量和电荷守恒。这个工作发表于前苏联联合核子研究所的预印本，编号为 JINR Preprint-1957-P-65，在学术界公开交流。

　　1957年7月，段一士先生向前苏联JETP杂志投寄论文《任意自旋场的广义协变方程》。这篇论文1958年在该杂志上发表，同年被翻译成英文。段一士先生首先采用半度规方法将平直时空中的任意自旋场方程推广到弯曲时空情况，在此基础上给出了任意自旋的广义协变场方程。这是国际上最早开展的引力规范的研究之一，是引力规范理论的基础。该文受到前苏联朗道院士的重视。1982年美国N.D. Birrel所著的"Quantum field in curved space"将此论文列为主要参考文献之一。

　　1957年夏季，段一士先生与在莫斯科苏联科学院物理化学研究所攻读硕士学位的黄友梅女士结婚。黄先生1953年毕业于南京大学化学系，1960年获前苏联科学院物理化学研究所副博士学位。后来，她任中国科学院兰州化学物理研究所研究员，曾担任多届研究室主任。黄先生的父亲黄瑞采是我国著名的土壤学家，南京农业大学终身教授，母亲林爱华操理家务。

- 1957—1958（30岁—31岁）

　　1957年，段一士先生服从国家安排，从前苏联回国到兰州大学任教。他创立并立志建好兰州大学理论物理专业，一直在粒子物理、广义相对论和拓扑场论等理论物理前沿进行科研工作，以此作为终身事业。

　　1958年，研究了核子间的相互作用，严格推导了两核子系统在束缚态情况下的格林函数积分方程和Bethe–Salpeter方程。相关研究工作发表于《兰州大学学报》。

　　1958年"大跃进"，为了能让学生继续学习物理学知识，段一士先生提出研制模拟计算机的计划，坚持带领学生学习电子学、电磁学和量子力学等课程。1958年8月1日，我国第一台电子计算机"103"在清华大学诞生。四个月之后，段一士先生和邝宇平等人研制的理论物理用大型模拟式电子计算机获得成功。 （注：这台模拟式计算机机柜长5米，高2米，装有200个电子管、6000个电阻及其他一些精密仪器，经费仅1万元。该计算机利用电子网络线路来模拟微分方程的近似差分方程式，用以求解各种类型的微分方程和积分微分方程，误差在1%左右。关于求解量子力学中薛丁格方程的设计方案的论文"解薛丁格方程模拟式电子计算机的几个方案"发表于《兰州大学学报》。省委第一书记等人于1959 年1月4日到兰州大学参观了这台计算机。）

- 1959（32岁）

三年自然灾害期间（1959—1961）， 在生活极端困难的情况下，段一 士先生依然坚持开展科学研究，包括基本粒子的广义协变理论、基于双时格林函数的铁磁性质和海森堡非线性场论等，取得了丰硕的成果。此外，还给学生讲授了基本粒子理论（1959—1960）、量子场论（1960—1961）、群论（1960—1961）、重整化理论（1960）等几门重要课程，并作了粒子理论中CPT定理讲座（1960）。

1959年，江隆基校长、徐躬耦教授和崔乃夫教务处长考虑到段一士先生在广义相对论、规范场和粒子物理方面的学术专长，经过讨论以后，决定由他担任理论物理教研室主任并指导研究生。于是，段一士先生成为兰州大学最年轻的教研室主任和最年轻的研究生导师。

- 1960（33岁）

1960年，段一士先生研究了基本粒子相互作用的手征对称性问题，探讨了手征对称与质量倒逆变换之间的内在联系。由此确定不同场之间的可能耦合形式，包括四费米子弱作用形式，电磁场与旋量场耦合形式，以及介子和核子间相互作用形式等。论文"质量倒逆变换和基本粒子的相互作用"发表于《兰州大学学报》。

1960年，段一士先生研究了广义相对论中的 Duffin–Kemmer 理论，从而导出在半度规表象中自旋为0和1的基本质点的广义协变方程式，并比较了不同玻色场和费米场与引力场耦合的不同形式。论文"广义相对论中的Duffin–Kemmer理论"发表于《兰州大学学报》（与学生邹鹏程合作）。

1960年，段一士先生研究了赝标量场的广义协变方程，得到了旋量场与赝标量场的广义协变相互作用拉格朗日函数，探讨了引力场对核子结构的影响。论文"赝标量场的广义协变方程"发表于《兰州大学学报》（与学生张敬业合作）。

- 1961（34岁）

    1961年，段一士先生利用双时格林函数方法研究了铁磁共振问题，求出了一般温度下的铁磁共振频率，在考虑外磁场足够强并取一些特定参数的情况下，得到了与实验符合的Oguchi铁磁共振频率公式。论文"关于铁磁共振理论（磁晶各向异性对铁磁共振频率的影响）"发表于《兰州大学学报》（与学生张敬业合作），这是国内利用双时格林函数方法开展凝聚态物理研究的早期文献之一。

    1961年，段一士先生开展了海森堡非线性场论的研究，讨论了基本粒子广义协变理论与海森堡理论之间的关系，对海森堡所提出的统一场理论从广义相对论的角度提供了理论根据。论文"基本粒子广义协变理论和海森堡理论"发表于《兰州大学学报》（与学生张敬业合作）。

    1961年，江隆基校长为落实教育部60条实施细则，加强基础课教学，要求由骨干教师承担基础课教学。这一年物理系普通物理课程（包括力学、分子物理、电磁学、原子物理和光学）的力学部分由段一士先生讲授。

    同年9月，段一士先生的大女儿段晓宁在南京鼓楼医院出生。

- 1962（35岁）

    1962年，段一士先生研究了广义相对论中的福克（V.A. Fock）谐和坐标条件。利用半度规表象的数学工具对福克的谐和坐标条件进行了详细的讨论，证明了该条件对惯性参考系具有充分性但却不具有唯一性。论文"关于广义相对论中福克的谐和坐标条件"发表于《物理学报》（与学生张敬业合作）。

    1962年，段一士先生对π介子衰变问题进行了研究。直接分析了Lehmann格林函数的解析性质，建立起了与前人不同的色散关系，得到了π介子衰变振幅的表达式。论文"π介子衰变"发表于《兰州大学学报》（与学生马中骐合作）。

    同年，段一士先生被评为兰州大学优秀教师。

- 1963（36岁）

    1963年，段一士先生阐明了在广义相对论中采用半度规表象的必要性，利用半度规数学工具对广义相对论中的守恒定律问题作了详细的讨论。从广义位移变换出发，得到了广义协变的能量动量守恒定律，具有严格的对黎曼坐标的广义协变性。这一新的协变形式的守恒定律是广义相对论中的一个重要发展，克服了爱因斯坦、朗道等人所提出的引力理论中的能量动量守恒定律只适用于伽利略坐标和其他方面存在的缺点。由于Møller的能量表达式不能解释Bondi平面波异于零的能量密度，该理论同样成功地克服了Møller 理论中存在的问题并能得到合理的引力辐射公式。论文"广义相对论中的能量动量守恒定律"发表于《物理学报》（与学生张敬业合作）。该成果1981 年获甘肃省高校科技进步一等奖，后来在刘辽教授著的《广义相对论》一书中做了专门的介绍，并称为"能量－动量守恒定律的段一士表述"。 （注：该工作与1957年得到的基本粒子的广义协变理论不同。1957 年的论文主要讨论引力场中基本粒子的守恒量问题。而在这里是采用了半度规表象，试图给出局域守恒的引力场的能量动量。）

- 1964（37岁）

    1964年，段一士先生再次被评为兰州大学优秀教师。

- 1965—1967（38岁—40岁）

    1965年，段一士先生通过强相互作用K-π中间态，利用色散关系理论研究了K介子衰变问题，计算了不同衰变道的分支比，得到了与实验一致的结果。论文"K_3介子衰变"发表于《物理学报》（与学生朱重远合作）。

    段一士先生研究了π-π共振态问题。从解析性和幺正性出发，对具有共振行为的π-π散射分波振幅的形式做了普遍的讨论。并计算了p波和d波的共振参数。两篇论文"关于π-π共振态"和"π-π共振态"均发表于《物理学报》（与学生葛墨林合作）。

- 1968（41岁）

    1968年12月，段一士先生的小女儿段晓静在南京军区医院出生。

- 1969—1973（42岁—46岁）

    1969年3月，珍宝岛事件后，段一士先生开展了国防研究。在兰州大学物理系成立了技术革新组，研究并领导完成了反坦克火箭非接触引信军工项目。参加这一项目研发的有聂崇礼、程麟生、齐耀刚、葛墨林、康宁、张承理、张春光、罗长荣等人。在珍宝岛自卫反击战时，我国的火箭弹都是接触引信，可前苏联坦克的装甲是斜面，我军的火箭弹一接触就跳弹。段一士先生用小晶体管做非接触引信，锡箔纸模拟坦克的装甲，小灯泡亮或不亮判断是否引爆，验证了电容感应式非接触式电子引信原理。以此为基础承担了军委炮兵解决防止反坦克火箭跳弹现象的科研任务。经过多次艰辛的试验，克服了跳弹现象。通过计算风速、海拔、气压、温度等因素对弹道的修正，极大提高了命中率。研制完成后，亲自去北京南口做打靶实验，实弹验证了该非接触式电子引信反坦克火箭弹确实具有极强的穿甲杀伤力。段一士先生曾回忆："他们（工作人员）拿白石灰画了个圆圈，我一炮就打到了这个圈里，立刻打穿了钢板，把里面的狗打死了。"叶剑英元帅亲自观摩打靶实验并高度赞扬。

1972年，程开甲院士途经兰州，到兰州大学看望段一士先生并进行了学术交流。（注：1951年，程开甲院士在南京大学物理系任副教授，段一士先生任助教。）

- 1974（47岁）

1974年，段一士先生在《兰州大学学报》发表论文"关于规范场理论"（与学生葛墨林合作）。从规范协变微商的普遍形式出发，得到了任意场函数ψ"平行位移"算符的积分表达式，并以此为基础结合李群和李代数的结构对规范场理论进行了较深入的讨论。对发展规范不变的物理理论提供了新思路。

段一士先生在《兰州大学学报》发表论文"电子—核子深度非弹的标度性（单内线光锥近似）"（与学生葛墨林合作）。用单内线光锥近似理论研究 电子—核子深度非弹所呈现的标度性。该理论具有严格的规范不变性，并易于 处理束缚态问题。

- 1975（48岁）

1975年，段一士先生系统研究了磁单极问题，在《兰州大学学报》发表论文"磁单极子与黑格斯场"（与学生葛墨林合作）。指出 G. 't Hooft关于磁单极子的理论的实质，是由于规范势可以部分地由 Higgs 场表达，并且磁单极子的存在密切与Higgs场在同位旋空间的拓扑性质有关，而不必由规范场方程求解。这一工作对于建立非阿贝尔规范理论中的磁单极理论具有重要意义。该理论与Arafune的类似理论几乎同时提出，但是观点更清晰和明确。这是国际上首次引入规范势可分解的思想。

段一士先生在《西北大学学报》发表论文"不可易规范场的规范不变分量的对偶荷"（与葛墨林和侯伯宇合作）。

- 1976（49岁）

1976年，段一士先生在《科学通报》发表论文"磁单极子与黑格斯场"（与葛墨林和侯伯宇合作），进一步阐述了磁单极与规范场的联系。

段一士先生在《物理学报》发表论文"不可易规范场的对偶荷"（与葛 墨林和侯伯宇合作）。研究了不可易SU(2)规范场的各种规范不变物理量—电 荷、对偶荷（磁荷）、电磁场以及有质量矢量粒子场的表达式与关系，特别是对 偶荷（磁荷）与电荷同位旋空间的大范围拓扑性质的关系。

- 1977（50岁）

  1977年，段一士先生在《兰州大学学报》发表论文"SU(2)规范理论与磁单极"和"磁单极理论在同位旋空间的几何表述"（与学生葛墨林合作）。

  同年夏天，著名理论物理学家杨振宁先生访问兰州大学，在听完段一士先生关于规范场和磁单极的报告后，称赞其理论"妙，真妙，妙极了！"。

  研究成果"反坦克火箭非接触电子引信"获军委炮兵国防科研一等奖。

- 1978（51岁）

  1978年，段一士先生在《高能物理与核物理》发表论文"非阿贝尔规范势分解的几何表述"（与学生葛墨林合作）。从无扰嘉当几何的观点出发讨论了非阿贝尔规范势的分解。借助γ-矩阵和洛仑兹群生成元的对易关系，得到相应的"同位旋空间"的规范势的表达式。

  1978年，段一士先生在《中国科学》发表论文"The decomposition and reduction of gauge field and dual charged solution of Abelianizable Field"（与侯伯宇和葛墨林合作），系统介绍了规范势可分解的思想。进一步将规范势可分解的观点和拓扑流守恒定律推广到半单纯群规范理论的范围，并提出不同Higgs机制使规范粒子获得质量的新理论。该工作受到国际同行的广泛关注，段一士先生赴美国讲学时，在伊利诺里州立大学、密苏里大学等五所高校作了与此相关的学术报告。

  研究成果"磁单极理论的研究"获1978年度全国科学大会自然科学重大成果奖，"非接触电子引信"获1978年度全国科学大会国防科研重大成果奖。

- 1979（52岁）

  1979年，段一士先生在《中国科学》发表论文"SU(2)规范理论与N个磁单极运动体系的电动力学"（与学生葛墨林合作）。阐明了关于规范势可分解和具有内部结构的基本观点，并导出了SU(2)规范理论中N个磁单极运动体系存在时的基本方程，给出磁荷量子化条件。论文后来成为该领域研究的经典文献。国际上称段一士先生与合作者提出的规范势可分解和具有内部结构的理论为"Duan–Ge–Cho decomposition"。 比国际著名物理学家L. Faddeev的工作早二十年（参看：L. Faddeev and A. Niemi, Phys. Rev. Lett. 82, 1624 (1999)）。规范势可分解的理论广泛应用于宇宙学、凝聚态物理、粒子物理、微分几何与拓扑等领域。

应邀并与葛墨林合作在《中山大学学报》发表综述论文"关于规范场理论"。在这一评述性报告中，从规范场理论的发展概况，论述研究规范场的重要性和必要性。希望促进我国高能物理学界对规范场理论的研究予以足够的重视。

在《自然杂志》发表关于广义相对论的评述文章"广义相对论及其实验基础"。

段一士先生在《中国科学》发表论文"规范场的分解约化及可Abel化场对偶荷解"（与侯伯宇和葛墨林合作）。根据荷算符的同位旋方向，将规范势与其场强规范协变地分解成以该荷及其对偶荷为源的场和带该荷的矢量粒子场。

建立学位制后，段一士先生任我国首批国务院学位委员会特批的博士生导师。

他根据研究生的基础及当时理论物理学科发展的前沿问题，从1979年到2002年段一士先生给研究生讲授了以下课程：广义相对论、量子场论、粒子物理、李群李代数、规范场论、引力规范理论、弱电统一理论、量子色动力学、近代微分几何及拓扑、拓扑场论、SU(N)规范势分解、非对易几何、抽象代数、现代宇宙论、超对称和超弦、时空缺陷理论、拓扑流理论及其应用、复流形与Calabi–Yau流形等。

任兰州大学物理系主任（1979年—1994年）。

任高能物理学会首届常务理事、中国引力和相对论天体物理学会首届理事、任甘肃省物理学会理事长。

7月，应邀赴瑞士日内瓦参加国际高能物理会议，并作了关于规范场理论的学术报告。会后访问了欧洲核子研究中心（CERN）。

12月，段一士先生受教育部委托，在兰州大学举办了全国高校理论物理专业的广义相对论讲习班，全国六十余位教师参加了讲习班。

- 1980（53岁）

1980年，段一士先生参加在广州从化举办的全国理论粒子物理会议，在会议文集上发表论文"Exceptional algebra and topological properties of isospace of constant crvature"（与学生葛墨林合作）。

在《高能物理与核物理》发表论文"常曲率空间与SU(2)对偶荷"（与学生葛墨林合作）。讨论了四维常曲率空间拓扑性质与SU(2)对偶荷间的联系，所得的结论是三维同位旋空间拓扑荷的自然推广。

在《高能物理与核物理》发表论文"SU(2)规范理论中的规范真空效

应"（与学生葛墨林合作）。

研究成果"反坦克火箭电子引信"获中国人民解放军科技成果二等奖。

应邀赴美国Mossouri（密苏里）大学讲学，内容有规范场理论等。

1980年—1990年，被聘任为国家教委物理学科教学指导委员会委员。

80年代初，中国举行的中美联合培养物理类研究生（CUSPEA）招生考试中，段一士先生与钱伯初教授、汪志诚教授等共同努力培养，使得兰州大学物理系学生成绩多次名列前茅，引起了国内外同行对兰州大学的重视。

- 1981（54岁）

1981年，研究成果"广义相对论中的守恒定律"获甘肃省高校科技成果一等奖。

兰州大学成立理论物理博士点，段一士先生任博士点负责人。

任甘肃省自然辩证法研究会理事长。

- 1982（55岁）

1982年，段一士先生在《理论物理通讯》上发表论文"Infrared behavior of the running coupling constant in QCD"（与学生蔡善皋合作）。基于Slavnov–Taylor恒等式，利用胶子传播子的Schwinger–Dyson方程研究了QCD跑动耦合常数的红外行为。

将以前国防项目的部分成果整理发表于《兵工学报》，题为"电容感应式非触发近炸引信理论研究"（与张承理、张重光合作）。

9月—12月，时任甘肃省团委书记的胡锦涛总书记邀请段一士先生给中学生作了数次科普报告。

1982年，与兄弟单位（复旦大学、中山大学、西北大学、中国科学院高能物理研究所、中国科学院理论物理研究所等）合作完成的研究成果"经典规范场理论研究"获国家自然科学三等奖。（注：段一士先生与葛墨林的贡献是：提出规范势分解和具有内部结构的理论，以及半单纯群规范场理论中拓扑流的一般理论。在理论物理中有关陈类和欧拉类等拓扑问题中得到应用，并进一步发展了拓扑流和拓扑粒子理论。）

段一士先生任在上海举办的"第三届国际格拉斯曼广义相对论会议（Marcel Grossmann Meetings）"组织委员，并作了题为"广义相对论中的能量动量守恒定律"的学术报告。

- 1983（56岁）

1983年，段一士先生在斯坦福直线加速器中心（SLAC）访问期间，完成论文"The running coupling constant in QCD"（预印本号SLAC-PUB-3261）。利用重整化群理论讨论了QCD跑动耦合常数的性质，给出了精确的跑动耦合常数与转移动量之间的关系。还完成论文"Harmonic maps and their application to general relativity"（预印本号SLAC-PUB-3265）。提出了一种基于调和映射研究欧拉方程解的新方法，并证明在二维情况下不同调和映射对应的欧拉方程的解之间满足共形变换。这种方法还被推广用于研究广义相对论中的Ernst方程。

1983年，利用广义协变Dirac方程，计算了电子反常磁矩的引力修正，完成论文"A quantum effect in general relativity"（与学生俞重远合作。预印本号SLAC-PUB-3158）。

段一士先生在《中国科学》上发表论文"广义相对论中的能量动量守恒定律"（与学生王友棠合作）。利用能量动量守恒定律给出了四极矩引力辐射公式，此公式被脉冲双星中的PSR1913+16的引力辐射衰减效应的数据证实是正确的。论文同时在中英文版《中国科学》发表。

在《理论物理通讯》上发表论文"The decomposition of the gauge potential and its quantization theory"（与学生赵书城合作）。从规范势分解的角度研究了Yang–Mills场的量子化和重整化问题。

任中国物理学会第三届理事。

任中国引力和相对论天体物理学会第二届理事。

任全国理论物理教材编委会副主任委员。

1983年—1984年，段一士先生应邀赴斯坦福直线加速器中心（SLAC）做访问教授。在此期间访问了斯坦福大学，并讲授引力规范理论，赴巴尔的摩（Baltimore）参加学术会议。

- 1984（57岁）

1984年，段一士先生在斯坦福大学访问期间，首次提出并建立了ϕ-映射拓扑流理论。这是拓扑量子力学和拓扑场论的一个重要进展。论文"The structure of the topological current"在斯坦福加速器中心预印本刊印，预印本号SLAC-PUB-3301。该理论后来被称为"Duan's topological current theory"，在玻色–爱因斯坦凝聚体、固体旋错和位错、宇宙弦、超导、螺旋波等方面得到广泛应用。

在《物理学报》发表论文"调和映射理论中Euler方程的解和它在广义

相对论中的应用"。讨论了调和映射理论中Euler方程的解，分析了解之间的关系。

段一士先生首次从规范场论的角度研究固体中的旋错和位错问题，与段祝平合作完成"Gauge field theory of dislocation and disclination continuum"（预印本号SLAC-PUB-3286）和"Gauge field theory and conservation laws in elastic dislocation and disclination continuum mechanics"（预印本号SLAC-PUB-3302）。

在《兰州大学学报》发表论文"一种新的有质量规范场理论"（与学生赵书城合作）。提出一种非自发破缺的有质量规范场理论。

在《兰州大学学报》发表论文"QCD中有效耦合常数的色散关系与红外行为"（与学生杨孔庆合作）。在QCD的有效耦合常数的形式理论的基础上，利用色散关系的非微扰方法，研究了函数的色散关系，得到了QCD中的有效耦合常数的红外行为。

段一士先生被评为全国优秀教师。

任国际霍普金斯粒子理论现代问题讨论会常任组织委员。

- 1985（58岁）

1985年，段一士先生在《理论物理通讯》发表论文"Gauge theories of gravitation"（Proceedings of symposium on Yang–Mills gauge theories）。

在《兰州大学学报》发表论文"QCD中有效耦合常数的Padé逼近"（与学生杨孔庆、王文福、陈成明合作）和"量子电动力学中的有效作用荷"（与学生罗春荣合作）。

研究成果"广义相对论中的能量动量守恒定律"获国家教委科学技术进步一等奖。

任中国高能物理学会第二届理事。

应中科院研究生院邀请，在北京玉泉路中科院研究生院开设广义相对论与宇宙学课程讲习班。

赴意大利佛罗伦萨参加国际会议。

- 1986（59岁）

1986年，段一士先生在《兰州大学学报》发表论文"有质量矢量介子场的Abel规范理论"（与学生赵书城合作）。

第三次被评为兰州大学优秀教师。

赴德国玻恩参加国际会议。

- 1987（60岁）

1987.1—1989.12，段一士先生主持国家自然科学基金项目"规范场理论的进一步发展和应用"（No. 18670133）。

1987.9—1989.9，段一士先生主持高等学校博士学科点专项科研基金"超统一理论守恒定律与拓扑流"（No. 32860393）。

1987年，段一士先生在《兰州大学学报》发表论文"Higgs机制与背景场方法"（与学生赵书城合作）。

在《物理学报》发表论文"Einstein–Cartan引力理论中的广义协变守恒定律"（与学生刘继承、董学耕合作）。从广义Noether 定理出发，对Einstein–Cartan引力理论中一般拉氏量所对应的守恒定律作了普遍的讨论，并由广义位移变换得到了包含挠率的一般拉氏密度所对应的广义协变能量动量守恒定律，论述了超势存在的必然性。

6月17号，在兰州大学成功举办了第11届国际约翰霍普金斯粒子物理前沿问题研讨会（The 11th Johns Hopkins workshop on current problem in particle theory）。会议上段一士先生作大会报告"ϕ-mapping method, topological current, and string theory"。详细阐述了ϕ-映射拓扑流理论。参加会议的有：欧洲核子研究中心的韦斯特教授（Peter West）、哈佛大学的斯多明格教授（A. Strominger）、霍普金斯大学多莫科西教授（G. Domokos）等。时任省委书记李子奇、省长贾志杰到会祝贺。段一士先生主持本次研讨会并赋诗："万物始于一，常道法自然。深观宇宙妙，穷极粒子玄。"

- 1988（61岁）

1988年，段一士先生、G. Domokos教授和S. Kovesi-Domokos教授主编的会议文集"Frontier in particle physics proceedings, 11th Johns Hopkins workshop on current problems in particle theory"由新加坡世界科技出版社出版。

段一士先生任中国物理学会第四届理事。

任第六、七届甘肃政协委员会委员（1988—1998）。

赴美国巴尔的摩（Baltimore）参加国际会议。

- 1989（62岁）

1989年，任甘肃省物理学会名誉理事长（1989—2007）。

段一士先生再次被评为全国优秀教师。

赴意大利佛罗伦萨参加国际会议。

- 1990（63岁）

1990.1—1992.12，段一士先生主持国家自然科学基金项目"拓扑场论与拓扑流理论"（No. 18975017）。

1990.9—1992.9，段一士先生主持高等学校博士学科点专项科研基金"拓扑规范场论和拓扑引力理论"（No. 9073004）。

1990年，段一士先生开展了位错拓扑结构的研究，建立了位错与旋错的规范场理论，引起了国际同行的关注。论文"Topological structure of dislocation in the gauge field theory of dislocations and disclinations continuum"发表于Int. J. Engng. Sci.（与学生张胜利合作）。

段一士先生被评为全国高等学校先进科技工作者。

- 1991（64岁）

1991年，获得"国务院表彰的发展我国高等教育突出贡献奖"，享受国务院特殊津贴。

完成了位错拓扑流结构的研究工作，发展了拓扑流理论。论文"Topological current structure of dislocations in the 4-dimensional gauge field theory of dislocation and disclination continuum"发表于Int. J. Engng. Sci.（与学生张胜利合作）。

在《兰州大学学报》发表论文"弯曲时空共形（Weyl）规范理论及引力荷、惯性因子和宇宙常数"（与学生赵书城合作）和"推广的调和映射方程和它的孤子解"。

5月，参加中国科学技术协会第四次全国代表大会。

8月，赴美国巴尔的摩参加国际会议。

- 1992（65岁）

1992.1—1993.12，段一士先生主持国家自然科学基金项目"规范场理论和拓扑理论在固体物理中的应用"（No. 19172030）。

1992.1—1994.9，段一士先生主持高等学校博士学科点专项科研基金"拓扑流理论和拓扑场论在材料科学中的应用"。

1992年，开展了量子群方面的研究工作，论文"The quantum group $SU_q(2)$ and the q-analogue of angular momenta in classical mechanics"发表于J. Phys. A（与学生张胜利合作）。

研究了广义相对论中的一类新的坐标条件，论文"A new coordinate

condition in general relativity"发表于Gen. Rel. Grav.（与学生张胜利合作）。

开展了共形场论的研究工作，论文"Conformal (Weyl) invariance and Higgs mechanism"发表于IL Nuovo Cimento（与学生赵书城合作）。

研究了旋错的拓扑流理论，论文"Topological current structure of disclinations in the 4-dimensional gauge field theory of dislocation and disclination continuum"，发表于Int. J. Engng. Sci.（与学生张胜利合作）。

（注：1984—1992年，段一士先生及其合作者在Int. J. Engng. Sci.及其它杂志上发表的几篇论文，将规范场理论、活动标架理论以及规范势分解和具有内部结构的理论应用于固体缺陷理论，提出了位错和旋错连续统的严格非线性规范理论，并提出用平行场和拓扑流理论建立位错和旋错的拓扑理论。由扰率和曲率的表达式直接得到位错和旋错的局部和整体拓扑性质的严格描述。美国力学和应用数学权威D.G.B. Edelen教授认为："这一研究工作是该领域的一个重要发展，它严格地从规范理论本身直接地统一了固体缺陷的几何与拓扑理论。段教授及其合作者在Int. J. Engng. Sci.上发表的有关文章将可成为该领域中最重要及可持久保留下去的论文"，并在其专著《Gauge theory and defects in solids》中把该工作作为重要参考文献。）

- 1993（66岁）

1993年，段一士先生研究了Gauss–Bonnet–Chern定理的拓扑结构和拓扑流，论文"Topological structure of Gauss–Bonnet–Chern density and its topological current"发表于J. Math. Phys.（与学生孟新河合作）。

11月，段一士先生发起物理学新效应讲座，目的是为了进一步提高兰州大学物理研究与教学人才的学术水平，使物理专业和相关学科的高年级学生和研究生对物理学的最新发展有所了解，活跃物理学师生的学术氛围，同时该讲座也为兰州地区物理学教学与科研人员的学术交流服务。其内容为：获得诺贝尔物理学奖的主要内容；对物理学发展起突出作用的新效应；对其他自然科学和技术应用起重大作用的物理学新效应。自创办以来，段一士先生共主持了八十多次讲座，他亲自做的报告有：脉冲双星引力辐射、夸克模型、Top夸克的发现、原子弹、近代宇宙学、Tau轻子的发现及其重要意义、大统一理论超弦与夸克、纳米科学与纳米技术、贫铀武器、近代宇宙学暗能量、渐进自由—量子色动力学、爱因斯坦与相对论、宇宙微波背景辐射和宇宙学、21世纪物理学前沿十大问题、Higgs粒子与产生基本粒子质量的机制等。该讲座的支持单位有：物理学新效应讲座基金会、兰州大学理论物理研究所、兰州大学物理科学与技术学院、中国科学院近代物理研究所和甘肃省物理学会等。

任兰州大学理论物理研究所所长（1993—2004）。

赴匈牙利布达佩斯参加国际会议。

- 1994（67岁）

1994年，段一士先生开展了规范势分解与时空缺陷拓扑方面的研究，论文"Gauge potential decomposition, space-time defects and Planck's constant"发表于J. Math. Phys.（与学生张胜利、冯世祥合作）。

任兰州大学物理系名誉系主任。

赴意大利佛罗伦萨参加国际会议。

- 1995（68岁）

1995.1—1997.12，段一士先生主持国家自然科学基金项目"拓扑流和拓扑荷分歧理论"（No. 19475018）。

1995.9—1997.9，段一士先生主持高等学校博士学科点专项科研基金"物质结构和场的拓扑与拓扑流理论"（No. 9573007）。

1995年，段一士先生在《物理学报》发表论文"广义相对论中广义协变的角动量守恒定律"（与学生冯世祥合作）。利用Einstein引力—物质系统的局域Lorentz不变性获得了广义协变的角动量守恒定律，并用两个典型的实例说明 了此定律的合理性。

1995年，段一士先生在Helv. Phys. Acta.上发表论文"General decomposition theory of spin connection structure of Gauss–Bonnet–Chern density and Morse theory"，研究了Gauss–Bonnet–Chern（GBC）密度和Morse理论的自旋联络的广义分解理论（与学生李希国合作）。

赴美国巴尔的摩参加国际会议。

- 1996（69岁）

1996年，论文"Conservative angular momentum as SU(2) charges in complex gravity"发表于Commun. Theor. Phys.，"About the energy of the universe"发表于Chin. Phys. Lett.（与学生冯世祥合作）。

赴德国海德堡参加国际会议。

- 1997（70岁）

1997年，段一士先生研究了时空缺陷的拓扑流理论，建立了完整的ϕ-

映射拓扑流分岔理论。论文"The origin and bifurcation of the space-time defects in the early universe"发表于Gen. Rel. Grav.（与学生杨国宏、姜颖合作）。

7月，国际国内同行齐聚兰州，参加段一士先生主持的第21届国际约翰霍普金斯粒子物理前沿问题研讨会，共同探讨粒子和场论中的拓扑。段一士先生赋诗一首："拓扑规范聚众仙，时逢从心所欲年。为学不随流俗转，超越创新常道间。"

- 1998（71岁）

1998.1—2000.12，段一士先生主持国家自然科学基金项目"规范场内部结构与新拓扑不变量"（No. 19775021）。

1998年，段一士先生研究了凝聚态系统中的拓扑问题，开辟了拓扑物态研究的新领域。论文"Topological structure of the London equation"发表于Phys. Rev. B（与学生张宏、李晟合作）。

开展了Gauss–Bonnet–Chern拓扑流分岔的研究以及与Morse理论的联系。论文"The bifurcation theory of the Gauss–Bonnet–Chern topological current and Morse function"发表于Nucl. Phys. B（与学生李晟、杨国宏合作）。

应邀在吴大猷教授九十寿辰纪念文集（Jingshin theoretical physics symposium in honor of professor Ta-You Wu）上发表论文"Decomposition theory of spin connection, topological structure of Gauss–Bonnet–Chern topological current and Morse theory"（与学生李晟合作）。

7月，与来兰州大学访问的国际著名数学家丘成桐教授讨论了两人感兴趣的数学问题。

赴瑞典哥德堡（Goteborg）参加国际会议。

- 1999（72岁）

1999.1—2001.12，段一士先生主持高等学校博士学科点专项科研基金"规范场分解和内部结构与新拓扑不变量"（No. 98073009），此项目被评为优秀成果。

1999年，段一士先生研究了三维矢量序参量场的点缺陷，论文"Point defects of a three-dimensional vector order parameter"发表于Phys. Rev. E（与学生张宏合作）。

研究了光束中波位错的分岔条件，论文"Branch conditions for wave dislocations in light beams"发表于J. Phys. A: Math. Gen.（与学生贾

广、张宏合作）。

- 2000（73岁）

2000年，段一士先生研究了Chern–Simons 涡旋的演化。论文 "The evolution of the Chern–Simons vortices" 发表于Phys. Rev. D（与学生傅立斌合作）。

研究了任意维拓扑缺陷的拓扑性质，论文"A new topological aspect of the arbitrary dimensional topological defects" 发表于 J. Math. Phys.（与学生姜颖合作）。

2000年，赴南京参加量子力学学会，并作了题为"拓扑量子力学的新发展以及在物理学前沿中的应用"的学术报告。

- 2001（74岁）

2001.1—2004.12，段一士先生主持国家自然科学基金项目"拓扑规范场理论的新发展和在物理前沿科学中的应用"（No. 10175028）。

2001年，段一士先生研究了两分量BEC的环流条件，论文 "Circulation condition of the two-component Bose–Einstein condensate"，论文发表在Phys. Lett. A（与学生贾多杰合作）。

研究了分数量子霍尔效应中的拓扑激发，论文"Topological excitation in the fractional quantum Hall system" 论文发表于 Phys. Lett. A（与学生张鹏鸣、张宏合作）。

在《高能物理与核物理》发表论文"$SO(n)$规范势的可分解理论"，研究了$SO(n)$规范势分解问题（与学生李希国、宋建军合作）。

2001年，应邀给青海师范大学教师举办了培训班，作了粒子物理基础、拓扑物理和宇宙学等讲座。

赴意大利佛罗伦萨参加国际会议。

- 2002（75岁）

2002.1—2004.12，段一士先生主持高等学校博士学科点专项科研基金"拓扑规范场理论和拓扑量子力学的新发展和应用"（No. 20010730007）。

2002年，段一士先生在Mod. Phys. Lett. A 发表论文 "Decomposition of the $SU(N)$ connection and the Wu–Yang potential"，研究$SU(N)$联络的分解和Wu–Yang势（与学生张鹏鸣合作）。

2002年，研究了液晶的拓扑性质，论文"Topological aspects of liquid crystals"发表于Int. J. Theor. Phys.（与学生杨国宏合作）。

2002年，研究了Riemann–Cartan流形的挠率结构以及位错的拓扑性质，论文"Torsion structure in Riemann–Cartan manifold and dislocation"发表于Gen. Rel. Grav.（与学生李希国合作）。

2002年，赴北京参加国际弦物理大会。

4月24日，应邀赴南开大学与著名数学家陈省身先生深入讨论了陈先生的GBC定理和段一士先生的新见解，以及GBC定理在拓扑场论中的应用等问题。段一士先生曾证明，由规范场张量乘积在GBC定理中构成的不变量可严格表述为相应黎曼流形上矢量场的ϕ-映射拓扑荷。

10月24日，应邀赴浙江大学，参加为庆祝浙江近代物理中心成立十周年而举办的"面对二十一世纪物理的挑战"学术研讨会，并作了题为"宇宙弦与暗能量"的报告。

- **2003（76岁）**

2003年，段一士先生研究了Chern–Simons场论中的扭结拓扑结构。论文"Many knots in Chern–Simons field theory"发表于Phys. Rev. D（与学生刘鑫、傅立斌合作）。

研究了三分量自旋玻色爱因斯坦凝聚系统中的磁单极和涡旋线的拓扑性质。论文"Mermin–Ho vortices and monopoles in three-component spinor BEC"发表于J. Phys. A（与学生刘鑫、张鹏鸣合作）。

提出时空相变复标量序参量场是暗能量候选者之一，该序参量场的拓扑缺陷是宇宙弦。论文"Cosmic strings and quintessence"发表于Chin. Phys. Lett.（与学生任继荣、杨捷合作）。

获兰州大学2003年在SCI和EI及2004年在权威杂志上发表论文前三名奖励。

段一士先生76周岁，仍然躬耕讲坛。这年给研究生讲授了弱电统一理论（1月）、同伦理论（3月—5月）、Riemann–Cartan流形与挠率（6月）、 Noether守恒流与自然守恒流、ϕ-映射拓扑流理论（6月—12月）、抽象代数、李群和李代数、经典规范场理论和拓扑场论（2003年9月—2004年3月）等课程和讲座。

应邀赴湖南师范大学邀请参加关于发展理论物理学科的研讨会，并作了题为"科研、学习、教学"的报告。在多年的教学实践和科学研究中，段一士先生凝练出"言简意赅，珠联璧合，集纳新说，返璞归真"的教学之道，以及"去其糟粕，取其精华，去伪存真，止于至善"的治学之道。

- 2004（77岁）

2004年，段一士先生在JHEP发表了论文"Knotlike cosmic strings in the early universe"。通过Chern–Simons作用量，研究了早期宇宙黎曼—嘉当流形中带有扭结结构的宇宙弦的拓扑性质（与学生刘鑫合作）。

基于ϕ-映射拓扑流理论研究了声学几何中运动涡旋的横向力。论文"Transverse force on a moving vortex with the acoustic geometry"发表于Phys. Lett. A（与学生张鹏鸣、曹利明、钟承奎合作）。

获兰州大学2004年发表SCI一区论文作者第一名奖励。

段一士先生被评为兰州大学资深教授。

给研究生讲授了SO(N)规范场理论（3月—6月）、矩阵与行列式（10月）、广义相对论和引力规范理论（2004年9月—2005年6月）等课程和讲座。

应邀赴青海师范大学作了近代宇宙学、暗能量、宇宙大爆炸等讲座。

任兰州大学理论物理研究所名誉所长（2004—2016）。

科研成果"凝聚态和宇宙时空缺陷的拓扑规范场理论"获甘肃省科技进步一等奖。

科研成果"拓扑规范场论新发展及其在物理前沿科学中的应用"获教育部提名国家自然科学二等奖。

段一士先生等在国际上最早提出来规范势可分解和具有内部结构的新观点，并提出了ϕ-映射拓扑流理论，以此为基础建立了新拓扑场论和新拓扑量子力学，并创立了研究固体缺陷的拓扑规范场理论，揭示了场的几何性质与拓扑结构的内在联系。九十年代以后解决了以下重要问题：

(1) 在黎曼—嘉当流形中以挠率为基础，建立了新阿贝尔规范场论和相应的拓扑不变量，创建了固体位错和旋错新的拓扑缺陷理论；

(2) 利用黎曼—嘉当流形中由挠率张量构成的新阿贝尔规范场，研究了宇宙时空缺陷，国际上首次提出了早期宇宙中宇宙弦产生的物理机制，确定了宇宙弦的拓扑结构，提出产生宇宙弦的序参量场就是使宇宙加速膨胀的暗能量；

(3) 用提出的规范势分解和ϕ-映射拓扑流理论，在国际上首次严格证明了超导理论中的London理论，得到了超导体缺陷拓扑量子数与序参量波函数零点映射度之间的定量关系；

(4) 以建立的拓扑量子力学为基础，在国际上首次严格证明了著名物理学家朗道和费曼提出的量子力学中速度矢量场的旋度不为零的重要设想，并具有严格的奇异性，指出涡旋的拓扑量子数完全是由波函数 ϕ-映射的零点指数决定。研究了玻色-爱因斯坦凝聚中涡旋的拓扑结构；

(5) 利用提出的 ϕ-映射拓扑流理论，建立了缺陷系统拓扑分岔的新理论，并分别应用于超导、玻色-爱因斯坦凝聚及宇宙弦系统，得出了相应缺陷分岔的临界条件；

(6) 用建立的U(1)规范势分解理论和 ϕ-映射拓扑流理论严格证明了Chern–Simons流是对应弦的拓扑流，并严格证明了Chern–Simons作用量是扭结弦的拓扑不变量，比国际上著名数学家Witten的证明简明严格；

(7) 利用规范势分解和 ϕ-映射拓扑流理论建立起来的拓扑场论，还研究了螺旋波、波位错、以及太阳黑子涡旋的拓扑结构。

- 2005（78岁）

  2005.1—2007.12，段一士先生主持国家自然科学基金项目"宇宙弦、暗能量、额外维流形和p-膜的拓扑规范场论"（No. 10475034）。

  2005年，段一士先生基于Abelian-Higgs模型获得了描述自对偶涡旋的广义Bogomol'nyi方程。论文"Fermionic zero modes in self-dual vortex background"发表于Mod. Phys. Lett. A（与学生王永强、司铁岩、刘玉孝合作）。

  给研究生讲授了粒子物理与现代宇宙学（2005年9月—2006年1 月）。

  应邀赴四川大学作了爱因斯坦和相对论、科研学习教学、从粒子到宇宙、宇宙学和粒子物理的重大问题、原子弹的原理等学术报告。

- 2006（79岁）

  2006年，段一士先生利用 ϕ-映射拓扑流理论研究了带电的两凝聚体玻色系统中的涡旋线和扭结结构瞬子的拓扑性质。论文"Knotted soliton in a charged two-condensate Bose system"发表于Phys. Rev. B（与学生张欣会、刘玉孝、赵力合作）。

  研究了局域非阿贝尔拓扑张量流中的高维扭结拓扑缺陷。论文"Higher-dimensional knotlike topological defects in local non-Abelian topological tensor currents"发表于J. Math.Phys.（与学生吴绍锋、张鹏鸣合作）。

  研究了双能隙超导体中的拓扑涡旋的性质。论文"Topological vortices in a two-gap superconductor"发表于Phys. Lett. A （与学生赵力、张欣会合作）。

给研究生讲授了额外维理论（3月—5月）、复流形与Calabi–Yau流形（9月—12月）等课程。

受聘为兰州大学物理学隆基班导师，并为隆基班作了题为"热爱物理、学习与创新"和"宇宙微波背景辐射、宇宙学"的讲座。

应邀出席了由兰州大学团委办主办、物理科学与技术学院承办的"聚理兰大——院士专家论坛"第五讲，演讲题目是《21世纪物理前沿十大问题》。

应邀参加厦门大学八十五周年校庆学术活动，并作了题为"相对论一百年"的讲座。

应丘成桐教授的邀请，举办国际弦理论会议兰州卫星会议。出席会议的有北京师范大学刘辽教授、哈佛大学斯多明格教授、犹他大学吴咏时教授、中国科学技术大学闫沐霖教授等。

9月，诺贝尔奖获得者丁肇中教授来兰州大学访问期间，段一士先生与他进行了学术交流。

- 2007（80岁）

2007.1—2009.12，段一士先生主持国家自然科学基金项目"额外维和p-膜的拓扑场论"（No. 10610301019）。

2007年，段一士先生研究了类弦缺陷上费米子的局域化问题。论文"Localization of fermions on a string-like defect"发表于JHEP（与学生刘玉孝、赵力合作）。

研究了Hopf不变量的内部拓扑结构。论文"Inner topological structure of Hopf invariant"发表于J. Math. Phys.（与学生任继荣、李然合作）。

研究了球对称黑洞的拓扑性质与费米子吸收问题。论文"Fermion absorption cross section and topology of spherically symmetric black holes"发表于Phys. Lett. B（与学生刘玉孝、赵力、曹贞斌合作）。

研究了类弦缺陷上自对偶涡旋背景中的费米子。论文"Fermions in self-dual vortex background on a string-like defect"发表于Nucl. Phys. B（与学生刘玉孝、赵力、张欣会合作）。

给研究生讲授了粒子反应阈能（3月）、超弦理论/共形场与超对称（3月—5月）等课程和讲座。

为兰州大学隆基班作了题为"我国杰出教育家——江隆基"、"爱因斯坦和相对论"、"数学与物理：同气连枝，同胞共哺"的报告。

5月，诺贝尔奖获得者David Gross教授等访问兰州大学，并与段一士先生交流学术问题。

7月，段一士先生的弟子和国际国内同行在兰州大学举办理论物理前沿研讨会，段一士先生受邀作了题为"八十感悟"的报告。段一士先生的学生认为他的科研工作"独树一帜，从不盲目随流，特别是有些观点处于国际领先地位"。中国科学院理论物理研究所敬贺"科研英才、育人楷模"。

- 2008（81岁）

  2008年，段一士先生在JHEP上发表论文"Localization of matters on pure geometrical thick branes"，研究了纯几何厚膜上的物质场局域化问题（与学生刘玉孝、张欣会、张力达合作）。

  在J. Phys. A上发表论文"Vortices-induced quantum Rontgen effect in BEC: a consistent approach"，研究了玻色–爱因斯坦凝聚体中涡旋诱导的量子Rontgen效应（与学生黄汝南合作）。

  在J. Phys. Condens. Mat.上发表论文"Magnetic monopoles in ferromagnetic spin-triplet superconductors"，研究了反铁磁自旋三重态超导体中的磁单极（与学生张力达、刘玉孝合作）。

  在JHEP上发表论文"Localization and mass spectrum of matters on Weyl thick branes"，研究了Weyl厚膜上各种物质场的局域化和质量谱问题（与学生刘玉孝、张力达、魏少文合作）。

  在Phys. Rev. D上发表论文"Fermions on thick branes in the background of sine-Gordon kinks"，研究了sine-Gordon kinks 背景下厚膜上的费米子（与学生刘玉孝、张力达、张丽杰合作）。

  7月，段一士先生应邀承办中国引力与相对论天体物理学会学术年会，并作了题为"引力规范理论"的报告。

  9月和10月，应邀给兰州大学隆基班作了题为"攀登科学高峰——学习与创新的智慧和哲学"、"对称性破缺"的报告。

- 2009（82岁）

  2009年，段一士先生在J. Cosmol. Astropart. Phys.上发表论文"Bulk matters on symmetric and asymmetric de Sitter thick branes"，研究了对称、反对称德西特厚膜上的物质场（与学生刘玉孝、赵振华、魏少文合作）。

  在Phys. Rev. D上发表论文"Fermion localization and resonances on

a de Sitter thick brane"，研究了在德西特厚膜上费米子的局域化和共振态（与学生刘玉孝、杨捷、赵振华、付春娥合作）。

在Phys. Lett. B发表论文"Localization of matters on the bent brane in AdS(5) bulk"，研究了反德西特时空中弯曲膜上的物质局域化问题（与学生梁钧合作）。

中央电视台著名主持人水均益主持兰州大学百年校庆晚会时，现场采访了段一士先生。段一士先生谈到：兰州大学培养了许多博导、院长、校长和院士，并期望兰大培养的学生今后能获诺贝尔奖。

年末，中央教育台记者采访段一士先生。

- 2010（83岁）

2010年，段一士先生在Int. J. Mod. Phys. A上发表论文"Multi-vortices evolution in weak coupled ABJM theory"，研究了弱耦合ABJM理论中的多涡旋演化问题（与学生杨捷、李阳、刘玉孝合作）。

3月，"南风窗"杂志记者家中采访了段一士先生。

5月，兰州大学时任党委书记王寒松联系安排段一士先生在天津眼科中心做白内障眼科手术，并亲自到医院看望段一士先生。手术后葛墨林院士邀请段一士先生到南开大学休养，康复后在南开大学数学研究所进行了学术交流。

7月1日，韩国著名物理学家Yongmin Cho教授来兰州参加会议期间，专程到段一士先生家中拜访。

- 2012（85岁）

2012年，段一士先生在Chin. Phys. B上发表论文"Vortex lines in a ferromagnetic spin-triplet superconductor"，研究了铁磁自旋三重态超导体中的涡旋线（与学生赵力、杨捷合作）。

- 2013（86岁）

2013年5月，兰州大学采访了段一士先生，并以"我的科学人生"为题编入兰州大学"萃英记忆工程"丛书——《我的兰大：人物访谈录1》。

11月，段一士先生给兰州大学师生作题为"Higgs粒子与产生基本粒子质量的机制"的物理学新效应讲座。

- 2015（88岁）

2015年12月，段一士先生的《量子场论》由北京高等教育出版社出版。这是段一士先生在兰州大学长期讲授量子场论课程的讲稿和经验总结，是我国最早的量子场论讲义之一。

  1957年从前苏联莫斯科大学回国后，段一士先生编写了量子场论讲义并在兰州大学开始主讲这门课程，同时还自编教材给本科生开设了群论和广义相对论课程，这一传统持续至今。该教材的特点是以作者发现的广义守恒定理I和II为基础，建立了各种场在在相应变换下的守恒定律和守恒量，使量子场论内容化难为易，言简意赅。

- 2016（89岁）

  "数学与人文"第21辑《数学百草园》刊载"为学不随流俗转，超越创新常道间——段一士教授的学术人生"文章（任继荣撰稿）。

  为了学习段先生严谨的科学精神和宝贵的教学经验，传承兰州大学理论物理学科的优良传统，进一步推进理论物理前沿学科发展，物理科学与技术学院于2016年7月17日至19日在兰州大学举行"理论物理前沿研讨会暨段一士教授执教65周年座谈会"。开幕式上，主持人孙昌璞院士在报告中提到："朗道院士说段一士先生是最聪明的中国青年"，"杨振宁教授说山沟沟里也可以做出规范场"。葛墨林院士在报告中说他整个科研的开拓和认识都是由段一士先生这个根生长起来的。南京大学王凡教授介绍了段一士先生在磁单极方面的工作，并说到英国帝国大学的教授首次提到了"段学派（Duan school）"并引用了段一士先生 1979—2002年四篇规范势分解的论文。中国科学院理论物理研究所蔡荣根研究员说段一士先生的能量动量守恒定律被写进了刘辽教授的《广义相对论》教科书。段一士先生在开幕式上作了总结发言，他鼓励大家"青出于蓝而胜于蓝"，希望老师同学们今后更加努力工作，为我们国家做出贡献。

Appendix C

List of Prof. Yi-Shi Duan's Publications

[1] Y.S. Duan, Generalization of regular solutions of Einstein's gravity equations and Maxwell's equations for point-like charge, Soviet Physics JETP **27** (1954) 756–758 [arXiv:1707.02216 [gr-qc]].

[2] Y.S. Duan, Generalized nonsingular solutions for the scalar meson field of a point charge in general relativity theory, Soviet Physics JETP **31** (1956) 1098–1099 [arXiv:1707.02216 [gr-qc]].

[3] Y.S. Duan, General covariant equations for fields of arbitrary spin, Soviet Physics JETP **34** (1958) 632–636 [arXiv:1707.02217 [gr-qc]].

[4] 段一士，任意自旋基本质点的场的广义协变方程式，兰州大学学报 **01** (1958) 29–34.

[5] 段一士，两个核子互相作用系统的Green函数和Bethe-Salpeter方程式，兰州大学学报 **01** (1958) 35–42.

[6] 段一士、邝宇平，解薛丁格尔方程模拟式电子计算机的几个方案，兰州大学学报 **01** (1959) 15–22.

[7] 段一士，质量倒逆变换和基本粒子的相互作用，兰州大学学报 **02** (1960) 1–4.

[8] 段一士、邹鹏程，广义相对论中的Duffin–Kemmer理论，兰州大学学报 **02** (1960) 5–9.

[9] 段一士、张敬业，赝标量场的广义协变方程，兰州大学学报 **02** (1960) 11–14.

[10] 段一士、张敬业，关于铁磁共振理论（磁晶各向异性对铁磁共振频率的影响），兰州大学学报 **02** (1961) 25–31.

[11] 段一士、张敬业，基本粒子广义协变理论和海森堡理论，兰州大学学报 **03** (1961) 119–125.

[12] 段一士、张敬业，关于广义相对论中福克的谐和坐标条件，物理学报 **04** (1962) 211–217.

[13] 段一士、马中骐，π介子衰变，兰州大学学报 **02** (1962) 25–29.

[14] 段一士、张敬业，广义相对论中的能量动量守恒定律，物理学报 **11** (1963) 689–704.

[15] 段一士，广义相对论的基础和近代发展，物理译丛 **12** (1964) 1–10.

[16] 段一士、朱重远，K_β介子衰变，物理学报 **01** (1965) 103–113.

[17] 葛墨林、段一士，关于π-π共振态，物理学报 **11** (1965) 1903–1912.

[18] 葛墨林、段一士，π-π共振态，物理学报 **06** (1966) 724–728.

[19] 段一士、葛墨林，关于规范场理论，兰州大学学报 **00** (1974) 29–37.

[20] 段一士、葛墨林，电子——核子深度非弹的标度性，兰州大学学报 **00** (1974) 38–47.

[21] 侯伯宇、段一士、葛墨林，不可易规范场的对偶荷，兰州大学学报 **02** (1975) 26–33.

[22] 段一士、葛墨林，磁单极子与黑格斯场，兰州大学学报 **04** (1975) 19–25.

[23] 段一士、葛墨林、侯伯宇，不可易规范场的规范不变分量的对偶荷，西北大学学报 **01** (1975) 38–43.

[24] 段一士、葛墨林，磁单极子与黑格斯场，科学通报 **06** (1976) 282–285.

[25] 侯伯宇、段一士、葛墨林，不可易规范场的对偶荷，物理学报 **06** (1976) 514–520.

[26] 段一士、葛墨林，SU(2) 规范理论与磁单极，兰州大学学报 **01** (1977) 33–49.

[27] 段一士、葛墨林，磁单极理论在同位旋空间的几何表述，兰州大学学报 **03** (1977) 1–5.

[28] B.Y. Hou, Y.S. Duan and M.L. Ge. The decomposition and reduction of gauge field and dual charged solution of Abelianizable field, Sci. Sinica **21** (1978) 446–456.

[29] 段一士、葛墨林，非阿贝尔规范势分解的几何表述，高能物理与核物理 **06** (1978) 511–514.

[30] 段一士、葛墨林，SU(2)规范理论与N个磁单极运动体系的电动力学，兰州大学学报 **02** (1978) 53–69.

[31] 侯伯宇、段一士、葛墨林，规范场的分解约化及可Abel化场对偶荷解，中国科学 **01** (1979) 45–54.

[32] Y.S. Duan and M.L. Ge, SU(2) gauge theory and electrodynamics with N magnetic monopoles, Sci. Sinica **9**(11) (1979) 1072–1081(Chinese).

[33] 段一士、葛墨林，关于规范场理论，中山大学学报 **01** (1979) 101–107.

[34] 段一士，广义相对论及其实验基础，自然杂志 **03** (1979) 144–154.

[35] 葛墨林、段一士，常曲率空间与SU(2)对偶荷，高能物理与核物理 **02** (1980) 173–178.

[36] 葛墨林、段一士，SU(2)规范理论中的规范真空效应，高能物理与核物理 **05** (1980) 539–544.

[37] Y.S. Duan and S.G. Cai, Infrared behavior of the running coupling constant in QCD, Commun. Theor. Phys. **1** (1982) 385–388.

[38] 段一士、张承理、张重光，电容感应式非触发近炸引信理论研究，兵工学报（引信分册） **02** (1982) 29–37.

[39] Y.S. Duan and Y.T. Wang, The conservation law of energy momentum in general relativity, Sci. China Ser. A **26** (1983) 961–971.

[40] Y.S. Duan and S.C. Zhao, The decomposition of the gauge potential and its quantization theory, Commun. Theor. Phys. **2** (1983) 1553–1562.

[41] 段一士，王友棠，广义相对论中的能量动量守恒定律，中国科学 **04** (1983) 343–351.

[42] Y.S. Duan and C.Y. Yu, A quantum effect in general relativity, SLAC-PUB-3158, 1983.

[43] Y.S. Duan, The running coupling in QCD, SLAC-PUB-3261 (1983).

[44] Y.S. Duan, Harmonic maps and their application to general relativity, SLAC-PUB-3265 (1983).

[45] Y.S. Duan and Z.P. Duan, Gauge field theory of dislocation and disclination continuum, SLAC-PUB-3286 (1984).

[46] Y.S. Duan, The structure of the topological current, SLAC-PUB-3301 (1984).

[47] Y.S. Duan and Z.P. Duan, Gauge field theory and conservation laws in elastic dislocation and disclination continuum mechanics, SLAC-PUB-3302 (1984).

[48] 赵书城、段一士，一种新的有质量规范场理论，兰州大学学报 **02** (1984) 10–16.

[49] 杨孔庆、段一士，QCD中有效耦合常数的色散关系与红外行为，兰州大学学报 **04** (1984) 26–32.

[50] 段一士，调和映射理论中Euler方程的解和它在广义相对论中的应用，物理学报 **06** (1984) 826–832.

[51] Y.S. Duan, Gauge theories of gravitation, Commun. Theor. Phys. **4** (1985) 661–674.

[52] 杨孔庆、王文福、陈成明、段一士，QCD中有效耦合常数的Padé逼近，兰州大学学报 **02** (1985) 18–23.

[53] 罗春荣、段一士，量子电动力学中的有效作用荷，兰州大学学报 **03** (1985) 29–39.

[54] 段一士、赵书城，有质量矢量介子场的Abel规范理论，兰州大学学报 **01** (1986) 49–53.

[55] 段一士、刘继承、董学耕，Einstein–Cartan引力理论中的广义协变守恒定律，物理学报 **06** (1987) 760–768.

[56] 赵书城、段一士，Higgs机制与背景场方法，兰州大学学报 **01** (1987) 35–41.

[57] Y.S. Duan, J.C. Liu and X.G. Dong, General covariant energy-momentum conservation law in general spacetime, Gen. Rel. Grav. **20** (1988) 485–496.

[58] Y.S. Duan and S.L. Zhang, Topological structure of dislocation in the gauge field theory of disclations and disclinations, Int. J. Engng Sci. **28** (1990) 689–695.

[59] Y.S. Duan and S.L. Zhang, Topological structure of disclinations in the gauge field theory of dislocation and disclination continuum, Int. J. Engng Sci. **29** (1991) 153–160.

[60] 赵书城、段一士，弯曲时空共形规范理论及引力荷、惯性因子和宇宙常数，兰州大学学报 **03** (1991) 30–37.

[61] 段一士，推广的调和映射方程和它的孤子解，兰州大学学报 S1 (1991) 9–12.

[62] Y.S. Duan, S.L. Zhang and L.C. Jiang, A new coordinate condition in general relativity, Gen. Rel. Grav. **24** (1992) 1033–1045.

[63] Y.S. Duan and S.L. Zhang, Topological current structure of disclinations in the 4-dimensional gauge field theory of dislocation and disclination continuum, Int. J. Engng Sci. **30** (1992) 153–159.

[64] Y.S. Duan and F.C. Ma, The implication of Kamiokande-II Sun neutrino data,

Commun. Theor. Phys. **18** (1992) 87–90.

[65] S.C. Zhao and Y.S. Duan, Conformal (Weyl) invariance and Higgs mechanism, Nuovo Cim. A **105** (1992) 1739–1743.

[66] 张胜利、段一士，位错连续统规范理论中的等价应变张量，兰州大学学报 **02** (1992) 71–75.

[67] Y.S. Duan and X.H. Meng, Morse theory and structure of Gauss–Bonnet–Chern topological current, Proc. ISATQP-Shangxi 1992, science press (Beijing 1993) 137–146.

[68] Y.S. Duan and X.H. Meng, Topological gauge theory of dislocation in general decomposition gauge potential formulation, Int. J. Engng Sci. **31** (1993) 1173–1176.

[69] Y.S. Duan and X.H. Meng, Topological structure of Gauss–Bonnet–Chern density and its topological current, J. Math. Phys. **34** (1993) 1149–1161.

[70] Y.G. Miao and Y.S. Duan, An alternative Lorentz type symmetry of the theory of chiral bosons with interactions, Commun. Theor. Phys. **20** (1993) 125–128.

[71] X.G. Li and Y.S. Duan, From topological structure in Gauss–Bonnet–Chern to Morse theory, IMP&NLHIAL Annual Report (1994) 126–127.

[72] Y.S. Duan, S.L. Zhang and S.S. Feng, Gauge potential decomposition, space-time defects and Planck's constant, J. Math. Phys. **35** (1994) 4463–4468.

[73] S.S. Feng and Y.S. Duan, Conservative quantities and their algebra in the self-dual gravity, Gen. Rel. Grav. **27** (1995) 887–903.

[74] Y.S. Duan and X.G. Lee, General decomposition theory of spin connections topological structure of Gauss–Bonnet–Chern density and the Morse theory, Helv. Phys. Acta **68** (1995) 513–530.

[75] 段一士、冯世祥，广义相对论中广义协变的角动量守恒定律，物理学报 **09** (1995) 1373–1381.

[76] Y.S. Duan and S.S. Feng, General covariant conservation law of angular momentum in general relativity, Commun. Theor. Phys. **25** (1996) 99–104.

[77] S.X. Feng and Y.S. Duan, About the energy of the universe, Chin. Phys. Lett. **13** (1996) 409–411.

[78] S.S. Feng and Y.S. Duan, Conservative angular momentum as SU(2) charges in complex gravity, Commun. Theor. Phys. **25** (1996) 485–490.

[79] 俞重远、陈陆群、段一士、杨国宏，耦合孤子的相互作用，中国激光 **05** (1996) 403–408.

[80] S.X. Feng and Y.S. Duan, About the energy of the universe-reply, Chin. Phys. Lett. **14** (1997) 400–400.

[81] Y.S. Duan, G.H. Yang and Y. Jiang, The quantization of the space-time defects in the early universe, Helv. Phys. Acta **70** (1997) 565–577.

[82] Y.S. Duan and S.S. Feng, General covariant conservative angular momentum for topologically massive gravity, Commun. Theor. Phys. **27** (1997) 343–348.

[83] Y.S. Duan, G.H. Yang and Y. Jiang, The origin and bifurcation of the space-time defects in the early universe, Gen. Rel. Grav. **29** (1997) 715–725.

[84] 段一士、俞重远、吴振森，双折射光纤中非线性耦合Schrodinger方程的小振幅孤波解，物理学报 12 (1997) 2359-2362.

[85] Y.S. Duan, S. Li and G.H. Yang, The bifurcation theory of the Gauss–Bonnet–Chern topological current and Morse function, Nucl. Phys. B **514** (1998) 705–720.

[86] Y.S. Duan and S. Li, The evolution of the topological structure of the integral quantum Hall effect, Phys. Lett. A **246** (1998) 172–180.

[87] Y.S. Duan, H. Zhang and S. Li, Topological structure of the London equation, Phys. Rev. B **58** (1998) 125–127.

[88] Y.S. Duan and L.B. Fu, The general decomposition theory of SU(2) gauge potential topological structure and bifurcation of SU(2) Chern density, J. Math. Phys. **39** (1998) 4343–4355.

[89] Y.S. Duan and S. Li, The topological structure of the space-time disclination, J. Math. Phys. **39** (1998) 6696–6705.

[90] Y.S. Duan, Y. Jiang and G.H. Yang, Topological quantization of linear defects, Chin. Phys. Lett. **15** (1998) 781–783.

[91] Y.S. Duan and X.G. Lee, General decomposition of spin connection and topological structure in Gauss–Bonnet–Chern density, Commun. Theor. Phys. **29** (1998) 237–244.

[92] G.H. Yang and Y.S. Duan, Fourth-order topological tensor current and Fock's coordinate condition, Mod. Phys. Lett. A **13** (1998) 745–752.

[93] Y.S. Duan, G.H. Yang and Y. Jiang, The quantization and production of the space-time defects in the early universe, Int. J. Mod. Phys. A **13** (1998) 2001–2012.

[94] G.H. Yang and Y.S. Duan, Topological quantization of magnetic monopoles and their bifurcation theory, Int. J. Theor. Phys. **37** (1998) 2371–2382.

[95] G.H. Yang, Y.S. Duan and Y.C. Huang, Topological invariant in Riemann–Cartan manifold and space-time defects, Int. J. Theor. Phys. **37** (1998) 2953-2964.

[96] Y.S. Duan and Y. Jiang, The generation of the (k-1)-dimensional defect, arXiv:hep-th/9809011 (1998).

[97] Y.S. Duan, Y. Jiang and G.H. Yang, The topological quantization and bifurcation of the topological linear defects, arXiv:hep-th/9810054 (1998).

[98] Y.S. Duan, Y. Jiang and G.H. Yang, The topological quantization and the branch process of the (k-1)-dimensional topological defects, arXiv:hep-th/9810111 (1998).

[99] S. Li and Y.S. Duan, SO(4) monopole as a new topological invariant and its topological structure, arXiv:hep-th/9811086 (1998).

[100] G.H. Yang, G.J. Ni and Y.S. Duan, The relations among two transversal submanifolds and global manifold, arXiv:gr-qc/9901078 (1999).

[101] S. Li and Y.S. Duan, Decomposition Theory of spin connection and topological structure of Gauss–Bonnet–Chern theorem on manifold with boundary,

arXiv:math-ph/9903020 (1999).

[102] Y.S. Duan, L.B. Fu and H. Zhang, Topological quantum mechanics and the first Chern class, arXiv:hep-th/9908168 (1999).

[103] Y.S. Duan, L.B. Fu and X. Liu, The second Chern class in spinning system, arXiv:hep-th/9909008 (1999).

[104] Y.S. Duan, H. Zhang and L.B. Fu, Point defects of a three-dimensional vector order parameter, Phys. Rev. E **59** (1999) 528–534.

[105] Y.S. Duan and H. Zhang, Line defects of a two-component vector order parameter, Phys. Rev. E **60** (1999) 2568–2576.

[106] Y.S. Duan, Y. Jiang and T. Xu, A new topological aspect in the O(n) symmetric time-dependent Ginzburg–Landau model, Phys. Lett. A **252** (1999) 307–315.

[107] Y.S. Duan, H. Zhang and G. Jia, A mathematical framework for wave dislocation lines, Phys. Lett. A **253** (1999) 57–65.

[108] S.S. Feng and Y.S. Duan, Generally covariant conservative energy-momentum for gravitational anyons, Class. Quant. Grav. **16** (1999) 3237–3244.

[109] Y.S. Duan, L.B. Fu and H. Zhang, Topological current of point defects and its bifurcation, Mod. Phys. Lett. A **14** (1999) 2011–2023.

[110] G.H. Yang and Y.S. Duan, The origin and bifurcation of disclinations in the gauge field theory of dislocation and disclination continuum, Int. J. Engng Sci. **37** (1999) 1037–1050.

[111] Y.S. Duan and Y. Jiang, Strings and the gauge theory of spacetime defects, Int. J. Theor. Phys. **38** (1999) 563–573.

[112] S. Li and Y.S. Duan, The bifurcation of the topological structure in the sunspot's electric topological current with locally gauge-invariant Maxwell–Chern–Simons term, Astrophys. Space Sci. **268** (1999) 455–468.

[113] Y.S. Duan, Y. Jiang and G.H. Yang, Evolution of the topological linear defects, Chin. Phys. Lett. **16** (1999) 157–159.

[114] Y.S. Duan and H. Zhang, The topological structure of vortex in BEC, Eur. Phys. J. D **5** (1999) 47–51.

[115] Y.S. Duan and Y. Jiang, Can torsion play a role in angular momentum conservation law? Gen. Rel. Grav. **31** (1999) 3–12.

[116] Y.S. Duan, T. Xu and L.B. Fu, Topological structure in the O(n) symmetric time-dependent-Ginzburg–Landau model, Prog. Theor. Phys. **101** (1999) 467–471.

[117] Y.S. Duan, G. Jia and H. Zhang, Branch conditions for wave dislocations in right beams, J. Phys. A: Math. Gen. **32** (1999) 4943–4953.

[118] 郑驻军、段一士，二步幂零李群上一类左不变微分算子亚椭圆的充要条件，应用数学学报 **03** (1999) 321–326.

[119] L.B. Fu, Y.S. Duan and H. Zhang, Evolution of the Chern–Simons vortices, Phys. Rev. D **61** (2000) 045004.

[120] G.X. Zhang, Z.B. Li and Y.S. Duan, Exact Soliton solutions of non-linear wave equations, Sci. China Ser. A **44** (2000) 1103–1108.

[121] Y.S. Duan, Y. Jiang and G.H. Yang, The topological quantization and bifurcation of strings in early universe, Commun. Theor. Phys. **33** (2000) 105–112.

[122] Y.S. Duan, L.B. Fu and H. Zhang, Topological structure of Chern–Simons vortex, Commun. Theor. Phys. **33** (2000) 693–696.

[123] Y. Jiang and Y.S. Duan, The branch process of cosmic strings, J. Math. Phys. **41** (2000) 2616–2628.

[124] Y.S. Duan, L.B. Fu and G. Jia, Topological tensor current of $\tilde{p}$-branes in the phi-mapping theory, J. Math. Phys. **41** (2000) 4379–4386.

[125] Y. Jiang and Y.S. Duan, A new topological aspect of the arbitrary dimensional topological defects, J. Math. Phys. **41** (2000) 6463–6476.

[126] H. Zhang, Y.S. Duan and B.B. Hu, Bifurcation of the London equation in anisotropic superconductors, Europhys. Lett. **52** (2000) 101–107.

[127] Y.S. Duan, P.M. Zhang and H. Zhang, Soliton evolution in the composite-boson field, Int. J. Theor. Phys. **39** (2000) 2837–2851.

[128] D.J. Jia, Y.S. Duan and X.G. Lee, Circulation condition of the two-component Bose–Einstein condensate, Phys. Lett. A **289** (2001) 245–250.

[129] Y.S. Duan and D.J. Jia, Effective gauge dynamics of the Bose–Einstein condensates, Phys. Lett. A **290** (2001) 120–126.

[130] Y.S. Duan, P.M. Zhang and H. Zhang, Topological excitation in the fractional quantum Hall system, Phys. Lett. A **291** (2001) 296–300.

[131] T. Xu, G.H. Yang and Y.S. Duan, The bifurcation of vortex current in the time-dependent Ginzburg–Landau model, Commun. Theor. Phys. **35** (2001) 157–158.

[132] G.X. Zhang, Z.B. Li and Y.S. Duan, Exact solitary wave solutions of nonlinear wave equations, Sci. China A **44** (2001) 396–401.

[133] Y.S. Duan and D.J. Jia, Integer and half-integer quantization conditions in quantum mechanics, Chin. Phys. Lett. **18** (2001) 473–475.

[134] G.H. Yang, Y. Jiang and Y.S. Duan, Topological quantization of k-dimensional topological defects and motion equations, Chin. Phys. Lett. **18** (2001) 631–633.

[135] G.H. Yang, S.X. Feng, G.J. Ni and Y.S. Duan, Relations of two transversal submanifolds and global manifold, Int. J. Mod. Phys. A **16** (2001) 3535–3551.

[136] D.J. Jia and Y.S. Duan, Topological effects of instanton due to defects, Mod. Phys. Lett. A **16** (2001) 1863–1869.

[137] Y.S. Duan and P.M. Zhang, Inner structure of Gauss–Bonnet–Chern theorem and the Morse theory, Mod. Phys. Lett. A **16** (2001) 2483–2493.

[138] Y.S. Duan, X. Liu and P.M. Zhang, Novel theory for topological structure of vortices in a Bose–Einstein condensate, arXiv:cond-mat/0106103 (2001).

[139] 李希国、段一士、宋建军，SO(n)规范势的可分解理论，高能物理与核物理 **04** (2001) 296–303.

[140] 段一士、张鹏鸣，推广的Gross-Pitaevskii理论中的涡旋问题(英文)，原子核物理评论 **04** (2001) 225–231.

[141] L.B. Fu, J. Liu, S.G. Chen and Y.S. Duan, The configuration of a topological

current and its physical structure: an application and paradigmatic evidence, J. Phys. A: Math. Gen. **35** (2002) L181–L187.

[142] Y.S. Duan, X. Liu and P.M. Zhang, Decomposition theory of the U(1) gauge potential and the London assumption in topological quantum mechanics, J. Phys.: Condens. Matter **14** (2002) 7941–7947.

[143] G.H. Yang, H. Zhang and Y.S. Duan, Topological quantization of disclination points in three-dimensional liquid crystals, Chin. Phys. **11** (2002) 415–418.

[144] G.H. Yang, H. Zhang and Y.S. Duan, Topological aspect and bifurcation of disclination lines in two-dimensional liquid crystals, Commun. Theor. Phys. **37** (2002) 513–518.

[145] G.H. Yang, H. Zhang and Y.S. Duan, Topological aspects of liquid crystals, Int. J. Theor. Phys. **41** (2002) 991–1005.

[146] X.G. Lee, M. Baldo and Y.S. Duan, Torsion structure in Riemann–Cartan manifold and dislocation, Gen. Rel. Grav. **34** (2002) 1569–1577.

[147] Y.S. Duan and P.M. Zhang, Decomposing the SU(n) connection and the Wu–Yang potential, Mod. Phys. Lett. A **17** (2002) 2283-2287.

[148] Y.S. Duan, X. Liu and L.B. Fu, Many knots in Chern–Simons field theory, Phys. Rev. D **67** (2003) 085022.

[149] Y.S. Duan, X. Liu and P.M. Zhang, Mermin – Ho vortices and monopoles in three-component spinor BEC, J. Phys. A: Math. Gen. **36** (2003) 563–571.

[150] Y.S. Duan, J.P. Wang, X. Liu and P.M. Zhang, Topological excitation in an antiferromagnetic Bose–Einstein condensate, Prog. Theor. Phys. **110** (2003) 1–8.

[151] Y.S. Duan and J.P. Wang, Topological vortices in the modified two-dimensional Gross-Pitaevskii theory, Int. J. Theor. Phys. **42** (2003) 1733–1744.

[152] Y.S. Duan, X. Liu and L.B. Fu, Spinor decomposition of SU(2) gauge potential and spinor structures of Chern–Simons and Chern density, Commun. Theor. Phys. **40** (2003) 447–450.

[153] Y.S. Duan, J.R. Ren and J. Yang, Cosmic strings and quintessence, Chin. Phys. Lett. **20** (2003) 2133–2136.

[154] P.M. Zhang, Y.S. Duan and L.M. Cao, The order parameter field for cosmic string, inflation and dark energy, Mod. Phys. Lett. A **18** (2003) 2587–2597.

[155] Y.S. Duan, X. Liu and H. Zhang, Velocity field vortices and Mermin–Ho vortices in two-component spinor BEC, Ann. Phys. **308** (2003) 493–501.

[156] Y.S. Duan and X. Liu, Knotlike cosmic strings in the early universe, JHEP **02** (2004) 028.

[157] Y.S. Duan, Y.X. Liu and L.J. Zhang, Spinor field realizations of non-critical W(2,s) strings, Nucl. Phys. B **699** (2004) 174–182.

[158] P.M. Zhang, L.M. Cao, Y.S. Duan and C.K. Zhong, Transverse force on a moving vortex with the acoustic geometry, Phys. Lett. A **326** (2004) 375–380.

[159] Y.S. Duan, X. Liu and J. Yang, Knotlike vortex lines in spinor BEC, Ann. Phys. **312** (2004) 84–91.

[160] P.M. Zhang, Y.S. Duan, X. Liu and C.K. Zhong, Knot-like Mermin–Ho vortices in the superfluid He-3-A, Ann. Phys. **313** (2004) 16–25.

[161] P.M. Zhang, Y.S. Duan and J.R. Ren, Decomposition of the semi-simple group connection and cosmic, Mod. Phys. Lett. A **19** (2004) 745–753.

[162] Y.S. Duan and X.G. Shi, ϕ-mapping topological theory in two-gap superconductor, Int. J. Mod. Phys. B **18** (2004) 375–380.

[163] G.H. Yang, Y.S. Wang and Y.S. Duan, Contribution of disclination lines to free energy of liquid crystals in single-elastic constant approximation, Commun. Theor. Phys. **42** (2004) 185–188.

[164] Y.S. Duan, Y.X. Liu and L.J. Zhang, Corrections to the nonrelativistic ground energy of a helium atom, Chin. Phys. Lett. **21** (2004) 1714–1716.

[165] P.M. Zhang, Y.S. Duan and X. Liu, Knots in the non-Abelian Chern–Simons theory, Mod. Phys. Lett. A **19** (2004) 2629–2636.

[166] J. Liang and Y.S. Duan, Localization and mass spectrum of spin-1/2 fermionic field on a thick brane with Poincare symmetry, Europhys. Lett. **87** (2004) 40005.

[167] Y.S. Duan and J.R. Ren, The Non-Abelian topological gauge field theory of $\tilde{p}$-branes, arXiv:hep-th/0504114 (2005).

[168] Y.Q. Wang, Y.X. Liu, Z.H. Zhao and Y.S. Duan, Topological quantization of self-dual Chern–Simons vortices on Riemann surfaces, arXiv:hep-th/0508063 (2005).

[169] Y.Q. Wang, Y.X. Liu and Y.S. Duan, A new form of self-duality equations with topological term, arXiv:hep-th/0508104 (2005).

[170] Y.S. Duan, L. Zhao, X.H. Zhang and T.Y. Si, The U(1) topological gauge field theory for topological defects in liquid crystals, arXiv:hep-th/0512316 (2005).

[171] Y.S. Duan and J. Yang, The topological inner structure of Chern–Simons tensor current and the world-sheet of strings, Chin. Phys. Lett. **22** (2005) 1079–1082.

[172] Y.S. Duan, X.M. Zhang and M. Tian, The branch process of Skyrmions in the fractional quantum Hall effect, Chin. Phys. Lett. **22** (2005) 2047–2050.

[173] Y.S. Duan, W.J. Zhong and T.Y. Si, Self-dual Chern–Simons vortices in Higgs field, Chin. Phys. Lett. **22** (2005) 2462–2464.

[174] X. Liu, Y.Z. Zhang, Y.S. Duan and L.B. Fu, Various topological excitations in the SO(4) gauge field in higher dimensions, Ann. Phys. **318** (2005) 419–431.

[175] J.P. Wang and Y.S. Duan, Topological vortices in superfluid films, Commun. Theor. Phys. **44** (2005) 160–164.

[176] G.H. Yang, H. Zhang, L.J. Tian, Y.S. Wang and Y.S. Duan, Effect of disclination lines on free energy of nematic liquid crystals, Commun. Theor. Phys. **44** (2005) 213–222.

[177] X. Liu, Y.S. Duan and P.M. Zhang, Multi-types of Skyrmions in SU(n) quantum Hall system, Commun. Theor. Phys. **44** (2005) 371–374.

[178] Y.S. Duan and X.M. Zhang, Inner structure of statistical gauge potential in Chern–Simons–Ginzburg–Landau theory, Commun. Theor. Phys. **44** (2005)

559–562.

[179] G.H. Yang, J.J. Yan, L.J. Tian and Y.S. Duan, Topological aspects of entropy and phase transition of Kerr black holes, Commun. Theor. Phys. **44** (2005) 631–637.

[180] T.Y. Si and Y.S. Duan, A novel theory of electroweak instantons, Mod. Phys. Lett. A **20** (2005) 2891–2902.

[181] Y.Q. Wang, T.Y. Si, Y.X. Liu and Y.S. Duan, Fermionic zero modes in self-dual vortex background, Mod. Phys. Lett. A **20** (2005) 3045–3053.

[182] Y.S. Duan, X.H. Zhang, Y.X. Liu and L. Zhao, Knotted solitons in a charged two-condensate Bose system, Phys. Rev. B **74** (2006) 144508.

[183] Y.S. Duan, W.J. Zhong, T.Y. Si and Y.Q. Wang, Knotted self-dual vortex in generalized Abelian Chern–Simons Higgs field, Int. J. Mod. Phys. A **21** (2006) 151–160.

[184] Y.S. Duan, S.F. Wu and P.M. Zhang, Topological structure and topological tensor current of Gauss–Bonnet–Chern theorem, Commun. Theor. Phys. **45** (2006) 793–798.

[185] Y.S. Wang, G.H. Yang, L.J. Tian and Y.S. Duan, Stability of disclinations in nematic liquid crystals, Commun. Theor. Phys. **45** (2006) 993–999.

[186] X.G. Shi and Y.S. Duan, Topological structure of phase vortex in resonating valence bond superconductivity, Commun. Theor. Phys. **46** (2006) 33–36.

[187] T.Y. Si and Y.S. Duan, Distribution of topological defects on axisymmetric surface, Commun. Theor. Phys. **46** (2006) 319–322.

[188] Y.S. Duan, Y.X. Liu and Y.Q. Wang, Fermionic zero modes in gauge and gravity backgrounds on T2, Mod. Phys. Lett. A **21** (2006) 2019–2025.

[189] Y.S. Duan and S. F. Wu, Magnetic branes from generalized 't Hooft tensor, Mod. Phys. Lett. A **21** (2006) 2599–2606.

[190] Y.S. Duan, S.F. Wu and P.M. Zhang, Higher-dimensional knotlike topological defects in local non-Abelian topological tensor currents, J. Math. Phys. **47** (2006) 092303.

[191] Y.S. Duan and S.F. Wu, Regular magnetic monopole from generalized 't Hooft tensor, Chin. Phys. Lett. **23** (2006) 2932–2935.

[192] Y.S. Duan, X.H. Zhang and L. Zhao, Topological aspect of black hole with Skyrme hair, Int. J. Mod. Phys. A **21** (2006) 5895–5903.

[193] Y.S. Duan, L. Zhao and X.H. Zhang, Topological vortices in a two-gap superconductor, Phys. Lett. A **360** (2006) 183–189.

[194] Y.X. Liu, L. Zhao and Y.S. Duan, Localization of fermions on a string-like defect, JHEP **04** (2007) 97.

[195] Y.X. Liu, L. Zhao, X.H. Zhang and Y.S. Duan, Fermions in self-dual vortex background on a string-like defect, Nucl. Phys. B **785** (2007) 234–245.

[196] Y.X. Liu, L. Zhao, Z.B. Cao and Y.S. Duan, Fermion absorption cross section and topology of spherically symmetric black holes, Phys. Lett. B **650** (2007) 286–292.

[197] Y.S. Duan, X.H. Zhang and Y.X. Liu, Topological excitation in Skyrme theory, Commun. Theor. Phys. **47** (2007) 85–88.

[198] Y.S. Duan and X.G. Shi, Evolution of Nielsen–Olesen's string from Chern–Simons field theory, Commun. Theor. Phys. **48** (2007) 147–150.

[199] Y.X. Liu, Y.Q. Wang and Y.S. Duan, Fermionic zero modes in self-dual vortex background on a torus, Commun. Theor. Phys. **48** (2007) 675–679.

[200] Y.S. Duan, J.R. Ren and R. Li, Vector and spinor decomposition of SU(2) gauge potential, their equivalence, and knot structure in SU(2) Chern–Simons theory, Commun. Theor. Phys. **47** (2007) 875–878.

[201] Y.S. Duan, L. Zhao, X.H. Zhang and T.Y. Si, Topological defects in liquid crystal films, Commun. Theor. Phys. **47** (2007) 1125–1128.

[202] Y.S. Duan, L. Zhao and X.H. Zhang, Topological structure of knotted vortex lines in liquid crystals, Commun. Theor. Phys. **47** (2007) 1129–1134.

[203] Y.S. Duan and L. Zhao, Topological structure and evolution of space-time dislocations and disclinations, Int. J. Mod. Phys. A **22** (2007) 1335–1351.

[204] Y.S. Duan, L. Zhao, Y.X. Liu and J.R. Ren, Topological aspect of Chern–Simons p-branes, Chin. Phys. **16** (2007) 881–886.

[205] H. Zhang, B.B. Hu, B.W. Li and Y.S. Duan, Topological constraints on scroll and spiral waves in excitable media, Chin. Phys. Lett. **24** (2007) 1618–1621.

[206] P.M. Zhang, X.G. Lee, S.F. Wu and Y.S. Duan, Knotted wave dislocation with the hopf invariant, Int. J. Theor. Phys. **46** (2007) 1753–1762.

[207] J.R. Ren, R. Li and Y.S. Duan, Inner topological structure of Hopf invariant, J. Math. Phys. **48** (2007) 073502.

[208] Y.X. Liu, L. Zhao, L.J. Zhang and Y.S. Duan, Fermionic zero modes in self-dual vortex background on a sphere, Int. J. Mod. Phys. A **22** (2007) 3643–3653.

[209] Y.S. Duan, X.H. Zhang and L. Zhao, Novel topological invariant in the U(1) gauge field theory, Mod. Phys. Lett. A **22** (2007) 2201–2208.

[210] P.M. Zhang, X.G. Lee, S.F. Wu and Y.S. Duan, Topological objects in the O(3) nonlinear sigma model, Mod. Phys. Lett. A **22** (2007) 2379–2386.

[211] Y.S. Duan and Z.B. Cao, Topological zero-thickness cosmic strings, Mod. Phys. Lett. A **22** (2007) 2471–2478.

[212] Y.X. Liu, Y.S. Duan and L.J. Zhang, Angular momentum conservation law for Randall-Sundrum models, Mod. Phys. Lett. A **22** (2007) 2855–2864.

[213] Y.X. Liu, X.H. Zhang, L.D. Zhang and Y.S. Duan, Localization of matters on pure geometrical thick branes, JHEP **02** (2008) 067.

[214] Y.X. Liu, L.D. Zhang, S.W. Wei and Y.S. Duan, Localization and mass spectrum of matters on Weyl thick branes, JHEP **08** (2008) 041.

[215] Y.X. Liu, L.D. Zhang, L.J. Zhang and Y.S. Duan, Fermions on thick branes in the background of sine-Gordon kinks, Phys. Rev. D **78** (2008) 065025.

[216] X. Liu, Y.S. Duan, W.L. Yang and Y.Z. Zhang, Inner structure of Spinc(4) gauge potential on 4-dimensional manifolds, Ann. Phys. **323** (2008) 2107–2114.

[217] L.D. Zhang, Y.S. Duan and Y.X. Liu, The modified London equation,

Abrikosov-like vortices and knot solitons in two-gap superconductors, J. Phys. A: Math. Theor. **41** (2008) 375201.

[218] J.R. Ren, T. Zhu and Y.S. Duan, Topological aspect of knotted vortex filaments in excitable media, Chin. Phys. Lett. **25** (2008) 353–356.

[219] J.R. Ren, T. Zhu and Y.S. Duan, Topological properties of spatial coherence function, Chin. Phys. Lett. **25** (2008) 367–370.

[220] W.J. Zhong and Y.S. Duan, Topological quantization of instantons in SU(2) Yang-Mills theory, Chin. Phys. Lett. **25** (2008) 1534–1537.

[221] Y.S. Duan and R.N. Huang, Vortices-induced quantum Rontgen effect in BEC: a consistent approach, J. Phys. A: Math. Theor. **41** (2008) 045304.

[222] Y.X. Liu, L.J. Zhang, Y.Q. Wang and Y.S. Duan, Energy momentum for Randall-Sundrum models, Mod. Phys. Lett. A **23** (2008) 769–779.

[223] L. Zhao, Y.X. Liu and Y.S. Duan, Fermions in gravity and gauge backgrounds on a brane world, Mod. Phys. Lett. A **23** (2008) 1129–1139.

[224] Y.X. Liu, X.H. Zhang and Y.S. Duan, Detecting extra dimension by helium-like ions, Mod. Phys. Lett. A **23** (2008) 1853–1860.

[225] Y.S. Duan, L.D. Zhang and Y.X. Liu, A new description of cosmic strings in brane world scenario, Mod. Phys. Lett. A **23** (2008) 2023–2030.

[226] X.H. Zhang, Y.X. Liu and Y.S. Duan, Localization of fermionic fields on braneworlds with bulk Tachyon matter, Mod. Phys. Lett. A **23** (2008) 2093–2101.

[227] Y.S. Duan, L.D. Zhang and Y.X. Liu, Self-dual vortices in the Abelian Chern–Simons model with two complex scalar fields, Mod. Phys. Lett. A **23** (2008) 2189–2198.

[228] Y.S. Duan and M. Tian, General spherical symmetry solution to Einstein equations and Newtonian gravity modification at small length scale, Int. J. Mod. Phys. A **23** (2008) 2649–2652.

[229] L.D. Zhang, Y.S. Duan and Y.X. Liu, Magnetic monopoles in ferromagnetic spin-triplet superconductors, J. Phys.: Condens. Matter **20** (2008) 235219.

[230] Y.X. Liu, L.J. Zhang and Y.S. Duan, Fermionic zero modes in gauge and gravity backgrounds on a sphere, Commun. Theor. Phys. **49** (2008) 1577–1579.

[231] J.R. Ren, D.H. Xu, X.H. Zhang and Y.S. Duan, Topological aspects of superconductors at dual point, Commun. Theor. Phys. **49** (2008) 1627–1630.

[232] J.R. Ren, R. Li and Y.S. Duan, A new route to the interpretation of Hopf invariant, Commun. Theor. Phys. **50** (2008) 53–58.

[233] J.R. Ren, T. Zhu and Y.S. Duan, Topology of knotted optical vortices, Commun. Theor. Phys. **50** (2008) 345–348.

[234] J.R. Ren, S.W. Wei, D.H. Xu and Y.S. Duan, Topological vortices in a spin-triplet superconductor, Commun. Theor. Phys. **50** (2008) 525–528.

[235] Y.S. Duan, M. Tian and X.H. Zhang, Self-dual vortices in Schrodinger–Chern–Simons model, Commun. Theor. Phys. **50** (2008) 757–762.

[236] J.R. Ren, J.B. Wang, R. Li, D.H. Xu and Y.S. Duan, Topological aspects of

solitons in ferromagnets, Commun. Theor. Phys. **50** (2008) 777–780.

[237] Z.B. Cao, Y.X. Liu and Y.S. Duan, Some Topological Properties of Anyons, Commun. Theor. Phys. **50** (2008) 1155–1158.

[238] Y.X. Liu, Z.H. Zhao, S.W. Wei and Y.S. Duan, Bulk matters on symmetric and asymmetric de Sitter thick branes, JCAP, **2** (2009) 003.

[239] Y.X. Liu, J. Yang, Z.H. Zhao, C.E. Fu and Y.S. Duan, Fermion localization and resonances on a de Sitter thick brane, Phys. Rev. D **80** (2009) 065019.

[240] Y.X. Liu, C.E. Fu, L. Zhao and Y.S. Duan, Localization and mass spectra of fermions on symmetric and asymmetric thick branes, Phys. Rev. D **80** (2009) 065020.

[241] J. Liang and Y.S. Duan, Localization of matters on thick branes, Phys. Lett. B **678** (2009) 491–496.

[242] J. Liang and Y.S. Duan, Localization of matters on the bent brane in AdS(5) bulk, Phys. Lett. B **680** (2009) 489–495.

[243] J. Liang and Y.S. Duan, Localization of matter and fermion resonances on double walls, Phys. Lett. B **681** (2009) 172–178.

[244] J.R. Ren, R. Li and Y.S. Duan, Decomposition of SU(2) connection and knot structure in Skyrme-Faddeev model, Int. J. Mod. Phys. A **23** (2009) 1447–1456.

[245] Z.B. Cao and Y.S. Duan, Generalized Dirichlet normal ordering in open Bosonic strings, Commun. Theor. Phys. **51** (2009) 110–114.

[246] Z.B. Cao and Y.S. Duan, Canonical analysis of noncommutativity of open bosonic string, Mod. Phys. Lett. A **24** (2009) 239–249.

[247] X.G. Shi, M. Yu and Y.S. Duan, Topological solution of Bohmian quantum mechanics, Mod. Phys. Lett. A **24** (2009) 453–461.

[248] Z.B. Cao and Y.S. Duan, Open string in pp-wave background with B-field, Mod. Phys. Lett. A **24** (2009) 759–768.

[249] Z.B. Cao and Y.S. Duan, Extra dimensions with nontrivial torsion and spins of cosmic strings, Mod. Phys. Lett. A **24** (2009) 1147–1157.

[250] M. Tian, X.H. Zhang and Y.S. Duan, Topological structure of Gauss–Bonnet–Chern theorem and p-branes, Chin. Phys. B **18** (2009) 1301–1305.

[251] X.G. Shi and Y.S. Duan, Topological structure of vortex lines of quantum electron plasmas in three- dimensional space, Int. J. Mod. Phys. B **23** (2009) 2439–2447.

[252] J. Yang, Y.S. Duan and Y.X. Liu, The generalization of Chern–Simons current and the topological tensor current of p-branes, Int. J. Theor. Phys. **48** (2009) 2889–2899.

[253] J. Liang and Y.S. Duan, Topological aspects of the proton vortex clusters in the core of a neutron star, Int. J. Mod. Phys. D **18** (2009) 1839–1849.

[254] X.H. Zhang, Y.S. Duan, Y.X. Liu and L. Zhao, Self-dual vortices in the fractional quantum hall system, Int. J. Mod. Phys. B **24** (2010) 423–433.

[255] J. Liang and Y.S. Duan, Localizsion of matters on a non-Z(2)-symmetric thick brane, Mod. Phys. Lett. A **25** (2010) 511–523.

[256] X.M. Zhang and Y.S. Duan, The topological structure of the SU(2) Chern–Simons topological current in the four-dimensional quantum Hall effect, Chin. Phys. Lett. **27** (2010) 077301.

[257] X.G. Shi, Y.S. Duan and Z.J. Qin, Topological effect of reduced quantum trajectory in coherence and decoherence, Commun. Theor. Phys. **54** (2010) 433–438.

[258] J. Yang, Y. Li, Y.X. Liu and Y.S. Duan, Multivortices evolution in weak coupled ABJM theory, Int. J. Mod. Phys. A **25** (2010) 5369–5381.

[259] X.G. Shi and Y.S. Duan, Reduced quantum topological trajectory in coherence and decoherence, Mod. Phys. Lett. A **26** (2011) 1363–1373.

[260] L. Zhao, J. Yang, Q.Y. Xie, M. Tian and Y.S. Duan, Vortex lines in a ferromagnetic spin-triplet superconductor, Chin. Phys. B **21** (2012) 057401.

[261] G. Ren, Y.S. Duan, Extended theory of harmonic maps connects general relativity to chaos and quantum mechanism, Chaos Soliton. Fract. **103** (2017) 567–570 [arXiv:1703.03920 [gr-qc]].

[262] G. Ren, Y.S. Duan, Extended harmonic map equations and the chaotic soliton solutions, [arXiv:1609.04429 [nlin.CD]].